Contents

Preface		vii
1	*Preliminaries*	1
1.1	Probability and Bayes' Theorem	1
1.2	Examples on Bayes' Theorem	10
1.3	Random variables	11
1.4	Several random variables	17
1.5	Means and variances	22
	Exercises on Chapter 1	30
2	*Bayesian Inference for the Normal Distribution*	33
2.1	Nature of Bayesian inference	33
2.2	Normal prior and likelihood	36
2.3	Several normal observations with a normal prior	40
2.4	Dominant likelihoods	44
2.5	Locally uniform priors	47
2.6	Highest density regions (HDRs)	50
2.7	Normal variance	52
2.8	HDRs for the normal variance	55
2.9	The role of sufficiency	57
2.10	Conjugate prior distributions	63
2.11	The exponential family	66
2.12	Normal mean and variance both unknown	68
2.13	Conjugate joint prior for the normal distribution	73
	Exercises on Chapter 2	77
3	*Some Other Common Distributions*	80
3.1	The binomial distribution	80
3.2	Reference prior for the binomial likelihood	86
3.3	Jeffreys' rule	90

3.4 The Poisson distribution 95
3.5 The uniform distribution 99
3.6 Reference prior for the uniform distribution 103
3.7 The tramcar problem 105
3.8 The first digit problem; invariant priors 106
3.9 The circular normal distribution 110
3.10 Approximations based on the likelihood 114
Exercises on Chapter 3 120

4 *Hypothesis Testing* 123

4.1 Hypothesis testing 123
4.2 One-sided hypothesis tests 128
4.3 Lindley's method 130
4.4 Point (or sharp) null hypotheses with prior information 131
4.5 Point null hypotheses for the normal distribution 134
4.6 The Doogian philosophy 142
Exercises on Chapter 4 143

5 *Two-sample Problems* 147

5.1 Two-sample problem—both variances known 147
5.2 Variances unknown but equal 150
5.3 Variance unknown and unequal (Behrens–Fisher Problem) 153
5.4 The Behrens–Fisher controversy 156
5.5 Inferences concerning a variance ratio 159
5.6 Comparison of two proportions; the 2×2 table 161
Exercises on Chapter 5 165

6 *Correlation, Regression and the Analysis of Variance* 168

6.1 Theory of the correlation coefficient 168
6.2 Examples on the use of the correlation coefficient 174
6.3 Regression and the bivariate normal model 175
6.4 Conjugate prior for the bivariate regression model 182
6.5 Comparison of several means—the one way model 185
6.6 The two way layout 193
6.7 The general linear model 197
Exercises on Chapter 6 201

7 *Other Topics* 205

7.1 The likelihood principle 205
7.2 The stopping rule principle 210
7.3 Informative stopping rules 214
7.4 The likelihood principle and reference priors 216
7.5 Bayesian decision theory 218
7.6 Decision theory and hypothesis testing 224
7.7 Empirical Bayes methods 226
Exercises on Chapter 7 228

Appendix—Common Statistical Distributions 233

A.1 Normal distribution 234
A.2 Chi-squared distribution 234
A.3 Normal approximation to chi-squared 235
A.4 Gamma distribution 235
A.5 Inverse chi-squared distribution 236
A.6 Inverse chi distribution 237
A.7 Log chi-squared distribution 238
A.8 Student's t distribution 239
A.9 Normal/chi-squared distribution 240
A.10 Beta distribution 241
A.11 Binomial distribution 241
A.12 Poisson distribution 242
A.13 Negative binomial distribution 243
A.14 Hypergeometric distribution 244
A.15 Uniform distribution 245
A.16 Pareto distribution 246
A.17 Circular normal distribution 247
A.18 Behrens' distribution 249
A.19 Snedecor's F distribution 250
A.20 Fisher's z distribution 251
A.21 Cauchy distribution 252
A.22 Probability one beta variable is greater than another 252
A.23 Distribution of the correlation coefficient 253

Tables 255

A.1 Percentage points of the Behrens–Fisher distribution 255
A.2 Highest density regions for the chi-squared distribution 257

A.3 Highest density regions for the inverse chi-squared distribution 259
A.4 Values of chi-squared corresponding to HDRs for log chi-squared 261
A.5 Values of F corresponding to HDRs for log F 263

Further Reading 282
References 284
Index 291

Note: The tables in the Appendix are intended for use in conjunction with a standard set of statistical tables, for example, Neave (1978). They extend the coverage of these tables so that they are roughly comparable with those of Isaacs *et al.* (1974) or with the tables in Appendix A of Novick and Jackson (1974). However, tables of values easily calculated with a pocket calculator have been omitted. The tables have been computed using NAG routines and algorithms described in Patil (1965) and Jackson (1974).

Bayesian Statistics:

an introduction

PETER M. LEE

Provost of Wentworth College, University of York, England

A CHARLES GRIFFIN BOOK

Edward Arnold
A member of the Hodder Headline Group
LONDON MELBOURNE AUCKLAND
Co published in the Americas by Halsted Press,
an imprint of John Wiley & Sons Inc.
New York — Toronto

To My Mother
and
to the Memory of My Father

First published in Great Britain 1989

Reprinted 1991, 1992, 1994

Co published in the Americas by Halsted Press, an imprint of John Wiley & Sons, Inc., 605 Third Avenue, New York, NY 10158-0012

British Library Cataloguing in Publication Data

Lee, P.M. (Peter M)
Bayesian statistics
1. Statistical analysis. Bayesian theories
I. Title
519.5′42

ISBN 0-85264-309-8 (paper)
ISBN 0-85264-298-9 (boards)

Library of Congress Cataloguing in Publication Data

Available upon request
Halsted Press ISBN 0 470 21961 0

Typeset in 10½/12 pt Times by Wearside Tradespools, Fulwell, Sunderland

Printed and bound in Great Britain for Edward Arnold, a division of Hodder Headline Plc, 338 Euston Road, London NW1 3BH by Athenæum Press Ltd, Newcastle upon Tyne

Preface

When I first learned a little statistics, I felt confused, and others I spoke to confessed that they had similar feelings. Not because the mathematics was difficult—most of that was a lot easier than pure mathematics—but because I found it difficult to follow the logic by which inferences were arrived at from data. It sounded as if the statement that a null hypothesis was rejected at the 5% level meant that there was only a 5% chance of that hypotheses was true, and yet the books warned me that this was not a permissible interpretation. Similarly, the statement that a 95% confidence interval for an unknown parameter ran from -2 to $+2$ sounded as if the parameter lay in that interval with 95% probability and yet I was warned that all I could say was that if I carried out similar procedures time after time then the unknown parameters would lie in the confidence intervals I constructed 95% of the time. It appeared that the books I looked at were not answering the questions that would naturally occur to a beginner, and that instead they answered some rather recondite questions which no-one was likely to want to ask.

Subsequently, I discovered that the whole theory had been worked out in very considerable detail in such books as Lehmann (1959, 1986). But attempts such as those that Lehmann describes to put everything on a firm foundation raised even more questions. I gathered that the usual t test could be justified as a procedure that was "uniformly most powerful unbiased", but I could only marvel at the ingenuity that led to the invention of such criteria for the justification of the procedure, while remaining unconvinced that they had anything sensible to say about a general theory of statistical inference. Of course, Lehmann and others with an equal degree of common sense were capable of developing more and more complicated constructions and exceptions so as to build up a theory that appeared to cover most problems without doing anything obviously silly, and yet the whole enterprise seemed reminiscent of the construction of epicycle upon epicycle in order to preserve a theory of planetary motion based on circular motion; there seemed to be an awful lot of "adhockery".

I was told that there was another theory of statistical inference, based ultimately on the work of the Rev. Thomas Bayes, a Presbyterian minister who lived from 1702 to 1761, whose key paper was published posthumously by his friend Richard Price as Bayes (1763) [more information about Bayes himself and his work can be found in Holland (1962), Todhunter (1865, 1949) and Stigler (1986)]. However, I was warned that there was something not quite proper about this theory, because it depended on your personal beliefs and so was not objective. More precisely, it depended on taking some expression of your beliefs about an unknown quantity before the data were available (your "prior probabilities") and modifying them in the light of the data (via the so-called "likelihood function") to arrive at your "posterior probabilities" using the formulation that "posterior is proportional to prior times likelihood". The standard, or "classical", theory of statistical inference, on the other hand, was said to be objective, because it does not refer to anything corresponding to the Bayesian notion of "prior beliefs". Of course, the fact that in this theory you sometimes looked for a 5% significance test and sometimes for a 0.1% significance test, depending on what you thought about the different situations involved, was said to be quite a different matter.

I went on to discover that this theory could lead to the sorts of conclusions that I had naïvely expected to get from statistics when I first learned about it. Indeed, some lecture notes of Lindley's [and subsequently his book, Lindley (1965)] and the pioneering book by Jeffreys (1939, 1948, 1961) showed that if the statistician had "personal probabilities" that were of a certain conventional type then conclusions very like those in the elementary books I had first looked at could be arrived at, with the difference that a 95% confidence interval really did mean an interval in which the statistician was justified in thinking that there was a 95% probability of finding the unknown parameter. On the other hand, there was the further freedom to adopt other initial choices of personal beliefs and thus to arrive at different conclusions.

Over a number of years I taught the standard, classical, theory of statistics to a large number of students, most of whom appeared to have similar difficulties to those I had myself encountered in understanding the nature of the conclusions that this theory comes to. However, the mere fact that students have difficulty with a theory does not prove it wrong. More importantly, I found the theory did not improve with better acquaintance, and I went on studying Bayesian theory. It turned out that there were real differences in the conclusions arrived at by classical and Bayesian statisticians, and so the former was not just a special case of the latter corresponding to a conventional choice of prior beliefs. On the contrary, there was a strong disagreement between statisticians as to the conclusions to be arrived at in certain standard situations, of which I will cite three examples. The first concerns a test of a sharp null hypothesis (for example, a test that the mean of a distribution is exactly equal to zero), especially when the sample size was large. The second concerns the Behrens–Fisher problem, that is, the inferences that can be made about the difference between the means of two populations when no assumption is made about their variances. The third is the likelihood principle, which asserts that you can only take account of the probability of events that have actually occurred under various hypotheses, and not of events that might have happened but did not; this principle follows from Bayesian statistics and is contradicted by the classical theory. A particular case concerns the relevance of stopping rules, that is to say whether or not you are entitled to take into account the fact that the experimenter decided when to stop experimenting depending on the results so far available rather than having decided to use a fixed sample size all along. The more I thought about all these controversies, the more I was convinced that the Bayesians were right on these disputed issues.

At long last, I decided to teach a third-year course on Bayesian statistics in the University of York, which I have now done for a few years. Most of the students who took the course did find the theory more coherent than the classical theory they had learned in the first course on mathematical statistics they had taken in their second year, and I became yet more clear in my own mind that this was the right way to view statistics. I do, however, admit that there are topics (such as nonparametric statistics) which are difficult to fit into a Bayesian framework.

A particular difficulty in teaching this course was the absence of a suitable book for students who were reasonably well-prepared mathematically and already knew some statistics, even if they knew nothing of Bayes apart from Bayes' Theorem. I wanted to teach them more, and to give more information about the incorporation of real as opposed to conventional prior information, than they could get from Lindley (1965), but I did not think they were well enough prepared to face books like Box and Tiao (1973) or Berger (1985), and so I found that in teaching the course I had to get together material from a large number of sources, and in the end found myself writing this book. It seems less and less likely that students in mathematics departments will be completely unfamiliar with the ideas of statistics, and yet they are not (so far) likely to have encountered Bayesian methods in their first course on statistics; this book is designed with these facts in mind. It is assumed that the reader has a knowledge of calculus of one and two variables and a fair degree of mathematical maturity, but most of the book does not assume a knowledge of linear algebra. The development of the text is self-contained, but from time to time the contrast between Bayesian and classical conclusions is pointed out, and it is supposed that in most cases the reader will have some idea as to the conclusion that a classical statistician would come to, although no very detailed knowledge of classical statistics is expected. It should be possible to use the book as a course text for final-year undergraduate or beginning graduate students or for self-study for those who want a concise account of the way in which the Bayesian approach to statistics develops and the contrast between it and the conventional approach. The theory is built up step by step, rather than doing everything in the greatest generality to start with, and important notions such as sufficiency are brought out of a discussion of the salient features of specific examples.

I am indebted to Professor R. A. Cooper for helpful comments on an earlier draft of this book, although of course he cannot be held responsible for any errors in the final version.

30 March 1988 PETER M. LEE

1

Preliminaries

1.1 Probability and Bayes' Theorem

Notation

The notation will be kept as simple as possible, but it is useful to express statements about probability in the language of set theory. You probably know most of the symbols below, but if you do not you will find it easy enough to get the hang of this useful shorthand. We consider sets $A, B, C, \ldots$ of elements $x, y, z, \ldots$ and we use the word "iff" to mean "if and only if". Then we write

$x \in A$ iff x is a member of A;

$x \notin A$ iff x is *not* a member of A;

$A = \{x, y, z\}$ iff A is the set whose only members are x, y, and z (and similarly for larger or smaller sets);

$A = \{x; S(x)\}$ iff A is the set of elements for which the statement $S(x)$ is true;

$\varnothing = \{x; x \neq x\}$ for the null set, i.e. the set with no elements;

$x \notin \varnothing$ for all x;

$A \subset B$ (i.e. A is a subset of B) iff $x \in A$ implies $x \in B$;

$A \supset B$ (i.e. A is a superset of B) iff $x \in A$ is implied by $x \in B$;

$\varnothing \subset A, A \subset A$ and $A \supset A$ for all A;

$A \cup B = \{x; x \in A \text{ or } x \in B\}$ (where "P or Q" means "P or Q or both") (referred to as the union of A and B or as A union B);

$AB = A \cap B = \{x; x \in A \text{ and } x \in B\}$ (referred to as the intersection of A and B or as A intersect B);

A and B are disjoint iff $AB = \varnothing$;

$A \backslash B = \{x; x \in A, \text{ but } x \notin B\}$. (referred to as the difference set A less B)

Let (A_n) be a sequence $A_1, A_2, A_3, \ldots$ of sets. Then

$$\bigcup_{n=1}^{\infty} A_n = \{x; x \in A_n \text{ for one or more } n\};$$

$$\bigcap_{n=1}^{\infty} A_n = \{x; x \in A_n \text{ for all } n\};$$

(A_n) exhausts B if $\bigcup_{n=1}^{\infty} A_n \supset B$;

(A_n) consists of exclusive sets if $A_m A_n = \varnothing$ for $m \neq n$;
(A_n) consists of exclusive sets given B if $A_m A_n B = \varnothing$ for $m \neq n$;
(A_n) is non-decreasing if $A_1 \subset A_2 \subset A_3 \subset \ldots$ i.e. $A_n \subset A_{n+1}$ for all n;
(A_n) is non-increasing if $A_1 \supset A_2 \supset A_3 \supset \ldots$ i.e. $A_n \supset A_{n+1}$ for all n.

We sometimes need a notation for intervals on the real line, viz.

$[a, b] = \{x; a \leqslant x \leqslant b\}$;
$(a, b) = \{x; a < x < b\}$;
$[a, b) = \{x; a \leqslant x < b\}$;
$(a, b] = \{x; a < x \leqslant b\}$

where a and b are real numbers or $+\infty$ or $-\infty$.

Axioms for probability

In the study of probability and statistics we refer to as complete a description of the situation as we need in a particular context as an *elementary event*. Thus if we are concerned with the tossing of a red die and a blue die, then a typical elementary event is "red three, blue five", or if we are concerned with the numbers of Labour and Conservative MPs in the next parliament, a typical elementary event is "Labour 350, Conservative 250". Often, however, we want to talk about one aspect of the situation. Thus in the case of the first example we might be interested in whether or not we get a red three, which possibility includes "red three, blue one", "red three, blue two", etc. Similarly, in the other example we could be interested in whether there is a Labour majority of at least 100, which can also be analysed into elementary events. With this in mind an *event* is defined as a set of elementary events (this has the slightly curious consequence that, if you are *very* pedantic, an elementary event is not an event since it is an element rather than a set). We find it useful to say that one event E *implies* another event F if E is contained in F. Sometimes it is useful to generalize this by saying that, given H, E implies F if EH is contained in F. For example, given a red three has been thrown, throwing a blue three implies throwing an even total.

Note that the definition of an elementary event depends on the context. If we were never going to consider the blue die, then we could perfectly well treat events such as "red three" as elementary events. In a particular context, the elementary events in terms of which it is sensible to work are usually clear enough.

Events are referred to above as possible future occurrences, but they can also describe present circumstances, known or unknown. Indeed, the relationship which probability attempts to describe is one between what you currently know and something else about which you are uncertain, both of them being referred to as events. In other words, for at least some pairs of events E and H there is a number $\mathrm{P}(E|H)$ defined which is called the probability of the event E given the hypothesis H. I might, for example, talk of the probability of the event E that I throw a red three given the hypothesis H that I have rolled two fair dice once, or the probability of the event E of a Labour majority of at least 100 given the hypothesis H which consists of my knowledge of the political situation to date. Notice that in this context the term "hypothesis" can be applied to a large class of events, although later on we will find that in statistical arguments we are usually concerned with hypotheses which are more like the hypotheses in the ordinary meaning of the word.

Various attempts have been made to define the notion of probability. Many early writers claimed that $\mathrm{P}(E|H)$ was m/n where there were n symmetrical and so equally likely possibilities given H of which m resulted in the occurrence of E. Others have argued that $\mathrm{P}(E|H)$ should be taken as the long-run frequency with which E happens when H holds. These notions can help our intuition in some cases, but I think they are impossible to turn into precise, rigorous definitions. The difficulty with the first lies in finding genuinely "symmetrical" possibilities—for example, real dice are only approximately symmetrical. In any case there is a danger of circularity in the definitions of symmetry and probability. The difficulty with the second is that we never know how long we have to go on trying before we are within, say, 1% of the true value of the probability. Of course, we may be able to give a value for the number of trials we need to be within 1% of the true value with, say, probability 0.99, but this is leading to another vicious circle of definitions. Another difficulty is that sometimes we talk of the probability of events (for example, nuclear war in the next five years) about which it is hard to believe in a large number of trials, some resulting in "success" and some in "failure". A good, brief discussion is to be found in Nagel (1939).

It seems to me, and to an increasing number of statisticians, that the only satisfactory way of thinking of $\mathrm{P}(E|H)$ is as a measure of my degree of belief in E given that I know that H is true. It seems reasonable that this measure should abide by the following axioms:

P1 $\quad \mathrm{P}(E|H) \geqslant 0 \quad$ for all E, H.

P2 $\quad \mathrm{P}(H|H) = 1 \quad$ for all H.

P3 $\quad \mathrm{P}(E \cup F|H) = \mathrm{P}(E|H) + \mathrm{P}(F|H) \quad$ whenever $\quad EFH = \varnothing$.

P4 $\quad \mathrm{P}(E\,|\,FH)\mathrm{P}(F|H) = \mathrm{P}(EF|H)$.

By taking $F = H \backslash E$ in P3 and using P1 and P2, it easily follows that

$$\mathrm{P}(E|H) \leqslant 1 \qquad \text{for all } E, H$$

so that $\mathrm{P}(E|H)$ is always between 0 and 1. Also, by taking $F = \varnothing$ in P3 it follows that

$$\mathrm{P}(\varnothing|H) = 0$$

Now intuitive notions about probability always seem to agree that it should be a quantity between 0 and 1 which falls to 0 when we talk of the probability of something we are certain will not happen and rises to 1 when we are certain it will happen (and we are certain that H is true given H is true). Further, the additive property in P3 seems highly reasonable—we would, for example, expect the probability that the red die lands three or four should be the sum of the probability that it lands three and the probability that it lands four.

Axiom P4 may seem less familiar. It is sometimes written as

$$\mathrm{P}(E|FH) = \frac{\mathrm{P}(EF|H)}{\mathrm{P}(F|H)}$$

although of course this form cannot be used if the denominator (and hence the numerator) on the right-hand side vanishes. To see that it is a reasonable thing to assume, consider the following data on criminality among the twin brothers or sisters of criminals [quoted in the famous book by Fisher (1925b)]. The twins were classified according to whether they had a criminal conviction (C) or not (N) and according to whether they were monozygotic (M) (which is more or less the same as identical—we will return to this in the next section) or dizygotic (D), resulting in the following table:

	C	N	Total
M	10	3	13
D	2	15	17
Total	12	18	30

If we denote by H the knowledge that an individual has been picked at random from this population, then it seems reasonable to say that

$$\mathrm{P}(C \mid H) = 12/30,$$
$$\mathrm{P}(MC \mid H) = 10/30.$$

If on the other hand we consider an individual picked at random from among the twins with a criminal conviction in the population, we see that

$$\mathrm{P}(M \mid CH) = 10/12.$$

and hence

$$\mathrm{P}(M \mid CH)\,\mathrm{P}(C \mid H) = \mathrm{P}(MC \mid H)$$

so that P4 holds in this case. It is easy to see that this relationship does not depend on the particular numbers that happen to appear in the data.

In many ways, the argument in the preceding paragraph is related to derivations of probabilities from symmetry considerations, so perhaps it should be stressed that while in certain circumstances we may believe in symmetries or in equally probable cases, we cannot base a general definition of probability on such arguments.

It is convenient to use a stronger form of axiom P3 in many contexts, viz.

P3* $$\mathrm{P}\left(\bigcup_{n=1}^{\infty} E_n \mid H \right) = \sum_{n=1}^{\infty} \mathrm{P}(E_n \mid H)$$

whenever the (E_n) are exclusive events given H. There is no doubt of the mathematical simplifications that result from this assumption, but we are supposed to be modelling our degrees of belief and it is questionable whether these have to obey this form of the axiom. Indeed, one of the greatest advocates of Bayesian theory, Bruno de Finetti, was strongly against the use of P3*. His views can be found in de Finetti (1972, Section 5.32) or in de Finetti (1974–75, Section 3.11.3).

There is certainly some arbitrariness about P3*, which is sometimes referred to as an assumption of *σ-additivity*, in that it allows additivity over some but not all infinite collections of events (technically over countable but not over uncountable collections). However, it is impossible in a lot of contexts to allow additivity over *any* (arbitrary) collection of events. Thus if we want a model for picking a point "completely at random" from the unit interval

$$[0, 1] = \{x; 0 \leqslant x \leqslant 1\}$$

it seems reasonable that the probability that the point picked is in any particular sub-interval of the unit interval should equal the length of that sub-interval. However, this clearly implies that the probability of picking *any* one particular x is zero (since any such x belongs to intervals of arbitrarily small lengths). But the probability that *some* x is picked is unity, and it is impossible to get one by adding a lot of zeroes.

Mainly because of its mathematical convenience, we shall assume P3* while being aware of the problems.

"Unconditional" probability

Strictly speaking, there is, in my view, no such thing as an unconditional probability. However, it often happens that many probability statements are made conditional on everything that is part of an individual's knowledge at a particular time, and when many statements are to be made conditional on the same event it makes for cumbersome notation to refer to this same conditioning event every time. There are some cases where we have so much experimental data in circumstances judged to be relevant to a particular situation that there is a fairly general agreement as to the probability of an event. Thus in tossing a coin you and I both have experience of tossing similar coins many times and so are likely to believe that "heads" is approximately as likely as not, so that the probability of "heads" is approximately $\frac{1}{2}$ given your knowledge or mine.

In these cases we write

$$\mathrm{P}(E) \text{ for } \mathrm{P}(E \mid \Omega),$$
$$\mathrm{P}(E \mid F) \text{ for } \mathrm{P}(E \mid F\Omega)$$

where Ω is the set of possibilities consistent with the sum total of data available to the individual or individuals concerned. We usually consider sets F for which $F \subset \Omega$, so that $F\Omega = F$. It easily follows from the axioms that

$$0 \leqslant \mathrm{P}(E) \leqslant 1,$$
$$\mathrm{P}(\Omega) = 1, \qquad \mathrm{P}(\varnothing) = 0,$$
$$\mathrm{P}\left(\bigcup_{n=1}^{\infty} E_n\right) = \sum_{n=1}^{\infty} \mathrm{P}(E_n)$$

whenever the (E_n) are exclusive events (or more properly whenever they are exclusive events given Ω), and

$$\mathrm{P}(E|F)\mathrm{P}(F) = \mathrm{P}(EF).$$

Many books begin by asserting that unconditional probability is an intuitive notion and use the latter formula in the form

$$\mathrm{P}(E|F) = \mathrm{P}(EF)/\mathrm{P}(F) \qquad \text{(provided } P(F) \neq 0\text{)}$$

to define conditional probability.

Odds

It is sometimes convenient to use a language more familiar to bookmakers to express probabilistic statements. We define the *odds on E against F given H* as the ratio

$$\mathrm{P}(E|H)/\mathrm{P}(F|H) \qquad \text{to} \qquad 1$$

or equivalently

$$\mathrm{P}(E|H) \qquad \text{to} \qquad \mathrm{P}(F|H).$$

A reference to the odds on E against F with no mention of H is to be interpreted as a reference to the odds on E against F given Ω where Ω is some set of background knowledge as above.

Odds do not usually have properties as simple as probabilities, but sometimes, for example in connection with Bayesian tests of hypotheses, they are more natural to consider than separate probabilities.

Independence

Two events E and F are said to be *independent* given H if

$$\mathrm{P}(EF|H) = \mathrm{P}(E|H)P(F|H).$$

From axiom P4 it follows that if $\mathrm{P}(F|H) \neq 0$ this condition is equivalent to

$$\mathrm{P}(E|FH) = \mathrm{P}(E|H)$$

so that if E is independent of F given H then the extra information that

F is true does not alter the probability of E from that given H alone, and this gives the best intuitive idea as to what independence means. However, the restriction of this interpretation to the case where $P(F|H) \neq 0$ makes the original equation slightly more general.

More generally, a sequence (E_n) of events is said to be *pairwise independent* given H if

$$P(E_m E_n | H) = P(E_m | H) P(E_n | H) \qquad \text{for } m \neq n$$

and is said to consist of *mutually independent* events given H if for every finite subset of them

$$P(E_{n_1} E_{n_2} \ldots E_{n_k} | H) = P(E_{n_1} | H) P(E_{n_2} | H) \ldots P(E_{n_k} | H).$$

We must note that pairwise independence does not imply mutual independence and that

$$P(E_1 E_2 \ldots E_n | H) = P(E_1 | H) P(E_2 | H) \ldots P(E_n | H)$$

is not enough to ensure that the finite sequence $E_1, E_2, \ldots, E_n$ consists of mutually independent events given H.

Naturally, if no conditioning event is explicitly mentioned, the probabilities concerned are conditional on Ω as defined above.

Some simple consequences of the axioms; Bayes' Theorem

We have already seen a few consequences of the axioms, but it is useful at this point to note a few more. We first note that it follows simply from P4 and P2 and the fact that $HH = H$ that

$$P(E|H) = P(EH|H)$$

and in particular

$$P(E) = P(E\Omega).$$

Next we note that if, given H, E implies F, that is, $EH \subset F$ and so $EFH = EH$, then by P4 and the above

$$P(E|FH) P(F|H) = P(E|H).$$

From this and the fact that $P(E|FH) \leqslant 1$ it follows that if, given H, E implies F, then

$$P(E|H) \leqslant P(F|H).$$

In particular if E implies F then

$$P(E|F)\, P(F) = P(E),$$
$$P(E) \leqslant P(F).$$

For the rest of this subsection we can work in terms of "unconditional"probabilities, although the results are easily generalized. Let (H_n) be a sequence of exclusive and exhaustive events, and let E be any event. Then

$$\mathrm{P}(E) = \sum_n \mathrm{P}(E|H_n)\mathrm{P}(H_n)$$

since by P4 the terms on the right-hand side are $\mathrm{P}(EH_n)$, allowing us to deduce the result from P3*. This result is sometimes called the *generalized addition law*.

The key result in the whole book is *Bayes' Theorem*. This is simply deduced as follows. Let (H_n) be a sequence of events. Then by P4

$$\mathrm{P}(H_n|E)\mathrm{P}(E) = \mathrm{P}(EH_n) = \mathrm{P}(H_n)\mathrm{P}(E|H_n)$$

so that provided $\mathrm{P}(E) \neq 0$

$$\mathrm{P}(H_n|E) \propto \mathrm{P}(H_n)\mathrm{P}(E|H_n).$$

This relationship is one of several ways of stating Bayes' Theorem, and is probably the best way in which to remember it. When we need the constant of proportionality, we can easily see from the above that it is $1/\mathrm{P}(E)$.

It should be clearly understood that there is nothing controversial about Bayes' Theorem as such. It is frequently used by probabilists and statisticians, whether or not they are Bayesians. The distinctive feature of Bayesian statistics is the application of the theorem in a wider range of circumstances than is usual in classical statistics. In particular, Bayesian statisticians are always willing to talk of the probability of a hypothesis, both unconditionally (its *prior probability*) and given some evidence (its *posterior probability*), whereas other statisticians will only talk of the probability of a hypothesis in restricted circumstances.

When (H_n) consists of exclusive and exhaustive events we can combine the last two results to see that

$$\mathrm{P}(H_n|E) = \frac{\mathrm{P}(H_n)\mathrm{P}(E|H_n)}{\sum_m \mathrm{P}(H_m)\mathrm{P}(E|H_m)}.$$

A final result we will find useful from time to time is the *generalized multiplication law*, which runs as follows. If $H_1, H_2, \ldots, H_n$ are any events then

$$\mathrm{P}(H_1H_2 \ldots H_n) = \mathrm{P}(H_1)\mathrm{P}(H_2|H_1)\mathrm{P}(H_3|H_1H_2) \ldots \mathrm{P}(H_n|H_1H_2 \ldots H_{n-1})$$

provided all the requisite conditional probabilities are defined, which in practice they will be provided $P(H_1H_2 \ldots H_{n-1}) \neq 0$. This result is easily proved by repeated application of P4.

1.2 Examples on Bayes' Theorem

The biology of twins

Twins can be either monozygotic (M) (that is, developed from a single egg) or dizygotic (D). Monozygotic twins often look very similar and then are referred to as identical twins, but it is not always the case that one finds very striking similarities between monozygotic twins, while some dizygotic twins can show marked resemblances. Whether twins are monozygotic or dizygotic is not, therefore, a matter which can be settled simply by inspection. However, it is *always* the case that monozygotic twins are of the same sex, whereas dizygotic twins can be of opposite sex. Hence if the sexes of a pair of twins are denoted GG, BB or GB (note GB is indistinguishable from BG)

$$P(GG|M) = P(BB|M) = \tfrac{1}{2}, \qquad P(GB|M) = 0,$$
$$P(GG|D) = P(BB|D) = \tfrac{1}{4}, \qquad P(GB|D) = \tfrac{1}{2}.$$

It follows that

$$\begin{aligned} P(GG) &= P(GG|M)P(M) + P(GG|D)P(D) \\ &= \tfrac{1}{2}P(M) + \tfrac{1}{4}\{1 - P(M)\} \end{aligned}$$

from which it can be seen that

$$P(M) = 4P(GG) - 1$$

so that although it is not easy to be certain whether a particular pair are monozygotic or not, it is easy to discover the *proportion* of monozygotic twins in the whole population of twins simply by observing the sex distribution among *all* twins.

A political example

The following example is a simplified version of the situation just before the time of the British national referendum as to whether the United Kingdom should remain part of the European Economic Community which was held in 1975. Suppose that at that date, which was shortly after an election which the Labour Party had won, the proportion of the electorate supporting Labour (L) stood at 52%, while the proportion supporting the Conservatives (C) stood at 48% (it being assumed for simplicity that support for all other parties was negligible, although this was far from being the case). There were many opinion polls taken at the time, so we can take it as known that 55% of Labour supporters and 85% of Conservative voters intended to vote "Yes" (Y) and the remainder intended to vote "No" (N). Suppose that knowing all this you met someone at the time who said that she intended to vote "Yes", and you were interested in knowing which political party she supported. If this information were all you had available, you could reason as follows.

$$\begin{aligned} \mathrm{P}(L \mid Y) &= \frac{\mathrm{P}(Y \mid L)\mathrm{P}(L)}{\mathrm{P}(Y \mid L)\mathrm{P}(L) + \mathrm{P}(Y \mid C)\mathrm{P}(C)} \\ &= \frac{(0.55)(0.52)}{(0.55)(0.52) + (0.85)(0.48)} = 0.41. \end{aligned}$$

A warning

In the case of Connecticut v. Teal [see DeGroot *et al.* (1986, p. 9)], a case of alleged discrimination on the basis of a test to determine eligibility for promotion was considered. It turned out that of those taking the test 48 were black (B) and 259 were white (W), so that if we consider a random person taking the test

$$\mathrm{P}(B) = 48/307 = 0.16, \qquad \mathrm{P}(W) = 259/307 = 0.84.$$

Of the blacks taking the test, 26 passed (P) and the rest failed (F), whereas of the whites, 206 passed and the rest failed, so that altogether 232 people passed. Hence

$$\mathrm{P}(B \mid P) = 26/232 = 0.11, \qquad \mathrm{P}(W \mid P) = 206/232 = 0.89.$$

There is a temptation to think that these are the figures which indicate the possibility of discrimination. Now there certainly is an argument for saying that there was discrimination in this case, *but* the figures that should be considered are

$$\mathrm{P}(P \mid B) = 26/48 = 0.54, \qquad \mathrm{P}(P \mid W) = 206/259 = 0.80.$$

It is easily checked that the probabilities here are related by Bayes' Theorem. It is worth while spending some time playing with hypothetical figures to convince oneself that the fact that $\mathrm{P}(B \mid P) < \mathrm{P}(W \mid P)$ is irrelevant to the real question as to whether $\mathrm{P}(P \mid B) < \mathrm{P}(P \mid W)$—it might or might not be, depending on the rest of the relevant information, that is, on $\mathrm{P}(B)$ and $\mathrm{P}(W)$.

1.3 Random variables

Discrete random variables

As explained in Section 1.1, there is usually a set Ω representing the possibilities consistent with the sum total of data available to the individual or individuals concerned. We now suppose that with each elementary event ω in Ω there is an integer $\tilde{m}(\omega)$ which may be positive, negative or zero. In the jargon of mathematics, we have a function $\tilde{m}$ mapping Ω to the set Z of all (signed) integers. We refer to the function as a *random variable* or an *r.v.*

An example in terms of the very first example we discussed, which was about tossing a red die and a blue die, is the integer representing the sum of the spots showing. In this case, ω might be "red three, blue two" and then $\tilde{m}(\omega)$ would be 5. Another example in terms of the second (political) example is the Labour majority (represented as a negative integer should the Conservatives happen to win), and here ω might be "Labour 350, Conservative 250", in which case $\tilde{m}(\omega)$ would be 100.

Rather naughtily, probabilists and statisticians tend not to mention the elementary event ω of which $\tilde{m}(\omega)$ is a function and instead just write $\tilde{m}$ for $\tilde{m}(\omega)$. The reason is that what usually matters is the value of $\tilde{m}$ rather than the nature of the elementary event ω, the definition of which is in any case dependent on the context, as noted earlier, in the discussion of elementary events. Thus we write

$$\mathrm{P}(\tilde{m} = m) \quad \text{for} \quad \mathrm{P}(\{\omega;\, \tilde{m}(\omega) = m\})$$

that is, for the probability that the random variable $\tilde{m}$ takes the particular value m. It is a useful convention to use the same letter without the accent to denote a particular value of the random variable which is denoted by the same letter with a tilde ($\tilde{}$) on top. The use of a tilde over a letter to denote a random variable should be distinguished from the use of a bar over a letter to denote an arithmetic mean. An alternative convention used by some statisticians is to use capital letters for random variables and corresponding lower-case letters for typical values of these random variables, but in a Bayesian context we have so many quantities that are regarded as random variables that this convention is too restrictive. Even worse than the habit of dropping mention of ω is the tendency to omit the tilde and so use the same notation for a random variable and for a typical value of it. While failure to mention ω rarely causes any confusion, the failure to distinguish between random variables and typical values of these random variables can, on occasion, result in real confusion. When there is any possibility of confusion, the tilde will be used in the text, but otherwise it will be omitted. Also, we will use

$$p(m) = \mathrm{P}(\tilde{m} = m) = \mathrm{P}(\{\omega;\, \tilde{m}(\omega) = m\})$$

for the probability that the random variable $\tilde{m}$ takes the value m. When we are talking about only one random variable, this abbreviation presents few problems, but when we have a second random variable $\tilde{n}$ and write

$$p(n) = \mathrm{P}(\tilde{n} = n) = \mathrm{P}(\{\omega; \tilde{n}(\omega) = n\})$$

then ambiguity can result. It is not clear in such a case what $p(5)$ would mean, or indeed what $p(i)$ would mean (unless it refers to $\mathrm{P}(\tilde{\imath} = i)$ where $\tilde{\imath}$ is yet a third random variable). When it is necessary to resolve such an ambiguity we will use

$$p_{\tilde{m}}(m) = \mathrm{P}(\tilde{m} = m) = \mathrm{P}(\{\omega; \tilde{m}(\omega) = m\})$$

so that, for example, $p_{\tilde{m}}(5)$ is the probability that $\tilde{m}$ is 5 and $p_{\tilde{n}}(i)$ is the probability that $\tilde{n}$ equals i. Again, all of this *seems* very much more confusing than it really is—it is usually possible to conduct arguments quite happily in terms of $p(m)$ and $p(n)$ and substitute numerical values at the end if and when necessary.

You could well object that you would prefer a notation that was free of ambiguity, and if you were to do so I should have a lot of sympathy. But the fact is that constant references to $\tilde{m}(\omega)$ and $p_{\tilde{m}}(m)$ rather than to m and $p(m)$ would clutter the page and be unhelpful in another way.

We refer to the sequence $(p(m))$ as the *(probability) density (function)* or *p.d.f.* of the random variable m (strictly $\tilde{m}$). The random variable is said to have a distribution (of probability) and one way of describing a distribution is by its p.d.f. Another is by its *(cumulative) distribution function*, or *c.d.f.* or *d.f.*, defined by

$$\begin{aligned} F(m) = F_{\tilde{m}}(m) &= \mathrm{P}(\tilde{m} \leqslant m) = \mathrm{P}(\{\omega; \tilde{m}(\omega) \leqslant m\}) \\ &= \sum_{k \leqslant m} p_{\tilde{m}}(k). \end{aligned}$$

Because the p.d.f. has the obvious properties

$$p(m) \geqslant 0, \qquad \sum_m p(m) = 1$$

the d.f. is (weakly) increasing, that is,

$$F(m) \leqslant F(m') \quad \text{if} \quad m \leqslant m'$$

and moreover

$$\lim_{m \to -\infty} F(m) = 0, \qquad \lim_{m \to \infty} F(m) = 1.$$

The binomial distribution

A simple example of such a distribution is the *binomial distribution* (see the Appendix on "Common Statistical Distributions"). Suppose we have a sequence of trials each of which, independently of the others, results in success (S) or failure (F), the probability of success being a constant π (such trials are sometimes called Bernoulli trials). Then the probability of any particular sequence of n trials in which k result in success is

$$\pi^k(1-\pi)^{n-k}$$

so that allowing for the $\binom{n}{k}$ ways in which k successes and $n-k$ failures can be ordered, the probability that a sequence of n trials results in k successes is

$$\binom{n}{k}\pi^k(1-\pi)^{n-k} \qquad (0\leqslant k\leqslant n).$$

If then k (strictly $\tilde{k}$) is a random variable defined as the number of successes in n trials, then

$$p(k)=\binom{n}{k}\pi^k(1-\pi)^{n-k}.$$

Such a distribution is said to be binomial of index n and parameter π, and we write

$$k\sim \mathrm{B}(n,\pi)$$

[or strictly $\tilde{k}\sim \mathrm{B}(n,\pi)$].

Continuous random variables

So far we have restricted ourselves to random variables which take only integer values. These are particular cases of *discrete* random variables. Other examples of discrete random variables occur (for example, a measurement to the nearest quarter-inch which is subject to a distribution of error), but these can nearly always be changed to integer-valued random variables (in the given example simply by multiplying by 4). More generally we can suppose that with each elementary event ω in Ω there is a real number $\tilde{x}(\omega)$. We can define the (cumulative) distribution, c.d.f. or d.f. of $\tilde{x}$ by

$$F(x)=\mathrm{P}(\tilde{x}\leqslant x)=\mathrm{P}(\{\omega;\,\tilde{x}(\omega)\leqslant x\}).$$

As in the discrete case the d.f. is (weakly) increasing, i.e.

$$F(x)\leqslant F(x') \quad \text{if} \quad x\leqslant x'$$

and moreover

$$\lim_{x \to -\infty} F(x) = 0, \qquad \lim_{x \to \infty} F(x) = 1.$$

It is usually the case that when $\tilde{x}$ is not discrete there exists a function $p(x)$, or more strictly $p_{\tilde{x}}(x)$, such that

$$F(x) = \int_{-\infty}^{x} p_{\tilde{x}}(\xi)\,\mathrm{d}\xi$$

in which case $p(x)$ is called a (probability) density (function) or p.d.f. When this is so, x (strictly $\tilde{x}$) is said to have a *continuous* distribution (or more strictly an absolutely continuous distribution). Of course, in the continuous case $p(x)$ is not itself interpretable directly as a probability, but for small δx

$$p(x)\delta x \cong \mathrm{P}(x < \tilde{x} \leqslant x + \delta x) = \mathrm{P}(\{\omega; x < \tilde{x}(\omega) \leqslant x + \delta x\}).$$

The quantity $p(x)\delta x$ is sometimes referred to as the *probability element.* Note that letting $\delta x \to 0$ this implies that

$$\mathrm{P}(\tilde{x} = x) = 0$$

for every particular value x, in sharp contrast to the discrete case. We can also use the above approximation if y is some strictly increasing function of x, for example,

$$y = g(x).$$

Then if values correspond in an obvious way

$$\mathrm{P}(y < \tilde{y} \leqslant y + \delta y) = \mathrm{P}(x < \tilde{x} \leqslant x + \delta x)$$

which on substituting in the above relationship gives in the limit

$$p(y)\,|\mathrm{d}y| = p(x)\,|\mathrm{d}x|$$

which is the rule for *change of variable* in probability densities. (It is not difficult to see that, because the modulus signs are there, the same result is true if F is a strictly decreasing function of x). Another way of obtaining this rule is by differentiating the obvious equation

$$F(y) = F(x)$$

[strictly $F_{\tilde{y}}(y) = F_{\tilde{x}}(x)$] which holds whenever y and x are corresponding values, that is, $y = g(x)$. We should, however, beware that these results need modification if g is *not* a one-to-one function.

In the continuous case, we can find the density from the d.f. by differentiation, namely,

$$p(x) = F'(x) = \mathrm{d}F(x)/\mathrm{d}x.$$

Although there are differences, there are many similarities between the discrete and the continuous cases, which we try to emphasize by using the same notation in both cases. We note that

$$F(x) = \sum_{\xi \leq x} p_{\tilde{x}}(\xi)$$

in the discrete case, but

$$F(x) = \int_{-\infty}^{x} p_{\tilde{x}}(\xi)\,\mathrm{d}\xi$$

in the continuous case. The discrete case is slightly simpler in one way, in that no complications arise over change of variable, so that

$$p(y) = p(x)$$

if y and x are corresponding values, that is, $y = g(x)$.

The normal distribution

The most important example of a continuous distribution is the so-called *normal* or Gaussian distribution. We say that z has a *standard normal* distribution if

$$p(z) = (2\pi)^{-\frac{1}{2}} \exp\left(-\tfrac{1}{2}z^2\right)$$

and when this is so we write

$$z \sim \mathrm{N}(0, 1).$$

The density of this distribution is the familiar bell-shaped curve, with about two-thirds of the area between -1 and 1, 95% of the area between -2 and 2, and almost all of it between -3 and 3. Its distribution function is

$$\Phi(z) = \int_{-\infty}^{z} (2\pi)^{-\frac{1}{2}} \exp\left(-\tfrac{1}{2}\zeta^2\right)\mathrm{d}\zeta$$

More generally, we say that x has a normal distribution, denoted

$$x \sim \mathrm{N}(\mu, \varphi)$$

if

$$x = \mu + z\sqrt{\varphi}$$

where z is as above, or equivalently if

$$p(x) = (2\pi\varphi)^{-\frac{1}{2}} \exp\{-\tfrac{1}{2}(x-\mu)^2/\varphi\}.$$

The normal distribution is encountered almost at every turn in statistics. Partly this is because (despite the fact that its density may seem somewhat barbaric at first sight) it is in many contexts the easiest distribution to work with, but this is not the whole story. The *central limit theorem* says (roughly) that if a random variable can be expressed as a sum of a large number of components no one of which is likely to be much bigger than the others, these components being approximately independent, then this sum will be approximately normally distributed. Because of this theorem, an observation which has an error contributed to by many minor causes is likely to be normally distributed. Similar reasons can be found for thinking that in many circumstances we would expect observations to be approximately normally distributed, and this turns out to be the case, although there are exceptions.

Mixed random variables

While most random variables are discrete or continuous, there are exceptions, for example, the time one has to wait in a queue until being served, which is zero with a positive probability (if the queue is empty when one arrives), but otherwise is spread over a continuous range of values. Such a random variable is said to have a *mixed* distribution.

1.4 Several random variables

Two discrete random variables

Suppose that with each elementary event ω in Ω we can associate a pair $(\tilde{m}(\omega), \tilde{n}(\omega))$ of integers. We write

$$p(m, n) = \mathrm{P}(\tilde{m} = m, \tilde{n} = n) = \mathrm{P}(\{\omega; \tilde{m}(\omega) = m, \tilde{n}(\omega) = n\}).$$

Strictly speaking, $p(m, n)$ should be written as $p_{\tilde{m},\tilde{n}}(m, n)$ for reasons discussed earlier, but this degree of pedantry in the notation is rarely necessary. Clearly

$$p(m, n) \geqslant 0, \qquad \sum_m \sum_n p(m, n) = 1.$$

The sequence $(p(m, n))$ is said to be a *bivariate (probability) density (function)* or *bivariate p.d.f.* and is called the *joint p.d.f.* of the random variables m and n (strictly $\tilde{m}$ and $\tilde{n}$). The corresponding *joint distribution function*, *joint c.d.f.* or *joint d.f.* is

$$F(m, n) = \sum_{k \leqslant m} \sum_{l \leqslant n} p_{\tilde{m}, \tilde{n}}(k, l).$$

Clearly the density of m (called its *marginal density*) is

$$p(m) = \sum_n p(m, n).$$

We can also define a conditional distribution for n given m (strictly for $\tilde{n}$ given $\tilde{m} = m$) by

$$p(n \mid m) = \mathrm{P}(\tilde{n} = n \mid \tilde{m} = m) = P(\{\omega; \tilde{n}(\omega) = n\} \mid \{\omega; \tilde{m}(\omega) = m\})$$
$$= p(m, n)/p(m)$$

to define the *conditional (probability) density (function)* or *conditional p.d.f.* This represents our judgement as to the chance that $\tilde{n}$ takes the value n given that $\tilde{m}$ is known to have the value m. If it is necessary to make our notation absolutely precise, we can always write

$$p_{\tilde{m}|\tilde{n}}(m \mid n)$$

so, for example, $p_{\tilde{m}|\tilde{n}}(4 \mid 3)$ is the probability that $\tilde{m}$ is 4 given $\tilde{n}$ is 3, but $p_{\tilde{n}|\tilde{m}}(4 \mid 3)$ is the probability that $\tilde{n}$ is 4 given that $\tilde{m}$ is 3, but it should be emphasized that we will not often need to use the subscripts. Evidently

$$p(n \mid m) \geqslant 0, \qquad \sum p(n \mid m) = 1$$

and

$$p(n) = \sum p(m, n) = \sum p(m)\, p(n \mid m).$$

We can also define a *conditional distribution function* or *conditional d.f.* by

$$F(n \mid m) = \mathrm{P}(\tilde{n} \leqslant n \mid \tilde{m} \leqslant m) = \mathrm{P}(\{\omega; \tilde{n}(\omega) \leqslant n\} \mid \{\omega; \tilde{m}(\omega) \leqslant m\})$$

$$= \sum_{k \leqslant n} p_{\tilde{n}|\tilde{m}}(k \mid m).$$

Two continuous random variables

As in the previous section, we have begun by restricting ourselves to integer values, which is more or less enough to deal with any discrete cases that arise. More generally, we can suppose that with each elementary event ω in Ω we can associate a pair $(\tilde{x}(\omega), \tilde{y}(\omega))$ of real numbers. In this case we define the *joint distribution function* or *joint d.f.* as

$$F(x, y) = \mathrm{P}(\tilde{x} \leqslant x, \tilde{y} \leqslant y) = \mathrm{P}(\{\omega; \tilde{x}(\omega) \leqslant x, \tilde{y}(\omega) \leqslant y\}).$$

Clearly the d.f. of x is

$$F(x, +\infty)$$

and that of y is

$$F(+\infty, y).$$

It is usually the case that when neither x nor y is discrete there is a function $p(x, y)$, or more strictly $F_{\tilde{x}, \tilde{y}}(x, y)$, such that

$$F(x, y) = \int_{-\infty}^{x} \int_{-\infty}^{y} p_{\tilde{x}, \tilde{y}}(\xi, \eta)\, \mathrm{d}\xi\, \mathrm{d}\eta$$

in which case $p(x, y)$ is called a *joint (probability) density (function)* or *joint p.d.f.* When this is so, the joint distribution is said to be *continuous* (or more strictly to be absolutely continuous). We can find the density from the d.f. by

$$p(x, y) = \partial^2 F(x, y) / \partial x\, \partial y.$$

Clearly

$$p(x, y) \geqslant 0, \qquad \int\int p(x, y)\, \mathrm{d}x\, \mathrm{d}y = 1$$

and

$$p(x) = \int p(x, y)\, \mathrm{d}y.$$

The last formula is the continuous analogue of

$$p(m) = \sum_{n} p(m, n)$$

in the discrete case.

By analogy with the discrete case, we define the *conditional density* of y given x (strictly of $\tilde{y}$ given $\tilde{x} = x$) as

$$p(y \mid x) = p(x, y) / p(x)$$

provided $p(x) \neq 0$. We can then define the *conditional distribution function* by

$$F(y \mid x) = \int_{-\infty}^{y} p(\eta \mid x)\, \mathrm{d}\eta.$$

There are difficulties in the notion of conditioning on the event that $\tilde{x} = x$ because this event has probability zero for every x in the continuous case, and it can help to regard the above distribution as in the limit of the distribution which results from conditioning on the event that $\tilde{x}$ is between x and $x + \delta x$, i.e.

$$\{\omega;\, x < \tilde{x}(\omega) \leqslant x + \delta x\}$$

as $\delta x \to 0$.

Bayes' Theorem for random variables

It is worth noting that conditioning the random variable y by the value of x does not change the *relative* sizes of the probabilities of those pairs (x, y) that can still occur. That is to say, the probability $p(y|x)$ is proportional to $p(x,y)$ and the constant of proportionality is just what is needed so that the conditional probabilities integrate to unity. Thus

$$p(y|x) \geqslant 0, \qquad \int p(y|x)\,\mathrm{d}y = 1.$$

Moreover

$$p(y) = \int p(x, y)\,\mathrm{d}x = \int p(x)p(y|x)\,\mathrm{d}x.$$

It is clear that

$$p(y|x) = p(x, y)/p(x) = p(y)p(x|y)/p(x)$$

so that

$$p(y|x) \propto p(y)p(x|y).$$

This is of course a form of *Bayes' Theorem*, and is in fact the commonest way in which it occurs in this book. We note that it applies equally well if the variables x and y are continuous or if they are discrete. The constant of proportionality is

$$1/p(x) = 1/\int p(y)p(x|y)\,\mathrm{d}y$$

in the continuous case or

$$1/p(x) = 1/\sum_y p(y)p(x|y)$$

in the discrete case.

Example

A somewhat artificial example of the use of this formula in the continuous case is as follows. Suppose y is the time before the first occurrence of a radioactive decay which is measured by an instrument, but that, because there is a delay built into the mechanism, the decay is recorded as having taken place at a time $x>y$. We actually have the value of x, but would like to say what we can about the value of y on the basis of this knowledge. We might, for example, have

$$\begin{aligned} p(y) &= \mathrm{e}^{-y} && (0<y<\infty), \\ p(x\,|\,y) &= k\mathrm{e}^{-k(x-y)} && (y<x<\infty). \end{aligned}$$

Then

$$\begin{aligned} p(y\,|\,x) &\propto p(y)p(x\,|\,y) \\ &\propto \mathrm{e}^{(k-1)y} && (0<y<x). \end{aligned}$$

Often we will find that it is enough to get a result up to a constant of proportionality, but if we need the constant it is very easy to find it because we know that the integral (or the sum in the discrete case) must be one. Thus in this case

$$p(y\,|\,x) = \frac{(k-1)\,\mathrm{e}^{(k-1)y}}{\mathrm{e}^{(k-1)x}-1} \qquad (0<y<x).$$

One discrete variable and one continuous variable

We also encounter cases where we have two random variables, one of which is continuous and one of which is discrete. All the above definitions and formulae extend in an obvious way to such a case provided we are careful, for example, to use integration for continuous variables but summation for discrete variables. In particular the formulation

$$p(y\,|\,x) \propto p(y)p(x\,|\,y)$$

for Bayes' Theorem is valid in such a case.

It may help to consider an example (again a somewhat artificial one). Suppose k is the number of successes in n Bernoulli trials, so $k\sim \mathrm{B}(n,\pi)$, but that the value of π is unknown, your beliefs about it being uniformly distributed over the interval $[0, 1]$ of possible values. Then

$$\begin{aligned} p(k\,|\,\pi) &= \tbinom{n}{k}\pi^k(1-\pi)^{n-k} && (k=0, 1, \dots, n), \\ p(\pi) &= 1 && (0\leqslant\pi\leqslant 1). \end{aligned}$$

so that

$$\begin{aligned} p(\pi\,|\,k) = p(\pi)p(k\,|\,\pi) &\propto \tbinom{n}{k}\pi^k(1-\pi)^{n-k} \\ &\propto \pi^k(1-\pi)^{n-k}. \end{aligned}$$

The constant can be found by integration if it is required. Alternatively, a glance at the Appendix will show that, given k, π has a beta distribution

$$\pi \mid k \sim \mathrm{Be}(k+1, n-k+1)$$

and that the constant of proportionality is the reciprocal of the beta function $\mathrm{B}(k+1, n-k+1)$. Thus this beta distribution should represent your beliefs about π *after* you have observed k successes in n trials. This example has a special importance in that it is the one which Bayes himself discussed.

Independent random variables

The idea of independence extends from independence of events to independence of random variables. The basic idea is that y is *independent* of x if being told that x has any particular value does not affect your beliefs about the value of y. Because of complications involving events of probability zero, it is best to adopt the formal definition that x and y are independent if

$$p(x, y) = p(x)p(y)$$

for all values x and y. This definition works equally well in the discrete and the continuous cases (and indeed in the case where one random variable is continuous and the other is discrete). It trivially suffices that $p(x, y)$ be a product of a function of x and a function of y.

All the above generalizes in a fairly obvious way to the case of *more* than two random variables, and the notions of pairwise and mutual independence go through from events to random variables easily enough. However, we will find that we do not often need such generalizations.

1.5 Means and variances

Expectations

Suppose that m is a discrete random variable and that the series

$$\sum mp(m)$$

is absolutely convergent. Then its sum is called the *mean* or *expectation* of the random variable, and we denote it

$$\mathsf{E}m \qquad \text{(strictly } \mathsf{E}\tilde{m}\text{)}.$$

A motivation for this definition is as follows. In a large number N of trials we would expect the value m to occur about $p(m)N$ times, so that the sum total of the values that would occur in these N trials (counted according to their multiplicity) would be about

$$\sum mp(m)N$$

so that the average value should be about

$$\sum mp(m)N/N = \mathsf{E}m.$$

Thus we can think of expectation as being, at least in some circumstances, a form of very long term average. On the other hand, there are circumstances in which it is difficult to believe in the possibility of arbitrarily large numbers of trials, so that interpretation is not always available. It can also be thought of as giving the position of the "centre of gravity" of the distribution imagined as a distribution of mass spread along the x-axis.

More generally, if $g(m)$ is a function of the random variable and the series $\sum g(m)p(m)$ is absolutely convergent, then its sum is the expectation of $g(m)$. Similarly, if $h(m, n)$ is a function of two random variables m and n and the series $\sum\sum h(m, n)p(m, n)$ is absolutely convergent, then its sum is the expectation of $h(m, n)$. These definitions are consistent in that if we consider $g(m)$ and $h(m, n)$ as random variables with densities of their own, then it is easily shown that we get these values for their expectations.

In the continuous case we define the expectation of a random variable x by

$$\mathsf{E}x = \int xp(x)\,\mathrm{d}x$$

provided that the integral is absolutely convergent, and more generally define the expectation of a function $g(x)$ of x by

$$\mathsf{E}g(x) = \int g(x)p(x)\,\mathrm{d}x$$

provided that the integral is absolutely convergent, and similarly for the expectation of a function $h(x, y)$ of two random variables. Note that the formulae in the discrete and continuous cases are, as usual, identical except for the use of summation in the one case and integration in the other.

The expectation of a sum and of a product

If x and y are any two random variables, independent or not, and a, b and c are constants, then in the continuous case

$$\begin{aligned}\mathsf{E}\{ax+by+c\} &= \int\int(ax+by+c)p(x, y)\,\mathrm{d}x\,\mathrm{d}y\\ &= a\int\int xp(x, y)\,\mathrm{d}x\,\mathrm{d}y + b\int\int yp(x, y)\,\mathrm{d}x\,\mathrm{d}y + c\int\int p(x, y)\,\mathrm{d}x\,\mathrm{d}y\\ &= a\int xp(x)\,\mathrm{d}x + b\int yp(y)\,\mathrm{d}y + c\\ &= a\mathsf{E}x + b\mathsf{E}y + c\end{aligned}$$

and similarly in the discrete case. Yet more generally, if $g(x)$ is a function of x and $h(y)$ a function of y, then

$$\mathsf{E}\{ag(x)+bh(y)+c\} = a\mathsf{E}g(x)+b\mathsf{E}h(y)+c.$$

We have already noted that the idea of independence is closely tied up with multiplication, and this is true when it comes to expectations as well. Thus if x and y are independent, then

$$\begin{aligned}\mathsf{E}xy &= \int\int xyp(x,y)\,\mathrm{d}x\,\mathrm{d}y\\ &= \int\int xyp(x)p(y)\,\mathrm{d}x\,\mathrm{d}y\\ &= \left(\int xp(x)\,\mathrm{d}x\right)\left(\int yp(y)\,\mathrm{d}y\right)\\ &= (\mathsf{E}x)(\mathsf{E}y)\end{aligned}$$

and more generally if $g(x)$ and $h(y)$ are functions of independent random variables x and y, then

$$\mathsf{E}g(x)h(y) = (\mathsf{E}g(x))(\mathsf{E}h(y)).$$

Variance, precision and standard deviation

We often need a measure of how spread out a distribution is, and for most purposes the most useful such measure is the *variance* $\mathcal{V}x$ of x, defined by

$$\mathcal{V}x = \mathsf{E}(x-\mathsf{E}x)^2.$$

Clearly if the distribution is very little spread out, then most values are close to one another and, thus, close to their mean, so that $(x-\mathsf{E}x)^2$ is small with high probability and hence $\mathcal{V}x$ is small. Conversely if the distribution is well spread out then $\mathcal{V}x$ is large. It is sometimes useful to refer to the reciprocal of the variance, which is called the *precision*. Further, because the variance is essentially quadratic, we sometimes work in terms of its positive square root, the *standard deviation*, especially in numerical work. It is often useful that

$$\begin{aligned}\mathcal{V}x &= \mathsf{E}(x-\mathsf{E}x)^2\\ &= \mathsf{E}(x^2-2(\mathsf{E}x)x+(\mathsf{E}x)^2)\\ &= \mathsf{E}x^2-(\mathsf{E}x)^2.\end{aligned}$$

The notion of a variance is analogous to that of a moment of inertia in mechanics, and this formula corresponds to the *parallel axes theorem* in mechanics. This analogy seldom carries much weight nowadays, because so many of those studying statistics took it up with the purpose of avoiding mechanics!

In discrete cases it is sometimes useful that

$$\mathcal{V}x = \mathsf{E}x(x-1) + \mathsf{E}x - (\mathsf{E}x)^2.$$

Examples

As an example, suppose that $k \sim \mathrm{B}(n, \pi)$. Then

$$\mathsf{E}k = \sum_{k=0}^{n} k\binom{n}{k}\pi^k(1-\pi)^{n-k}.$$

After a little manipulation, this can be expressed as

$$\mathsf{E}k = n\pi \sum_{j=0}^{n-1} \binom{n-1}{j}\pi^j(1-\pi)^{n-1-j}.$$

Because the sum is a sum of binomial $\mathrm{B}(n-1, \pi)$ probabilities, this expression reduces to $n\pi$, and so

$$\mathsf{E}k = n\pi.$$

Similarly

$$\mathsf{E}k(k-1) = n(n-1)\pi^2$$

and so

$$\begin{aligned}\mathcal{V}k &= n(n-1)\pi^2 + n\pi - (n\pi)^2 \\ &= n\pi(1-\pi).\end{aligned}$$

For a second example, suppose $x \sim \mathrm{N}(\mu, \phi)$. Then

$$\begin{aligned}\mathsf{E}x &= \int x(2\pi\phi)^{-\frac{1}{2}} \exp\{-\tfrac{1}{2}(x-\mu)^2/\phi\}\mathrm{d}x \\ &= \mu + \int (x-\mu)(2\pi\phi)^{-\frac{1}{2}} \exp\{-\tfrac{1}{2}(x-\mu)^2/\phi\}\mathrm{d}x.\end{aligned}$$

The integrand in the last expression is an odd function of $x-\mu$ and so the integral vanishes, so that

$$\mathsf{E}x = \mu.$$

Moreover

$$\mathcal{V}x = \int (x-\mu)^2(2\pi\phi)^{-\frac{1}{2}} \exp\{-\tfrac{1}{2}(x-\mu)^2/\phi\}\,\mathrm{d}x$$

so that on writing $z = (x-\mu)/\sqrt{\phi}$

$$\mathcal{V}x = \phi\int z^2(2\pi)^{-\frac{1}{2}} \exp(-\tfrac{1}{2}z^2)\,\mathrm{d}z.$$

Integrating by parts (using z as the part to differentiate) we get

$$\begin{aligned}\mathcal{V}x &= \phi\int (2\pi)^{-\frac{1}{2}} \exp(-\tfrac{1}{2}z^2)\,\mathrm{d}z \\ &= \phi.\end{aligned}$$

Variance of a sum; covariance and correlation

Sometimes we need to find the *variance of a sum* of random variables. To do this, we note that

$$\begin{aligned}\mathscr{V}(x+y) &= \mathsf{E}\{x+y-\mathsf{E}(x+y)\}^2\\ &= \mathsf{E}\{(x-\mathsf{E}x)+(y-\mathsf{E}y)\}^2\\ &= \mathsf{E}(x-\mathsf{E}x)^2+\mathsf{E}(y-\mathsf{E}y)^2+2\mathsf{E}(x-\mathsf{E}x)(y-\mathsf{E}y)\\ &= \mathscr{V}x+\mathscr{V}y+\mathscr{C}(x, y)\end{aligned}$$

where the *covariance* $\mathscr{C}(x, y)$ of x and y is defined by

$$\begin{aligned}\mathscr{C}(x, y) &= \mathsf{E}(x-\mathsf{E}x)(y-\mathsf{E}y)\\ &= \mathsf{E}xy-(\mathsf{E}x)(\mathsf{E}y).\end{aligned}$$

More generally

$$\mathscr{V}(ax+by+c) = a^2\mathscr{V}x+b^2\mathscr{V}y+2ab\mathscr{C}(x, y)$$

for any constants a, b and c. By considering this expression as a quadratic in a for fixed b or vice versa and noting that (because its value is always positive) this quadratic cannot have two unequal real roots, we see that

$$(\mathscr{C}(x, y))^2 \leqslant (\mathscr{V}x)(\mathscr{V}y).$$

We define the *correlation coefficient* $\rho(x, y)$ between x and y by

$$\rho(x, y) = \frac{\mathscr{C}(x, y)}{\sqrt{(\mathscr{V}x)(\mathscr{V}y)}}.$$

It follows that

$$-1 \leqslant \rho(x, y) \leqslant 1$$

and indeed a little further thought shows that $\rho(x, y) = 1$ if and only if

$$ax+by+c = 0$$

with probability 1 for some constants a, b and c with a and b having opposite signs, while $\rho(x, y) = -1$ if and only if the same thing happens except that a and b have the same sign. If $\rho(x, y) = 0$ we say that x and y are *uncorrelated*.

It is easily seen that if x and y are independent then

$$\mathscr{C}(x, y) = \mathsf{E}xy-(\mathsf{E}x)(\mathsf{E}y) = 0$$

from which it follows that *independent random variables are uncorrelated*.

The converse is *not* in general true, but it can be shown that *if* x and y have a bivariate normal distribution (as described in Section 6.1), then they are independent if and only if they are uncorrelated.

It should be noted that if x and y are uncorrelated, and in particular if they are independent

$$\mathscr{V}(x \pm y) = \mathscr{V}x + \mathscr{V}y$$

(we observe that there is a plus sign on the right-hand side even if there is a minus sign on the left).

Approximations to the mean and variance of a function of a random variable

Very occasionally it will be useful to have an approximation to the mean and variance of a function of a random variable. Suppose that

$$z = g(x).$$

Then if g is a reasonably smooth function and x is not too far from its expectation, Taylor's theorem implies that

$$z \cong g(\mathsf{E}x) + (x - \mathsf{E}x)g'(\mathsf{E}x).$$

It therefore seems reasonable that a fair approximation to the expectation of z is given by

$$\mathsf{E}z = g(\mathsf{E}x)$$

and if this is so, then a reasonable approximation to $\mathscr{V}z$ may well be given by

$$\mathscr{V}z = \mathscr{V}x\{g'(\mathsf{E}x)\}^2.$$

As an example, we suppose that

$$x \sim \mathrm{B}(n, \pi)$$

and that $z = g(x)$, where

$$g(x) = \sin^{-1}\sqrt{(x/n)}$$

so that

$$g'(x) = \tfrac{1}{2}n^{-1}[(x/n)\{1 - (x/n)\}]^{-\frac{1}{2}}.$$

and thus $g'(\mathsf{E}x) = g'(n\pi) = \tfrac{1}{2}n^{-1}[\pi(1-\pi)]^{-\frac{1}{2}}$. The argument above then implies that

$$\mathsf{E}z \cong \sin^{-1}\sqrt{\pi}$$
$$\mathscr{V}z \cong 1/4n.$$

The interesting thing about this transformation, which has a long history [see Eisenhart *et al.* (1947, Chapter 16) and Fisher (1954)] is that, to the extent to which the approximation is valid, the variance of z does not depend on the parameter π. It is accordingly known as a *variance-stabilizing transformation.* We will return to this transformation in Section 3.2 on the "Reference prior for the binomial likelihood".

Conditional expectations and variances

If the reader wishes, the following may be omitted on a first reading and returned to as needed.

We define the *conditional expectation* of y given x by

$$\mathsf{E}(y \mid x) = \int y p(y \mid x) \, \mathrm{d}y$$

in the continuous case and by the corresponding sum in the discrete case. If we wish to be pedantic, it can occasionally be useful to indicate what we are averaging over by writing

$$\mathsf{E}_{\tilde{y} \mid \tilde{x}}(\tilde{y} \mid x)$$

just as we can write $p_{\tilde{y} \mid \tilde{x}}(y \mid x)$, but this is rarely necessary (though it can slightly clarify a proof on occasion). More generally, the conditional expectation of a function $g(y)$ of y given x is

$$\mathsf{E}(g(y) \mid x) = \int g(y) p(y \mid x) \, \mathrm{d}y.$$

We can also define a conditional variance as

$$\begin{aligned} \mathscr{V}(y \mid x) &= \mathsf{E}[\{y - \mathsf{E}(y \mid x)\}^2 \mid x] \\ &= \mathsf{E}(y^2 \mid x) - \{\mathsf{E}(y \mid x)\}^2. \end{aligned}$$

Despite some notational complexity, this is easy enough to find since a conditional distribution is simply a particular case of a probability distribution. If we are really pedantic, then $\mathsf{E}(\tilde{y} \mid x)$ is a real number which is a function of the real number x, while $\mathsf{E}(\tilde{y} \mid \tilde{x})$ is a random variable which is a function of the random variable $\tilde{x}$, which takes the value $\mathsf{E}(\tilde{y} \mid x)$ when $\tilde{x}$ takes the value x. However, the distinction, which is hard to grasp in the first place, is usually unimportant.

We may note that the formula

$$p(y) = \int p(y \mid x) p(x) \, \mathrm{d}x$$

could be written

$$p(y) = \mathsf{E} p(y \mid \tilde{x})$$

but we must be careful that it is an expectation over values of $\tilde{x}$ (i.e. $\mathsf{E}_{\tilde{x}}$) that occurs here.

Very occasionally we make use of results such as

$$\begin{aligned}\mathsf{E}\tilde{y} &= \mathsf{E}_{\tilde{x}}\{\mathsf{E}_{\tilde{y}|\tilde{x}}(\tilde{y}\,|\,\tilde{x})\},\\ \mathcal{V}\tilde{y} &= \mathsf{E}_{\tilde{x}}\mathcal{V}_{\tilde{y}|\tilde{x}}(\tilde{y}\,|\,\tilde{x}) + \mathcal{V}_{\tilde{x}}\mathsf{E}_{\tilde{y}|\tilde{x}}(\tilde{y}\,|\,\tilde{x}).\end{aligned}$$

The proofs are possibly more confusing than helpful. They run as follows:

$$\begin{aligned}\mathsf{E}\{\mathsf{E}(\tilde{y}\,|\,\tilde{x})\} &= \int \mathsf{E}(\tilde{y}\,|\,x)p(x)\,\mathrm{d}x\\ &= \int\left(\int yp(y\,|\,x)\,\mathrm{d}y\right)p(x)\,\mathrm{d}x\\ &= \int\int yp(x,y)\,\mathrm{d}y\,\mathrm{d}x\\ &= \int yp(y)\,\mathrm{d}y\\ &= \mathsf{E}\tilde{y}.\end{aligned}$$

Similarly, we get the generalization

$$\mathsf{E}\{\mathsf{E}(g(\tilde{y})\,|\,\tilde{x})\} = \mathsf{E}g(\tilde{y})$$

and in particular

$$\mathsf{E}\{\mathsf{E}(\tilde{y}^2\,|\,\tilde{x})\} = \mathsf{E}\tilde{y}^2$$

hence

$$\begin{aligned}\mathsf{E}\mathcal{V}(\tilde{y}\,|\,\tilde{x}) &= \mathsf{E}\{\mathsf{E}(\tilde{y}^2\,|\,\tilde{x})\} - \mathsf{E}\{\mathsf{E}(\tilde{y}\,|\,\tilde{x})\}^2\\ &= \mathsf{E}\tilde{y}^2 - \mathsf{E}\{\mathsf{E}(\tilde{y}\,|\,\tilde{x})\}^2\end{aligned}$$

while

$$\begin{aligned}\mathcal{V}\mathsf{E}(\tilde{y}\,|\,\tilde{x}) &= \mathsf{E}\{\mathsf{E}(\tilde{y}\,|\,\tilde{x})\}^2 - [\mathsf{E}\{\mathsf{E}(\tilde{y}\,|\,\tilde{x})\}]^2\\ &= \mathsf{E}\{\mathsf{E}(\tilde{y}\,|\,\tilde{x})\}^2 - (\mathsf{E}\tilde{y})^2\end{aligned}$$

from which it follows that

$$\mathsf{E}\mathcal{V}(\tilde{y}\,|\,\tilde{x}) + \mathcal{V}\mathsf{E}(\tilde{y}\,|\,\tilde{x}) = \mathsf{E}\tilde{y}^2 - (\mathsf{E}\tilde{y})^2 = \mathcal{V}\tilde{y}.$$

Medians and modes

The mean is not the only measure of the centre of a distribution. We may also need to consider the *median*, which is defined as any value x_0 such that

$$\mathrm{P}(\tilde{x} \geqslant x_0) \leqslant \tfrac{1}{2} \qquad \text{and} \qquad \mathrm{P}(\tilde{x} \leqslant x_0) \geqslant \tfrac{1}{2}.$$

In the case of most continuous random variables there is a unique median such that

$$\mathrm{P}(\tilde{x} \geqslant x_0) = \mathrm{P}(\tilde{x} \leqslant x_0) = \tfrac{1}{2}.$$

We occasionally refer also to the *mode*, defined as that value at which the probability density function is a maximum. One important use we shall have for the mode will be in methods for finding the median based on the approximation

$$\text{mean} - \text{mode} = 3\,(\text{mean} - \text{median})$$

or equivalently

$$\text{median} = (2\,\text{mean} + \text{mode})/3$$

(see the Appendix).

Exercises on Chapter 1

1. A card came is played with 52 cards divided equally between four players, North, South, East and West, all arrangements being equally likely. Thirteen of the cards are referred to as trumps. If you know that North and South have ten trumps between them, what is the probability that all three remaining trumps are in the same hand?

2. Whether certain mice are black or brown depends on a pair of genes, each of which is either B or b. If both members of the pair are alike, the mouse is said to be homozygous, and if they are different it is said to be heterozygous. The mouse is brown only if it is homozygous bb. The offspring of a pair of mice have two such genes, one from each parent, and if the parent is heterozygous, the inherited gene is equally likely to be B or b. Suppose that a black mouse results from a mating between two heterozygotes.
   (a) What are the probabilities that this mouse is homozygous and that it is heterozygous?

   Now suppose that this mouse is mated with a brown mouse, resulting in seven offspring, all of which turn out to be black.
   (b) Use Bayes' Theorem to find the probability that the black mouse was homozygous BB.
   (c) Recalculate the same probability by regarding the seven offspring as seven observations made sequentially, treating the posterior after each observation as the prior for the next (cf. Fisher, 1956, 1959, Section II.2).

3. The example on Bayes' Theorem in Section 1.2 concerning the

biology of twins was based on the assumption that births of boys and girls occur equally frequently, and yet it has been known for a very long time that fewer girls are born than boys (cf. Arbuthnot, 1710). Suppose that the probability of a girl being born is p, so that

$$\begin{array}{lll} \mathrm{P}(GG\,|\,M)=p, & \mathrm{P}(BB\,|\,M)=1-p, & \mathrm{P}(GB\,|\,M)=0, \\ \mathrm{P}(GG\,|\,D)=p^2, & \mathrm{P}(BB\,|\,D)=(1-p)^2, & \mathrm{P}(GB\,|\,D)=2p(1-p). \end{array}$$

Find the proportion of monozygotic twins in the whole population of twins in terms of p and the sex distribution among all twins.

4. Suppose a red and a blue die are tossed. Let x be the sum of the number showing on the red die and twice the number showing on the blue die. Find the density function and the distribution function of x.

5. Modify the formula for the density of a one-to-one function $g(x)$ of a random variable x to find an expression for the density of x^2 in terms of that of x, in both the continuous and discrete case. Hence show that the square of a standard normal density has a chi-squared density on one degree of freedom as defined in the Appendix.

6. Suppose that $x_1, x_2, \ldots, x_n$ are independent and all have the same continuous distribution, with density $f(x)$ and distribution function $F(x)$. Find the distribution functions of

$$M = \max\{x_1, x_2, \ldots, x_n\} \quad \text{and} \quad m = \min\{x_1, x_2, \ldots, x_n\}$$

in terms of $F(x)$, and hence find expressions for the density functions of M and m.

7. Suppose that u and v are independently uniformly distributed on the interval $[0, 1]$, so that they divide the interval into three sub-intervals. Find the joint density function of the lengths of the first two sub-intervals.

8. Suppose that the random variable x has a negative binomial distribution $\mathrm{NB}(n, \pi)$ of index n and parameter π, so that

$$p(x) = \binom{n+x-1}{x}\pi^n(1-\pi)^x$$

Find the mean and variance of x and check that the answer agrees with that given in the Appendix.

9. The *skewness* of a random variable x is defined as $\gamma_1 = \mu_3/(\mu_2)^{3/2}$, where

$$\mu_n = \mathsf{E}(x - \mathsf{E}x)^n$$

(although some authors work in terms of $\beta_1 = \gamma_1{}^2$). Find the skewness of a random variable X with a binomial distribution $B(n, \pi)$ of index n and parameter π.

10. Let x and y have a bivariate normal distribution and suppose that x and y both have mean 0 and variance 1, so that their marginal distributions are standard normal and their joint density is

$$p(x, y) = \{2\pi\sqrt{(1-\rho^2)}\}^{-1} \exp\{-\tfrac{1}{2}(x^2 - 2\rho xy + y^2)/(1-\rho^2)\}.$$

Show that if the correlation coefficient between x and y is ρ, then that between x^2 and y^2 is ρ^2.

11. Suppose that x has a Poisson distribution $\mathrm{P}(\lambda)$ of mean λ and that, for given x, y has a binomial distribution $\mathrm{B}(n, \pi)$ of index n and parameter π.

(a) Show that the unconditional distribution of y is a Poisson distribution of mean $\lambda\pi = \mathsf{E}_{\tilde{x}}\mathsf{E}_{\tilde{y}|\tilde{x}}(\tilde{y}\,|\,\tilde{x})$.

(b) Verify that the formula

$$\mathcal{V}\tilde{y} = \mathsf{E}_{\tilde{x}}\mathcal{V}_{\tilde{y}|\tilde{x}}(\tilde{y}\,|\,\tilde{x}) + \mathcal{V}_{\tilde{x}}\mathsf{E}_{\tilde{y}|\tilde{x}}(\tilde{y}\,|\,\tilde{x})$$

derived in Section 1.5 holds in this case.

2

Bayesian Inference for the Normal Distribution

2.1 Nature of Bayesian inference

Preliminary remarks

In this section a general framework for Bayesian statistical inference will be provided. In broad outline we take prior beliefs about various possible hypotheses and then modify these prior beliefs in the light of relevant data which we have collected in order to arrive at posterior beliefs. (The reader may prefer to return to this section after reading the next section, which deals with one of the simplest special cases of Bayesian inference.)

Post is prior times likelihood

Almost all of the situations we will think of in this book fit into the following pattern. Suppose that you are interested in the values of k unknown quantities

$$\boldsymbol{\theta} = (\theta_1, \theta_2, \ldots, \theta_k)$$

(where k can be one or more than one) and that you have some *a priori* beliefs about their values which you can express in terms of the p.d.f.

$$p(\boldsymbol{\theta}).$$

Now suppose that you then obtain some data relevant to their values. More precisely, suppose that we have n observations

$$\boldsymbol{X} = (X_1, X_2, \ldots, X_n)$$

which have a probability distribution that depends on these k unknown quantities as parameters, so that the p.d.f. (continuous or discrete) of the vector $\boldsymbol{X}$ depends on the vector $\boldsymbol{\theta}$ in a known way. Usually the components of $\boldsymbol{\theta}$ and $\boldsymbol{X}$ will be integers or real numbers, so that the components of $\boldsymbol{X}$ are random variables, and so the dependence of $\boldsymbol{X}$ on $\boldsymbol{\theta}$ can be expressed in terms of a p.d.f.

$$p(\boldsymbol{X}|\boldsymbol{\theta}).$$

You then want to find a way of expressing your beliefs about $\boldsymbol{\theta}$ taking into account both your prior beliefs and the data. Of course, it is possible that your prior beliefs about $\boldsymbol{\theta}$ may differ from mine, but very often we will agree on the way in which the data are related to $\boldsymbol{\theta}$ (that is, on the form of $p(\boldsymbol{X}|\boldsymbol{\theta})$). If this is so, we will differ in our posterior beliefs (i.e. in our beliefs after we have obtained the data), but it will turn out that if we can collect enough data, then our posterior beliefs will usually become very close.

The basic tool we need is Bayes' Theorem for random variables (generalized to deal with random vectors). From this theorem we know that

$$p(\boldsymbol{\theta}|\boldsymbol{X}) \propto p(\boldsymbol{\theta})p(\boldsymbol{X}|\boldsymbol{\theta}).$$

Now we know that $p(\boldsymbol{X}|\boldsymbol{\theta})$ considered as a function of $\boldsymbol{X}$ for fixed $\boldsymbol{\theta}$ is a density, but we will find that we often want to think of it as a function of $\boldsymbol{\theta}$ for fixed $\boldsymbol{X}$. When we think of it in that way it does not have quite the same properties—for example, there is no reason why it should sum (or integrate) to unity. Thus in the extreme case where $p(\boldsymbol{X}|\boldsymbol{\theta})$ turns out not to depend on $\boldsymbol{\theta}$, then it is easily seen that it can quite well sum (or integrate) to ∞. When we are thinking of $p(\boldsymbol{X}|\boldsymbol{\theta})$ as a function of $\boldsymbol{\theta}$ we call it the *likelihood* function. We sometimes write

$$l(\boldsymbol{\theta}|\boldsymbol{X}) = p(\boldsymbol{X}|\boldsymbol{\theta}).$$

Just as we sometimes write $p_{\boldsymbol{X}|\boldsymbol{\theta}}(\boldsymbol{X}|\boldsymbol{\theta})$ to avoid ambiguity, if we really need to avoid ambiguity we write

$$l_{\boldsymbol{\theta}|\boldsymbol{X}}(\boldsymbol{\theta}|\boldsymbol{X})$$

but this will not usually be necessary. Sometimes it is more natural to consider the *log-likelihood* function

$$L(\boldsymbol{\theta}|\boldsymbol{X}) = \log l(\boldsymbol{\theta}|\boldsymbol{X}).$$

With this definition and the definition of $p(\boldsymbol{\theta})$ as the prior p.d.f. for $\boldsymbol{\theta}$ and of $p(\boldsymbol{\theta}|\boldsymbol{X})$ as the posterior p.d.f. for $\boldsymbol{\theta}$ given $\boldsymbol{X}$, we may think of Bayes' Theorem in the more memorable form

$$\text{Posterior} \propto \text{Prior} \times \text{Likelihood}.$$

This relationship summarizes the way in which we should modify our beliefs in order to take into account the data we have available.

Likelihood can be multiplied by any constant

We note that, because of the way we write Bayes' Theorem with a proportionality sign, it does not alter the result if we multiply $l(\boldsymbol{\theta}|\boldsymbol{X})$ by any constant or indeed more generally by anything which is a function of $\boldsymbol{X}$ alone. Accordingly, we can regard the definition of the likelihood as being *any constant multiple* of $p(\boldsymbol{X}|\boldsymbol{\theta})$ rather than necessarily equalling $p(\boldsymbol{X}|\boldsymbol{\theta})$ (and similarly the log-likelihood is undetermined up to an additive constant). Sometimes the integral

$$\int l(\boldsymbol{\theta}|\boldsymbol{X})\,\mathrm{d}\boldsymbol{\theta}$$

(interpreted as a multiple integral $\int\int \ldots \int \ldots \mathrm{d}\theta_1 \mathrm{d}\theta_2 \ldots \mathrm{d}\theta_k$ if $k>1$ and interpreted as a summation or multiple summation in the discrete case), taken over the admissible range of $\boldsymbol{\theta}$, is finite, although we have already noted that this is not always the case. When it is, it is occasionally convenient to refer to the quantity

$$\frac{l(\boldsymbol{\theta}|\boldsymbol{X})}{\int l(\boldsymbol{\theta}|\boldsymbol{X})\,\mathrm{d}\boldsymbol{\theta}}.$$

We shall call this the *standardized likelihood*, that is, the likelihood scaled so that the area, volume or hypervolume under the curve, surface or hypersurface is unity.

Sequential use of Bayes' Theorem

We should also note that the method can be applied *sequentially*. Thus, if you have an initial sample of observations $\boldsymbol{X}$, you have

$$p(\boldsymbol{\theta}|\boldsymbol{X}) \propto p(\boldsymbol{\theta})\,l(\boldsymbol{\theta}|\boldsymbol{X}).$$

Now suppose that you have a second set of observations $\boldsymbol{Y}$ distributed independently of the first sample. Then

$$p(\boldsymbol{\theta}|\boldsymbol{X},\boldsymbol{Y}) \propto p(\boldsymbol{\theta})\,l(\boldsymbol{\theta}|\boldsymbol{X},\boldsymbol{Y}).$$

But independence implies

$$p(\boldsymbol{X},\boldsymbol{Y}|\boldsymbol{\theta}) = p(\boldsymbol{X}|\boldsymbol{\theta})p(\boldsymbol{Y}|\boldsymbol{\theta})$$

from which it is obvious that

$$l(\boldsymbol{\theta}|\boldsymbol{X},\boldsymbol{Y}) \propto l(\boldsymbol{\theta}|\boldsymbol{X})\,l(\boldsymbol{\theta}|\boldsymbol{Y})$$

and hence

$$\begin{aligned} p(\boldsymbol{\theta}|\boldsymbol{X},\boldsymbol{Y}) &\propto p(\boldsymbol{\theta})\,l(\boldsymbol{\theta}|\boldsymbol{X})\;l(\boldsymbol{\theta}|\boldsymbol{Y}) \\ &\propto p(\boldsymbol{\theta}|\boldsymbol{X})\,l(\boldsymbol{\theta}|\boldsymbol{Y}). \end{aligned}$$

So we can find your posterior for $\boldsymbol{\theta}$ given $\boldsymbol{X}$ and $\boldsymbol{Y}$ by treating your posterior given $\boldsymbol{X}$ as the prior for the observation $\boldsymbol{Y}$. This formula will work *irrespective of the temporal order* in which $\boldsymbol{X}$ and $\boldsymbol{Y}$ are observed.

The predictive distribution

Occasionally (for example, when we come to consider Bayesian decision theory and empirical Bayes methods) we need to consider the marginal distribution

$$p(\boldsymbol{X}) = \int p(\boldsymbol{X} \mid \boldsymbol{\theta}) p(\boldsymbol{\theta}) \, d\boldsymbol{\theta}$$

which is called the *predictive distribution of* $\boldsymbol{X}$, since it represents our current predictions of the value of $\boldsymbol{X}$ taking into account both the uncertainty about the value of $\boldsymbol{\theta}$ and the residual uncertainty about $\boldsymbol{X}$ when $\boldsymbol{\theta}$ is known.

One valuable use of the predictive distribution is in checking your underlying assumptions. If, for example, $p(\boldsymbol{X})$ turns out to be small (in some sense) for the observed value of $\boldsymbol{X}$, it might suggest that the form of the likelihood you have adopted was suspect. Some people have suggested that another thing you might re-examine in such a case is the prior distribution you have adopted, although there are logical difficulties about this if $p(\boldsymbol{\theta})$ simply represents your prior beliefs. It might, however, be the case that seeing an observation the possibility of which you had neglected causes you to think more fully and thus bring out beliefs which had previously been subconscious.

A warning

The theory described above relies on the possibility of specifying the likelihood as a function, or equivalently on being able to specify the density $p(\boldsymbol{X} \mid \boldsymbol{\theta})$ of the observations $\boldsymbol{X}$ save for the fact that the k parameters $\theta_1, \theta_2, \ldots, \theta_k$ are unknown. It should be borne in mind that these assumptions about the form of the likelihood may be unjustified, and a blind following of the procedure described above can never lead to their being challenged (although the point made above in connection with the predictive distribution can be of help). It is all too easy to adopt a model because of its convenience and to neglect the absence of evidence for it.

2.2 Normal prior and likelihood

Posterior from a normal prior and likelihood

We say that x is normal of mean θ and variance ϕ and write

$$x \sim \mathrm{N}(\theta, \phi)$$

when

$$p(x) = (2\pi\phi)^{-\frac{1}{2}} \exp\{-\tfrac{1}{2}(x-\theta)^2/\phi\}.$$

Suppose that you have an unknown parameter θ for which your prior beliefs can be expressed in terms of a normal distribution, so that

$$\theta \sim N(\theta_0, \varphi_0)$$

and suppose also that you have an observation x which is normally distributed with mean θ equal to the parameter of interest, that is,

$$x \sim N(\theta, \varphi)$$

where θ_0, φ_0, and φ are known. As mentioned in Section 1.3, there are often grounds for suspecting that an observation might be normally distributed, usually related to the Central Limit Theorem, so this assumption is not implausible. If these assumptions are valid

$$p(\theta) = (2\pi\varphi_0)^{-\frac{1}{2}} \exp\{-\tfrac{1}{2}(\theta-\theta_0)^2/\varphi_0\}$$
$$p(x\,|\,\theta) = (2\pi\varphi)^{-\frac{1}{2}} \exp\{-\tfrac{1}{2}(x-\theta)^2/\varphi\}$$

and hence

$$\begin{aligned} p(\theta\,|\,x) &\propto p(\theta)p(x\,|\,\theta) \\ &= (2\pi\varphi_0)^{-\frac{1}{2}} \exp\{-\tfrac{1}{2}(\theta-\theta_0)^2/\varphi_0\} \times (2\pi\varphi)^{-\frac{1}{2}} \exp\{-\tfrac{1}{2}(x-\theta)^2/\varphi\} \\ &\propto \exp\{-\tfrac{1}{2}\theta^2(\varphi_0^{-1}+\varphi^{-1}) + \theta(\theta_0/\varphi_0 + x/\varphi)\} \end{aligned}$$

regarding $p(\theta\,|\,x)$ as a function of θ.

It is now convenient to write

$$\varphi_1 = \frac{1}{\varphi_0^{-1}+\varphi^{-1}}$$
$$\theta_1 = \varphi_1(\theta_0/\varphi_0 + x/\varphi)$$

so that

$$\varphi_0^{-1}+\varphi^{-1} = \varphi_1^{-1}$$
$$\theta_0/\varphi_0 + x/\varphi = \theta_1/\varphi_1$$

and hence

$$p(\theta\,|\,x) \propto \exp\{-\tfrac{1}{2}\theta^2/\varphi_1 + \theta\theta_1/\varphi_1\}.$$

Adding into the exponent

$$-\tfrac{1}{2}\theta_1{}^2/\varphi_1$$

which is constant as far as θ is concerned, we see that

$$p(\theta\,|\,x) \propto \exp\{-\tfrac{1}{2}(\theta-\theta_1)^2/\varphi_1\}$$

from which it follows that as a density must integrate to unity

$$p(\theta\,|\,x) = (2\pi\varphi_1)^{-\frac{1}{2}} \exp\{-\tfrac{1}{2}(\theta-\theta_1)^2/\varphi_1\}$$

that is, that the posterior density is

$$\theta \mid x \sim \mathrm{N}(\theta_1, \varphi_1).$$

In terms of the precision, which we recall can be defined as the inverse of the variance, the relationship $\varphi_0^{-1} + \varphi^{-1} = \varphi_1^{-1}$ can be remembered as

Posterior precision = Prior precision + Datum precision.

(It should be noted that this relationship has been derived *assuming* a normal prior and a normal likelihood.)

The relation for the posterior mean, θ_1, is only slightly more complicated. We have

$$\theta_1 = \theta_0 \frac{\varphi_0^{-1}}{\varphi_0^{-1} + \varphi^{-1}} + x \frac{\varphi^{-1}}{\varphi_0^{-1} + \varphi^{-1}}$$

which can be remembered as

Posterior mean = Weighted mean of prior mean and datum value, the weights being proportional to their respective precisions.

Example

According to Kennett and Ross (1983), the first apparently reliable datings for the age of Ennerdale granophyre were obtained from the K/Ar method (which depends on observing the relative proportions of potassium-40 and argon-40 in the rock) in the 1960s and early 1970s, and these resulted in an estimate of 370 ± 20 million years. Later in the 1970s, measurements based on the Rb/Sr method (depending on the relative proportions of rubidium-87 and strontium-87) gave an age of 421 ± 8 million years. It appears that the errors marked are meant to be standard deviations, and it seems plausible that the errors are normally distributed. If then a scientist S had the K/Ar measurements available in the early 1970s, it could be said that (before the Rb/Sr measurements came in), S's prior beliefs about the age of these rocks were represented by

$$\theta \sim \mathrm{N}(370, 20^2).$$

We could then suppose that the investigations using the Rb/Sr method result in a measurement

$$x \sim \mathrm{N}(\theta, 8^2)$$

We shall suppose for simplicity that the precisions of these measurements are known to be exactly those quoted, although this is not quite true (methods which take more of the uncertainty into account will be discussed later). If we use the method above, then, noting that the observation x turned out to be 421, we see that S's posterior beliefs about θ should be represented by

$$\theta \mid x \sim \mathrm{N}(\theta_1, \varphi_1)$$

where

$$\varphi_1 = (20^{-2} + 8^{-2})^{-1} = 55 = 7^2,$$
$$\theta_1 = 55(370/20^2 + 421/8^2) = 413.$$

Thus the posterior for the age of the rocks is

$$\varphi \mid x \sim \mathrm{N}(413, 7^2)$$

that is, 413 ± 7 million years.

Of course, all this assumes that the K/Ar measurements were available. If the Rb/Sr measurements were considered by another scientist $\hat{\mathrm{S}}$ who had no knowledge of these, but had a vague idea (in the light of knowledge of similar rocks) that their age was likely to be 400 ± 50 million years, that is,

$$\theta \sim \mathrm{N}(400, 50^2)$$

then $\hat{\mathrm{S}}$ would have a posterior variance

$$\varphi_1 = (50^{-2} + 8^{-2})^{-1} = 62 = 8^2$$

and a posterior mean of

$$\varphi_1 = 62(400/50^2 + 421/8^2) = 418$$

so that $\hat{\mathrm{S}}$'s posterior distribution is

$$\theta \mid x \sim \mathrm{N}(418, 8^2)$$

that is, 418 ± 8 million years. We note that this calculation has been carried out assuming that the prior information available is rather vague, and that this is reflected in the fact that the posterior is almost entirely determined by the data.

The situation can be summarized thus:

	Prior distribution	Likelihood from data	Posterior distribution
S	$\mathrm{N}(370, 20^2)$	$\mathrm{N}(421, 8^2)$	$\mathrm{N}(413, 7^2)$
$\hat{\mathrm{S}}$	$\mathrm{N}(400, 50^2)$		$\mathrm{N}(418, 8^2)$

We note that in numerical work it is usually more meaningful to think in terms of the standard deviation $\sqrt{\varphi}$, whereas in theoretical work it is usually easier to work in terms of the variance φ itself.

We see that after this single observation the ideas of S and Ŝ about θ as represented by their posterior distributions are much closer than before, although they still differ considerably.

Predictive distribution

In the case discussed in this section it is easy to find the predictive distribution, since

$$\tilde{x} = (\tilde{x} - \theta) + \theta$$

and, independently of one another,

$$\tilde{x} - \theta \sim \mathrm{N}(0, \varphi),$$
$$\theta \sim \mathrm{N}(\theta_0, \varphi_0)$$

from which it follows that

$$\tilde{x} \sim \mathrm{N}(\theta_0, \varphi + \varphi_0)$$

using the standard fact that the sum of independent normal variates has a normal distribution. (The fact that the mean is the sum of the means and the variance the sum of the variances is of course true more generally, as proved in Section 1.4 on "Several random variables".)

The nature of the assumptions made

Although the above example is very simple, it does exhibit the main features of Bayesian inference as outlined in the previous section. We have assumed that the distribution of the observation x is *known to be normal* but that there is *an unknown parameter* θ, in this case the mean of the normal distribution. The assumption that the variance is known is unlikely to be fully justified in a practical example, but it may provide a reasonable approximation. We should, however, realize that it is all too easy to concentrate on the parameters of a well-known family, in this case the normal family, and to forget that the assumption that the density is in that family for *any* values of the parameters may not be valid. The fact that the normal distribution is easy to handle, as witnessed by the way in which normal prior and normal likelihood combine to give normal posterior, is a good reason for looking for a normal model when it does provide a fair approximation, but there can easily be cases where it does not.

2.3 Several normal observations with a normal prior

Posterior distribution

We can generalize the situation in the previous section by supposing that *a priori*

$$\theta \sim N(\theta_0, \varphi_0)$$

but that instead of having just one observation we have *n independent* observations $\boldsymbol{x} = (x_1, x_2, \ldots, x_n)$ such that

$$x_i \sim N(\theta, \varphi).$$

We sometimes refer to $\boldsymbol{x}$ as an n-sample from $N(\theta, \varphi)$. Then

$$\begin{aligned} p(\theta \mid \boldsymbol{x}) &\propto p(\theta)p(\boldsymbol{x} \mid \theta) = p(x_1 \mid \theta)p(x_2 \mid \theta) \ldots p(x_n \mid \theta) \\ &= (2\pi\varphi_0)^{-\frac{1}{2}} \exp\{-\tfrac{1}{2}(\theta - \theta_0)^2/\varphi_0\} \\ &\quad \times (2\pi\varphi)^{-\frac{1}{2}} \exp\{-\tfrac{1}{2}(x_1 - \theta)^2/\varphi\} \\ &\quad \times (2\pi\varphi)^{-\frac{1}{2}} \exp\{-\tfrac{1}{2}(x_2 - \theta)^2/\varphi\} \\ &\quad \times \ldots \times (2\pi\varphi)^{-\frac{1}{2}} \exp\{-\tfrac{1}{2}(x_n - \theta)^2/\varphi\} \\ &\propto \exp\{-\tfrac{1}{2}\theta^2(1/\varphi_0 + n/\varphi) + \theta(\theta_0/\varphi_0 + \Sigma x_i/\varphi)\} \end{aligned}$$

Proceeding just as we did in the previous section when we had only one observation, we see that the posterior distribution is

$$\theta \sim N(\theta_1, \varphi_1)$$

where

$$\begin{aligned} \varphi_1 &= (1/\varphi_0 + n/\varphi)^{-1} \\ \theta_1 &= \varphi_1(\theta_0/\varphi_0 + \textstyle\sum x_i/\varphi). \end{aligned}$$

We could alternatively write these formulae as

$$\begin{aligned} \varphi_1 &= \{\varphi_0^{-1} + (\varphi/n)^{-1}\}^{-1} \\ \theta_1 &= \varphi_1\{\theta_0/\varphi_0 + \bar{x}/(\varphi/n)\} \end{aligned}$$

which shows that, assuming a normal prior and likelihood, the result is just the same as the posterior distribution obtained from the *single* observation of the mean $\bar{x}$, since we know that

$$\bar{x} \sim N(\theta, \varphi/n)$$

and the above formulae are the ones we had before with φ replaced by φ/n and x by $\bar{x}$. (Note that the use of a bar over the x here to denote a mean is unrelated to the use of a tilde over x to denote a random variable).

We would of course obtain the same result by proceeding sequentially from $p(\theta)$ to $p(\theta \mid x_1)$ and then treating $p(\theta \mid x_1)$ as prior and x_2 as data to obtain $p(\theta \mid x_1, x_2)$ and so on. This is in accordance with the general result mentioned in Section 2.1 on "Nature of Bayesian inference".

Example

We now consider a numerical example. The basic assumption in this section is that the variance is *known*, and yet in most practical cases it has to be estimated. There are a few circumstances in which the variance could be known, for example, when we are using a measuring instrument which has been used so often that its measurement errors are well known, but there are not many. Later in the book we will discover two factors which mitigate this assumption—first, that the numerical results are not very different when we do take into account the uncertainty about the variance, and, second, that the larger the sample size is, the less difference it makes.

The data we will consider are quoted by Whittaker and Robinson (1940, Section 97). They consider chest measurements of 10,000 men. Now, based on memories of my experience as an assistant in a gentlemen's outfitters in my university vacations I would suggest a prior

$$\theta \sim \mathrm{N}(38, 9).$$

Of course it is open to question whether these men form a random sample from the whole population, but unless I am given information to the contrary I would retain the prior quoted above, except that I might perhaps increase the variance. Whitaker and Robinson's data show that the mean turned out to be 39.8 with a standard deviation of 2.0 for their sample of 10,000. If we put the two together, we end with a posterior mean for the chest measurements of men in this population which is normal with variance

$$\varphi_1 = \{9^{-1} + (2^2/10{,}000)^{-1}\}^{-1} = 1/2500$$

and mean

$$\theta_1 = (1/2500)\{38/9 + 39.8/(2^2/10{,}000)\} = 39.8.$$

Thus for all practical purposes we have ended up with the distribution

$$\theta \sim \mathrm{N}(39.8,\ 2^2/10{,}000)$$

suggested by the data. We should note that this distribution is

$$\theta \sim \mathrm{N}(\bar{x},\ \varphi/n),$$

the distribution we referred to in Section 2.1 on "Nature of Bayesian inference" as the *standardized likelihood*. Naturally, the closeness of the posterior to the standardized likelihood results from the large sample size, and whatever my prior had been, unless it were very very extreme, I would have obtained very much the same result. More formally, the posterior will be close to the standardized likelihood insofar as the weight

$$\varphi_1/\varphi_0$$

associated with the prior mean is small, that is, in so far as φ_0 is large compared with φ/n. This is reassuring in cases where the prior is not very easy to specify, although of course there are cases where the amount of data available is not enough to get to this comforting position.

Predictive distribution

If we consider taking another single observation x_{n+1}, then the predictive distribution can be found just as in the previous section by writing

$$x_{n+1} = (x_{n+1} - \theta) + \theta$$

and noting that, independently of one another,

$$(x_{n+1} - \theta) \sim \mathrm{N}(0, \varphi),$$
$$\theta \sim \mathrm{N}(\theta_1, \varphi_1)$$

so that

$$x_{n+1} \sim \mathrm{N}(\theta_1, \varphi + \varphi_1).$$

It is easy enough to adapt this argument to find the predictive distribution of an m-vector $\boldsymbol{y} = (y_1, y_2, \ldots, y_m)$ where

$$y_i \sim \mathrm{N}(\theta, \psi)$$

by writing

$$\boldsymbol{y} = (\boldsymbol{y} - \theta\mathbf{1}) + \theta\mathbf{1}$$

where $\mathbf{1}$ is the constant vector

$$\mathbf{1} = (1, 1, \ldots, 1).$$

Then θ has its posterior distribution $\theta\,|\,\boldsymbol{x}$ and the components of the vector $\boldsymbol{y} - \theta\mathbf{1}$ are $\mathrm{N}(0, \psi)$ variates independent of θ and of one another, so that $\boldsymbol{y}$ has a multivariate normal distribution, although its components are not independent of one another.

Robustness

It should be noted that *any* statement of a posterior distribution and *any* inference are conditional not merely on the data, but also on the assumptions made about the likelihood. So, in this section, the posterior distribution ends up being normal as a consequence partly of the prior but also of the assumption that the data were distributed normally, albeit with an unknown mean. We say that an inference is *robust* if it is not seriously affected by changes in the assumptions on which it is based. The notion of robustness is not one which can be pinned down into a more precise definition, and its meaning depends on the context, but nevertheless the concept is of great importance, and increasing attention is paid in statistics to investigations of the robustness of various techniques. We can immediately say that the conclusion that the nature of the posterior is robust against changes in the prior is valid provided that the sample size is large and the prior is not-too-extreme normal distribution or nearly so. Some detailed exploration of the notion of robustness can be found in Kadane (1984).

2.4 Dominant likelihoods

Improper priors

We recall from the previous section that, when we have several normal observations with a normal prior and the variances are known, the posterior for the mean is

$$\theta \sim \mathrm{N}(\theta_1, \phi_1)$$

where θ_1 and ϕ_1 are given by the appropriate formulae and that this approaches the standardized likelihood

$$\theta \sim \mathrm{N}(\bar{x}, \phi/n)$$

in so far as ϕ_0 is large compared with ϕ/n, although this result is only approximate unless ϕ_0 is infinite. However, this would mean a prior density $\mathrm{N}(\theta_0, \infty)$ which, whatever θ_0 were, would have to be uniform over the whole real line, and clearly could not be represented by any *proper* density function. It is basic to the concept of a probability density that it integrates to 1 so, for example,

$$p(\theta) = \kappa \qquad (-\infty < \theta < \infty)$$

cannot possibly represent a probability density whatever κ is, and in particular $p(\theta) = 0$, which results from substituting $\phi = \infty$ into the normal density, cannot be a density. Nevertheless, we shall sometimes find it useful to extend the concept of a probability density to some cases like this where

$$\int_{-\infty}^{\infty} p(\theta)\,\mathrm{d}\theta = \infty$$

which we shall call *improper* "densities". The density $p(\theta) = \kappa$ can then be regarded as representing a normal density of infinite variance. Another example of an improper density we will have use for later on is

$$p(\theta) = \kappa/\theta \qquad (0<\theta<\infty).$$

It turns out that sometimes when we take an improper prior density then it can combine with an ordinary likelihood to give a posterior which is proper. Thus if we use the uniform distribution on the whole real line $p(\theta) = \kappa$ for some $\kappa \neq 0$, it is easy to see that it combines with a normal likelihood to give the standardized likelihood as posterior; it follows that the dominant feature of the posterior is the likelihood. The best way to think of an improper density is as an approximation which is valid for some large range of values, but is not to be regarded as truly valid throughout its range. In the case of a physical constant which you are about to measure, you may be very unclear what its value is likely to be, which would suggest the use of a prior that was uniform or nearly so over a large range, but it seems unlikely that you would regard values in the region of, say, 10^{100} as being as likely as, say, values in the region of 10^{-100}. But if you have a prior which is approximately uniform over some (possibly very long) interval and is never very large outside it, then the posterior is close to the standardized likelihoood, and so to the posterior which would have resulted from taking an improper prior uniform over the whole real line. [It is possible to formalize the notion of an improper density as part of probability theory—for details, see Rényi (1970).]

Approximation of proper priors by improper priors

This result can be made more precise. The following theorem is proved by Lindley (1965, Section 5.2); the proof is omitted.

Theorem. A random sample $\boldsymbol{x} = (x_1, x_2, \ldots, x_n)$ of size n is taken from $\mathrm{N}(\theta, \varphi)$ where φ is known. Suppose that there exist positive constants α, ε, M and c depending on $\boldsymbol{x}$ (small values of α and ε are of interest), such that in the interval I_α defined by

$$\bar{x} - \lambda_\alpha \surd(\varphi/n) \leqslant \theta \leqslant \bar{x} + \lambda_\alpha \surd(\varphi/n)$$

where

$$2\Phi(-\lambda_\alpha) = \alpha$$

the prior density of θ lies between $c(1-\varepsilon)$ and $c(1+\varepsilon)$ and outside I_α it is bounded by Mc. Then the posterior density $p(\theta\,|\,\boldsymbol{x})$ satisfies

$$\frac{(1-\varepsilon)}{(1+\varepsilon)(1-\alpha)+M\alpha}\,(2\pi\varphi/n)^{-\frac{1}{2}}\exp\{-\tfrac{1}{2}(\bar{x}-\theta)^2/(\varphi/n)\}$$

$$\leqslant p(\theta\,|\,\boldsymbol{x})$$

$$\leqslant\frac{(1+\varepsilon)}{(1-\varepsilon)(1-\alpha)}\,(2\pi\varphi/n)^{-\frac{1}{2}}\exp\{-\tfrac{1}{2}(\bar{x}-\theta)^2/(\varphi/n)\}$$

inside I_α, and

$$0\leqslant p(\theta\,|\,\boldsymbol{x})\leqslant\frac{M}{(1-\varepsilon)(1-\alpha)}\,(2\pi\varphi/n)^{-\frac{1}{2}}\exp(-\tfrac{1}{2}\lambda_\alpha{}^2)$$

outside it.

While we are not going to prove the theorem, it is worth while to give some idea of the sorts of bounds which it implies. Anyone who has worked with the normal distribution is likely to remember that the 1% point is 2.58, that is, that if $\alpha = 0.01$ then $\lambda_\alpha = 2.58$, so that I_α extends 2.58 standard deviations $[2.58\sqrt{(\varphi/n)}]$ on each side of the sample mean $\bar{x}$. Suppose then that *before* you had obtained any data you believed all values in some interval to be equally likely, and that there were no values that we believed to be more than three times as probable as the values in this interval. If then it turns out when you obtain the data the range I_α lies entirely in this interval, then you can apply the theorem with $\alpha = 0.01$, $\varepsilon = 0$, and $M = 3$, to deduce that within I_α the true density lies within multiples $(1-\alpha+M\alpha)^{-1} = 0.98$ and $(1-\alpha)^{-1} = 1.01$ of the normal density. We can regard this theorem as demonstrating how robust the posterior is to changes in the prior. Similar results hold for distributions other than the normal.

It is often sensible to analyse scientific data on the assumption that the likelihood dominates the prior. There are several reasons for this, of which two important ones are as follows. First, even if you and I both have strong prior beliefs about the value of some unknown quantity, we might not agree, and it seems sensible to use a neutral *reference prior* which is dominated by the likelihood and could be said to represent the views of someone who (unlike ourselves) had no strong beliefs *a priori.* The difficulties of public discourse in a world where different individuals have different prior beliefs constitute one reason why a few people have argued that, in the absence of agreed prior information, we should simply quote the likelihood function [see, for example, Edwards (1972)], but there are considerable difficulties in the way of this (see also Section 7.1 on "The likelihood principle"). Second, in many scientific contexts we would not bother to carry out an experiment unless we thought it was going to increase our knowledge significantly, and if that is the case then presumably the likelihood will dominate the prior.

2.5 Locally uniform priors

Bayes' postulate

We have already seen that it seems useful to have a reference prior to aid public discourse in situations where prior opinions differ or are not strong. A prior which does not change very much over the region in which the likelihood is appreciable and does not take very large values outside that region is said to be *locally uniform*. For such a prior

$$p(\theta \mid x) \propto p(x \mid \theta) = l(\theta \mid x)$$

so that on normalizing the posterior must equal the standardized likelihood.

Bayes himself appears to have thought that, at least in the case where θ is an unknown probability between 0 and 1, the situation where we "know nothing" should be represented by taking a *uniform prior* and this is sometimes known as *Bayes' postulate* (as distinct from his theorem).

However, it should be noted that if, for example

$$p(\theta) = 1 \qquad (0 < \theta < 1)$$

then on writing

$$\varphi = 1/\theta$$

we have according to the usual change-of-variable rule

$$p(\varphi) \mid \mathrm{d}\varphi \mid = p(\theta) \mid \mathrm{d}\theta \mid$$

or

$$\begin{aligned} p(\varphi) &= p(\theta) \mid \mathrm{d}\theta/\mathrm{d}\varphi \mid \\ &= 1/\varphi^2 \qquad (1 < \varphi < \infty) \end{aligned}$$

(as a check, this density does integrate to unity). Now it has been argued that if we "know nothing" about θ then we equally "know nothing" about φ, which should surely be represented by the improper prior

$$p(\varphi) = \text{constant} \qquad (1 < \varphi < \infty)$$

[although there are also arguments for a prior proportional to φ^{-1} or to $(\varphi - 1)^{-1}$], so that the idea that a uniform prior can be used to represent ignorance is not self-consistent. It cannot be denied that this is a serious objection, but it is perhaps not quite as serious as it seems at first sight. With most transformations the density of the transformed variable will not change very fast over a reasonably short interval. For example, while $1/\varphi^2$ changes quite considerably over long intervals of φ, it is sufficiently close to constancy over any moderately short interval that a posterior based on a uniform prior is unlikely to differ greatly from one based on the prior with density $1/\varphi^2$, provided that the amount of data available is not *very* small. This argument would not necessarily work if we were to consider a very extreme transformation, e.g. $\varphi = \exp(\exp(\theta))$, but it could be argued that the mere fact that such an extreme transformation even crossed your mind would suggest that you really had some prior information which made it sensible, and you should accordingly make use of your prior information.

Data translated likelihoods

Even though it may not make a great deal of difference within broad limits what we treat as our reference prior, provided that it is reasonably flat, there is still a natural urge to look for the "right" scale of measurement in which to have a uniform prior, from which the prior in any other scale of measurement can be deduced. One answer to this is to look for a scale of measurement in which the likelihood is *data translated.* The likelihood is said to be in such a form if

$$l(\theta \mid \boldsymbol{x}) = g(\theta - t(\boldsymbol{x}))$$

for some function t (which we will later note is a *sufficient statistic*). In looking to see whether the likelihood can be expressed in this way, we should bear in mind that the definition of the likelihood function allows us to multiply it by any function of the data $\boldsymbol{x}$ alone.

For example, if we have an n-sample from a normal distribution of unknown mean θ and known variance φ, we know that

$$l(\theta \mid x) = \exp\{-\tfrac{1}{2}(\theta - \bar{x})^2/(\varphi/n)\}$$

which is clearly of this form. On the other hand, if k has a binomial distribution of index n and parameter π, so that

$$l(\pi \mid k) = \pi^k(1-\pi)^{n-k}$$

then the likelihood cannot be put into the form $g(\pi - t(k))$.

If the likelihood *is* in data translated form, then different values of the data will give rise to the same functional form for the likelihood except for a shift in location. Thus in the case of the normal mean, if we consider two experiments, one of which results in a value of $\bar{x}$ which is, say, 5 larger than the other, then we get the same likelihood function in both cases except that corresponding values of θ differ by 5. This would seem to suggest that the main function of the data is to determine the location of the likelihood. Now if a uniform prior is taken for θ, the posteriors are also the same except that corresponding values differ by 5, so that the inferences made do seem simply to represent a determination of location. It is because of this that it seems sensible to adopt a uniform prior when the likelihood is data translated.

Transformation of unknown parameters

The next question is what we should do when it is not. Sometimes it turns out that there is a function

$$\psi = \psi(\theta)$$

which is such that we can write

$$l(\theta \mid \boldsymbol{x}) = g(\psi(\theta) - t(\boldsymbol{x}))$$

in which case the obvious thing to do is to take a prior uniform in ψ rather than in θ, implying a prior for the parameter θ given by the usual change-of-variable rule. If, for example, x has an exponential distribution, that is, $x \sim \mathrm{E}(\theta)$ (see under the gamma distribution in the Appendix), then

$$p(x \mid \theta) = \theta^{-1} \exp(-x/\theta)$$

so that (after multiplying by x as we are entitled to) we may write

$$\begin{aligned} l(\theta \mid x) &= (x/\theta) \exp(-x/\theta) \\ &= \exp\{(\log x - \log \theta) - \exp(\log x - \log \theta)\}. \end{aligned}$$

This is in the above form with

$$\begin{aligned} g(y) &= \exp\{-y - \exp(-y)\} \\ t(x) &= \log x \\ \psi(\theta) &= \log \theta. \end{aligned}$$

Unfortunately, it is often difficult to see how to express a likelihood function in this form even when it is possible, and it is not always possible. We shall find another case of this when we come to investigate the normal variance in Section 2.7 below, and a further one when we try to find a reference prior for the uniform distribution in Section 3.6 below. Sometimes there is a function $\psi(\theta)$ such that the likelihood is approximately of this form, for example when we have a binomial distribution of known index and unknown parameter (this case will be considered when the reference prior for the binomial parameter is discussed in Section 3.2 below).

For the moment, we can reflect that this argument strengthens the case for using a uniform (improper) prior for the mean of a normal distribution

$$p(\theta) \propto c \qquad (-\infty < \theta < \infty).$$

One way of thinking of the uniform distribution is as a normal distribution of infinite variance or equivalently zero precision. The equations for the case of normal mean and variance

Posterior precision = Prior precision + Datum precision
Posterior mean = Weighted mean of prior mean and datum value

then become

Posterior precision = Datum precision
Posterior mean = Datum value

which accords with the result

$$p(\theta \mid \boldsymbol{x}) \propto l(\theta \mid \boldsymbol{x})$$

for a locally uniform prior. An interesting defence of the notion of a uniform prior can be found in Savage *et al.* (1962, p. 20).

2.6 Highest density regions (HDRs)

Need for summaries of posterior information

In the case of our example on Ennerdale granophyre all the information available after the experiment is contained in the posterior distribution. One of the best ways of conveying this information would be to sketch the posterior density (though this procedure is more difficult in cases where we have several parameters to estimate, so that θ is multi-dimensional). It is less trouble to the statistician to say simply that

$$\theta \sim \mathrm{N}(413, 7^2)$$

although those without experience may need tables to appreciate what this assertion means.

Sometimes the probability that the parameter lies in a particular interval may be of interest. Thus there might be geological reasons why, in the above example, we wanted to know the chance that the rocks were less than 400 million years old. If this is the case, the probability required is easily found by use of tables of the normal distribution. More commonly, there are no limits of any special interest, but it seems reasonable to specify an interval in which "most of the distribution" lies. It would appear sensible to look for an interval which is such that the density at any point inside it is greater than the density at any point outside it, and it would also appear sensible to seek (for a given probability level) an interval that is as short as possible (in several dimensions, this means that it should occupy as small a volume as possible). Fortunately, it is clear that these conditions are equivalent. In most common cases there is one such interval for each probability level.

We shall refer to such an interval as a *highest (posterior) density region* or an *HDR*. Although this terminology is used by several authors, there are other terms in use, for example, *Bayesian confidence interval* (cf. Lindley, 1965, Section 5.2) and *credible interval* (cf. Edwards *et al.*, 1963, Section 5). In the particular example referred to above we could use the well-known fact that 95% of the area of a normal distribution lies within ± 1.96 standard deviations of the mean to say that $413 \pm 1.96 \times 7$, that is (399, 427) is a 95% HDR for the age θ given the data.

Relation to classical statistics

The traditional approach, sometimes called *classical statistics* or *sampling theory statistics*, would lead to similar conclusions in this case. From either standpoint

$$(\theta - x)/\sqrt{\varphi} \sim \mathrm{N}(0, 1)$$

and in either case the interval (399, 427) is used at a 95% level. However, in the classical approach it is x that is regarded as random and giving rise to a random interval which has a probability 0.95 of containing the fixed (but unknown) value θ. By contrast, the Bayesian approach regards θ as random in the sense that we have certain beliefs about its value, and think of the interval as fixed once the datum is available. Perhaps the tilde notation for random variables helps. With this, the classical approach amounts to saying that

$$|(\theta - \tilde{x})/\sqrt{\varphi}| < 1.96$$

with probability 0.95, while the Bayesian approach amounts to saying that

$$|(\tilde{\theta}-x)/\sqrt{\varphi}|<1.96$$

with probability 0.95.

Although there is a simple relationship between the conclusions that classical and Bayesian statisticians would arrive at in this case, there will be cases later on in which there is no great similarity between the conclusions arrived at.

2.7 Normal variance

A suitable prior for the normal variance

Suppose that we have an n-sample $\boldsymbol{x}=(x_1, x_2, \ldots, x_n)$ from $\mathrm{N}(\mu, \varphi)$ where the variance φ is *unknown* but the mean μ is *known*. Then clearly

$$\begin{aligned} p(\boldsymbol{x}|\varphi) &= (2\pi\varphi)^{-\frac{1}{2}} \exp\{-\tfrac{1}{2}(x_1-\mu)^2/\varphi\} \\ &\qquad \times \ldots \times (2\pi\varphi)^{-\frac{1}{2}} \exp\{-\tfrac{1}{2}(x_n-\mu)^2/\varphi\} \\ &\propto \varphi^{-n/2} \exp\{-\tfrac{1}{2}\sum(x_i-\mu)^2/\varphi\}. \end{aligned}$$

On writing

$$S=\sum(x_i-\mu)^2$$

(remember that μ is known; we shall use a slightly different notation when it is not) we see that

$$p(\boldsymbol{x}|\varphi) \propto \varphi^{-n/2} \exp(-\tfrac{1}{2}S/\varphi).$$

In principle, we might have any form of prior distribution for the variance φ. However, if we are to be able to deal easily with the posterior distribution (and, for example, to be able to find HDRs easily from tables), it helps if the posterior distribution is of a "nice" form. This will certainly happen if the prior is of a similar form to the likelihood, namely,

$$p(\varphi) \propto \varphi^{-\kappa/2} \exp(-\tfrac{1}{2}S_0/\varphi)$$

where κ and S_0 are suitable constants. For reasons which will emerge, it is convenient to write $\kappa=\nu+2$ so that

$$p(\varphi) \propto \varphi^{-\nu/2-1} \exp(-\tfrac{1}{2}S_0/\varphi)$$

leading to the posterior distribution

$$\begin{aligned} p(\varphi|\boldsymbol{x}) &\propto p(\varphi)p(\boldsymbol{x}|\theta) \\ &\propto \varphi^{-(\nu+n)/2-1} \exp\{-\tfrac{1}{2}(S_0+S)/\varphi\}. \end{aligned}$$

This distribution is in fact closely related to one of the best-known continuous distributions in statistics (after the normal distribution), viz. the *chi-squared distribution* or χ^2 *distribution*. This is seen more clearly if we work in terms of the precision $\lambda = 1/\varphi$ instead of in terms of the variance φ, since

$$\begin{aligned} p(\lambda|\boldsymbol{x}) &\propto p(\varphi|\boldsymbol{x})\,|\mathrm{d}\varphi/\mathrm{d}\lambda| \\ &\propto \lambda^{(\nu+n)/2+1} \exp\{-\tfrac{1}{2}(S_0+S)\lambda\} \times \lambda^{-2} \\ &\propto \lambda^{(\nu+n)/2-1} \exp\{-\tfrac{1}{2}(S_0+S)\lambda\}. \end{aligned}$$

This is now very close to the form of the chi-squared distribution (see the Appendix, or indeed any elementary textbook on statistics). It is in fact a very simple further step (left as an exercise!) to check that (for given $\boldsymbol{x}$)

$$(S_0+S)\lambda \sim \chi^2_{\nu+n}$$

that is, $(S_0+S)\lambda$ has a chi-squared distribution on $\nu+n$ degrees of freedom. [The term "degrees of freedom" (d.f.) is hallowed by tradition, but is just a name for a parameter]. We usually indicate this by writing

$$\varphi \sim (S_0+S)\chi^{-2}_{\nu+n}.$$

Clearly the same argument can be applied to the prior distribution, so our prior assumption is that

$$\varphi \sim S_0\chi_\nu^{-2}.$$

It may be that you cannot find suitable values of the parameters ν and S_0 so that a distribution of this type represents your prior beliefs, but clearly if values can be chosen so that they are reasonably well approximated, it is convenient. Usually, the approximation need not be too close since, after all, the chances are that the likelihood will dominate the prior. In fitting a plausible prior one possible approach is to consider the mean and variance of our prior distribution and then choose ν and S_0 so that

$$\mathsf{E}\varphi = S_0/(\nu-2),$$

$$\mathscr{V}\varphi = \frac{2S_0^{\,2}}{(\nu-2)^2(\nu-4)}.$$

We refer to this distribution as (a multiple of) an *inverse* chi-squared distribution. It is very important in Bayesian statistics, although the inverse chi-squared (as opposed to the chi-squared) distribution rarely occurs in classical statistics. Some of its properties are described in the Appendix, and its density for typical values of ν (and $S_1 = 1$) is illustrated in Fig. 2.7.

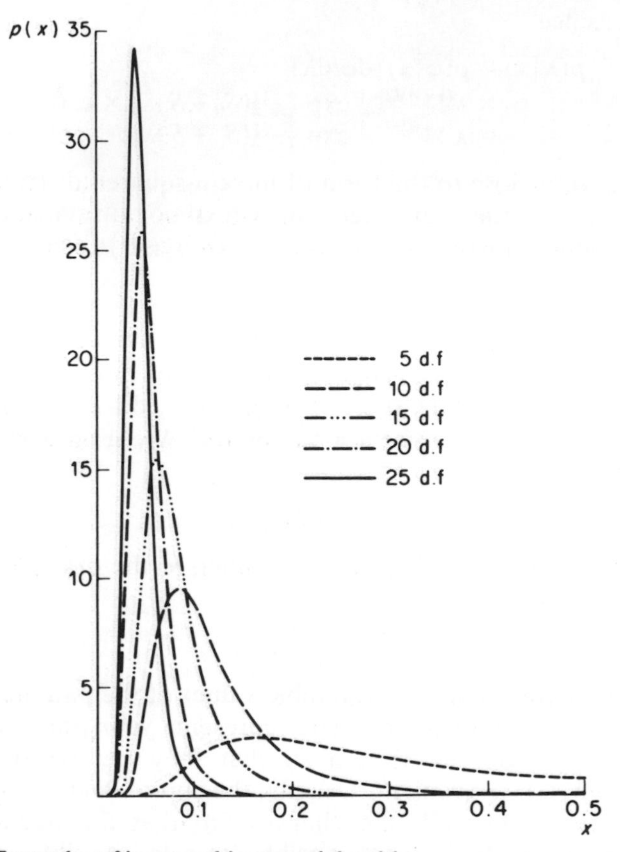

Fig. 2.7 Examples of inverse chi-squared densities

Reference prior for the normal variance

The next thing to investigate is whether there is something which we can regard as a reference prior by finding a scale of measurement in which the likelihood is data translated. For this purpose (and others) it is convenient to define the sample standard deviation s by

$$s^2 = S/n = \sum(x_i - \mu)^2/n$$

(again we shall use a different definition when μ is unknown). Then

$$\begin{aligned} l(\varphi|\boldsymbol{x}) &\propto p(\boldsymbol{x}|\varphi) \propto \varphi^{-n/2} \exp(-\tfrac{1}{2}S/\varphi) \\ &\propto s^n \varphi^{-n/2} \exp(-\tfrac{1}{2}ns^2/\varphi) \\ &= \exp[-\tfrac{1}{2}n(\log\varphi - \log s^2) - \tfrac{1}{2}ns^2/\varphi] \\ &= \exp[-\tfrac{1}{2}n(\log\varphi - \log s^2) - \tfrac{1}{2}n \exp\{-(\log\varphi - \log s^2)\}]. \end{aligned}$$

This is of data translated form

$$l(\varphi|\boldsymbol{x}) = g(\psi(\varphi) - t(\boldsymbol{x}))$$

with

$$\begin{aligned} g(y) &= \exp[-\tfrac{1}{2}ny - \tfrac{1}{2}n \exp(-y)] \\ t(\boldsymbol{x}) &= \log s^2 \\ \psi(\varphi) &= \log\varphi. \end{aligned}$$

The general argument about data translated likelihoods now suggests that we take as reference prior an improper density which is locally uniform in $\psi = \log\varphi$, that is, $p(\psi) \propto c$, which, in terms of φ corresponds to

$$p(\varphi) = p(\psi)\,|\mathrm{d}\psi/\mathrm{d}\varphi| \propto \mathrm{d}\log\varphi/\mathrm{d}\varphi$$

and so to

$$p(\varphi) \propto 1/\varphi.$$

(Although the above argument is complicated, and a similarly complicated example will occur in the case of the uniform distribution in Section 3.6, there will be no other difficult arguments about data translated likelihoods.)

This prior (which was first mentioned in Section 2.4 on dominant likelihoods) is in fact a particular case of the priors of the form $S_0\chi_\nu^{-2}$ which we were considering earlier, in which $\nu = 0$ and $S_0 = 0$. Use of the reference prior results in a posterior distribution

$$\varphi \sim S\chi_n^{-2}$$

which again is a particular case of the distribution found before, and is quite easy to use.

It is important to note that inferences about variances are not as robust as inferences about means if the underlying distribution turns out to be only approximately normal.

2.8 HDRs for the normal variance

What distribution should we be considering?

It might be thought that as the normal variance has (under the assumptions we are making) a distribution which is a multiple of the inverse chi-squared distribution we should be using tables of HDRs for the inverse chi-squared distribution to give intervals in which most of the posterior distribution lies. This procedure is, indeed, recommended by, for example, Novick and Jackson (1974, Section 7.3) and Schmitt (1969, Section 6.3). However, there is another procedure which seems to be marginally preferable.

The point is that we chose a reference prior which was uniform in $\log \varphi$ so that the density of $\log \varphi$ was constant and no value of $\log \varphi$ was more likely than any other *a priori*. Because of this it seems natural to use $\log \varphi$ in the posterior distribution and thus to look for an interval inside which the *posterior density of* $\log \varphi$ is higher than anywhere outside. It might seem that this implies the use of tables of HDRs of log chi-squared, but in practice it is more convenient to use tables of the corresponding values of chi-squared, and such tables can be found in the Appendix. In fact it does not make much difference whether we look for regions of highest density of the inverse chi-squared distribution or of the log chi-squared distribution, but in so far as there is a difference it seems preferable to base inferences on the log chi-squared distribution.

Example

When we considered the normal distribution with unknown mean but known variance, we had to admit that this was a situation which rarely occurred in real-life examples. This is even more true when it comes to the case where the mean is known and the variance unknown, and it should really be thought of principally as a building block towards the structure we shall erect to deal with the more realistic case where both mean and variance are unknown.

We shall therefore consider an example in which the variance was in fact unknown, but treat it *as if* the variance were known. The following numbers give the uterine weight (in mg) of 20 rats drawn at random from a large stock:

9	18	21	26
14	18	22	27
15	19	22	29
15	19	24	30
16	20	24	32

It is easily checked that $n = 20$, $\sum x_i = 420$, $\sum x_i^2 = 9484$, so that $\bar{x} = 21.0$ and

$$S = \sum (x_i - \bar{x})^2 = \sum x_i^2 - (\sum x_i)^2/n = 664.$$

In such a case we do not *know* that the mean is 21.0 (or at least it is difficult to imagine circumstances in which we could have this information). However, we shall exemplify the methodology for the case where the mean is known by analysing this data *as if* we knew that the mean were $\mu = 21.0$. If this were so, then we would be able to assert that

$$\varphi \sim 664\chi_{20}^{-2}.$$

All the information we have about the variance φ is contained in this statement, but of course it is not necessarily easy to interpret from the point of view of someone inexperienced with the use of statistical methods (or even of someone who is but does not know about the inverse chi-squared distribution). Accordingly, it may be useful to give some idea of the distribution if we look for an HDR. From the tables in the Appendix, we see that the values of chi-squared corresponding to a 95% HDR for log chi-squared are 9.958 and 35.227, so that the interval for φ is from 664/35.227 to 664/9.958, that is, is the interval (19, 67). (We note that it is foolish to quote too many significant figures in our conclusions, though it may be sensible to carry through extra significant figures in intermediate calculations.) It may be worth comparing this with the results from looking at HDRs for the inverse chi-squared distribution itself. From the tables in the Appendix a 95% HDR for the inverse chi-squared distribution on 20 degrees of freedom lies between 0.025 and 0.094, so that the interval for φ is from 664×0.025 to 664×0.094, that is, is the interval (17, 62). It follows that the two methods do not give notably different answers.

2.9 The role of sufficiency

Definition of sufficiency

When we considered the normal variance with known mean, we found that the posterior distribution depended on the data only through the single number S. It often turns out that the data can be reduced in a similar way to one or two numbers, and as long as we know them we can forget the rest of the data. It is this notion that underlies the formal definition of sufficiency.

Suppose observations $\boldsymbol{x} = (x_1, x_2, \ldots, x_n)$ are made with a view to gaining knowledge about a parameter θ, and that

$$t = t(\boldsymbol{x})$$

is a function of the observations. We call such a function a *statistic*. We often suppose that t is real-valued, but it is sometimes vector-valued.

Using the formulae in Section 1.4 on "Several random variables" and the fact that once we know $\boldsymbol{x}$ we automatically know the value of t, we see that for any statistic t

$$p(\boldsymbol{x}|\theta) = p(\boldsymbol{x}, t|\theta) = p(t|\theta)p(\boldsymbol{x}|t, \theta).$$

However, it sometimes happens that

$$p(\boldsymbol{x}|t, \theta)$$

does not depend on θ, so that

$$p(\boldsymbol{x}|\theta) = p(t|\theta)p(\boldsymbol{x}|t).$$

If this happens, we say that t is a *sufficient statistic* for θ given $\boldsymbol{x}$, often abbreviated by saying that t is sufficient for θ. It is occasionally useful to have a further definition as follows: a statistic $u = u(\boldsymbol{x})$ whose density $p(u|\theta) = p(u)$ does not depend on θ is said to be *ancillary* for θ.

Neyman's Factorization Theorem

The following theorem is frequently used in finding sufficient statistics:

Theorem. A statistic t is sufficient for θ given $\boldsymbol{x}$ if and only if there are functions f and g such that

$$p(\boldsymbol{x}|\theta) = f(t, \theta)g(\boldsymbol{x})$$

where $t = t(\boldsymbol{x})$.

Proof. If t is sufficient for θ given $\boldsymbol{x}$ we may take

$$f(t, \theta) = p(t|\theta) \quad \text{and} \quad g(\boldsymbol{x}) = p(\boldsymbol{x}|t).$$

Conversely, if the condition holds, we integrate or sum both sides of the equation over all values of $\boldsymbol{x}$ such that $t(\boldsymbol{x}) = t$. Then the left-hand side will, from the basic properties of a density, be the density of t at that particular value, and so we get

$$p(t|\theta) = f(t, \theta)G(t)$$

where $G(t)$ is obtained by summing or integrating $g(\boldsymbol{x})$ over all these values of $\boldsymbol{x}$. We then have

$$f(t, \theta) = p(t|\theta)/G(t).$$

Considering now any one value of $\boldsymbol{x}$ such that $t(\boldsymbol{x}) = t$ and substituting in the equation in the statement of the theorem

$$p(\boldsymbol{x}|\theta) = p(t|\theta)g(\boldsymbol{x})/G(t).$$

Since whether t is sufficient or not

$$p(\boldsymbol{x}|t, \theta) = p(\boldsymbol{x}, t|\theta)/p(t|\theta) = p(\boldsymbol{x}|\theta)/p(t|\theta)$$

we see that

$$p(\boldsymbol{x}|t, \theta) = g(\boldsymbol{x})/G(t).$$

Since the right-hand side does not depend on θ, it follows that t is indeed sufficient, and the theorem is proved.

Sufficiency Principle

Theorem. A statistic t is sufficient for θ given $\boldsymbol{x}$ if and only if

$$l(\theta|\boldsymbol{x}) \propto l(\theta|t)$$

whenever $t = t(\boldsymbol{x})$ (where the constant of proportionality does not, of course, depend on θ).

Proof. If t is sufficient for θ given $\boldsymbol{x}$ then

$$l(\theta|\boldsymbol{x}) \propto p(\boldsymbol{x}|\theta) = p(t|\theta)p(\boldsymbol{x}|t) \propto p(t|\theta) \propto l(\theta|t).$$

Conversely, if the condition holds then

$$p(\boldsymbol{x}|\theta) \propto l(\theta|\boldsymbol{x}) \propto l(\theta|t) \propto p(t|\theta)$$

so that for some function $g(\boldsymbol{x})$

$$p(\boldsymbol{x}|\theta) = p(t|\theta)g(\boldsymbol{x}).$$

The theorem now follows from the Factorization Theorem.

Corollary 1. For *any* prior distribution, the posterior distribution of θ given $\boldsymbol{x}$ is the same as the posterior distribution of θ given a sufficient statistic t.

Proof. From Bayes' Theorem $p(\theta|\boldsymbol{x})$ is proportional to $p(\theta|t)$; they must then be equal as they both integrate or sum to unity.

Corollary 2. If a statistic $t = t(\boldsymbol{x})$ is such that $l(\theta|\boldsymbol{x}) \propto l(\theta|\boldsymbol{x}')$ whenever $t(\boldsymbol{x}) = t(\boldsymbol{x}')$, then it is sufficient for θ given $\boldsymbol{x}$.

Proof. By summing or integrating it follows that

$$l(\theta|t) \propto p(t|\theta) = \sum p(\boldsymbol{x}'|\theta) \propto \sum l(\theta|\boldsymbol{x}') \propto l(\theta|\boldsymbol{x})$$

the summations being over all $\boldsymbol{x}'$ such that $t(\boldsymbol{x}') = t = t(\boldsymbol{x})$. The result now follows from the theorem.

Examples

Normal variance. In the case where the x_i are normal of known mean μ and unknown variance φ, we noted that

$$p(\boldsymbol{x}\,|\,\varphi) \propto \varphi^{-n/2} \exp(-\tfrac{1}{2}S/\varphi)$$

where $S = \sum(x_i - \mu)^2$. It follows from the Factorization Theorem that S is sufficient for φ given $\boldsymbol{x}$. Moreover, we can verify the Sufficiency Principle as follows. If we had simply been given the value of S without being told the values of $x_1, x_2, \ldots, x_n$ separately we could have noted that for each i

$$(x_i - \mu)/\sqrt{\varphi} \sim \mathrm{N}(0, 1)$$

so that S/φ is a sum of squares of n independent N(0, 1) variables. Now a χ_n^2 distribution is often *defined* as being the distribution of the sum of squares of n random variables with an N(0, 1) distribution, and the density of χ_n^2 can be deduced from this. It follows that $S/\varphi \sim \chi_n^2$ and hence if $y = S/\varphi$ then

$$p(y) \propto y^{n/2-1} \exp(-\tfrac{1}{2}y).$$

Using the change of variable rule it is then easily seen that

$$p(S\,|\,\varphi) \propto S^{n/2-1} \varphi^{n/2} \exp(-\tfrac{1}{2}S/\varphi)$$

We can thus verify the Sufficiency Principle in this particular case because

$$l(\varphi\,|\,\boldsymbol{x}) \propto p(\boldsymbol{x}\,|\,\varphi) \propto S^{-n/2+1} p(S\,|\,\varphi) \propto p(S\,|\,\varphi) \propto l(\varphi\,|\,S).$$

Poisson case. We recall that the integer-valued random variable x is said to have a Poisson distribution of mean λ (denoted $x \sim \mathrm{P}(\lambda)$) if

$$p(\lambda) = (\lambda^x/x!) \exp(-\lambda) \qquad (x = 0, 1, 2, \ldots)$$

We shall consider the Poisson distribution in more detail later. For the moment, all that matters is that it often serves as a model for the number of occurrences of a rare event, for example, for the number of times the King's Arms on the riverbank at York is flooded in a year. Then if $x_1, x_2, \ldots, x_n$ have independent Poisson distributions with the same mean (so could, for example, represent the numbers of floods in several successive years), it is easily seen that

$$p(\boldsymbol{x}\,|\,\lambda) \sim \lambda^T \exp(-n\lambda)$$

where

$$T = \sum x_i.$$

It follows from the Factorization Theorem that T is sufficient for λ given $\boldsymbol{x}$. Moreover we can verify the Sufficiency Principle as follows. If we had simply been given the value of T without being given the values of the x_i separately, we could have noted that a sum of independent Poisson distributions has a Poisson distribution with mean the sum of the means, so that

$$T \sim \mathrm{P}(n\lambda)$$

and hence

$$l(\lambda \mid T) \propto p(T \mid \lambda) = \{(n\lambda)^T/T!\} \exp(-n\lambda) \propto p(\boldsymbol{x} \mid \lambda) \propto l(\lambda \mid \boldsymbol{x})$$

in accordance with the sufficiency principle.

Order statistics and minimal sufficient statistics

It may be noted that it is easy to see that whenever $\boldsymbol{x} = (x_1, x_2, \ldots, x_n)$ consists of independently identically distributed observations whose distribution depends on a parameter θ, then the *order statistic*

$$\boldsymbol{x}_{(\mathrm{O})} = (x_{(1)}, x_{(2)}, \ldots, x_{(n)})$$

which consists of the values of the x_i arranged in increasing order, so that

$$x_{(1)} \leqslant x_{(2)} \leqslant \ldots \leqslant x_{(n)}$$

is sufficient for θ given $\boldsymbol{x}$.

This helps to underline the fact that there is, in general, no such thing as a *unique* sufficient statistic. Indeed, if t is sufficient for θ given $\boldsymbol{x}$, then so is (t, u) for *any* statistic $u(\boldsymbol{x})$. If t is a function of all other sufficient statistics that can be constructed, so that no further reduction is possible, then t is said to be *minimal sufficient*. Even a minimal sufficient statistic is not unique, since any one–one function of such a statistic is itself minimal sufficient.

It is not obvious that a minimal sufficient statistic always exists, but in fact it does. Although the result is more important than the proof, we shall now prove this. We define a statistic $u(\boldsymbol{x})$ which is a *set*, rather than a real number or a vector, by

$$u(\boldsymbol{x}) = \{\boldsymbol{x}'; \, l(\theta \mid \boldsymbol{x}') \propto l(\theta \mid \boldsymbol{x})\}.$$

Then it follows from Corollary 2 to the Sufficiency Principle that u is sufficient. Further, if $v = v(\boldsymbol{x})$ is any other sufficient statistic, then by the same principle whenever $v(\boldsymbol{x}') = v(\boldsymbol{x})$ we have

$$l(\theta \mid \boldsymbol{x}') \propto l(\theta \mid v) \propto l(\theta \mid \boldsymbol{x})$$

and hence $u(\boldsymbol{x}') = u(\boldsymbol{x})$, so that u is a function of v. It follows that u is minimal sufficient. We can now conclude that the condition that

$$t(\boldsymbol{x}') = t(\boldsymbol{x})$$

if and only if

$$l(\theta \mid \boldsymbol{x}') \propto l(\theta \mid \boldsymbol{x})$$

is equivalent to the condition that t is minimal sufficient.

Examples on minimal sufficiency

Normal variance. In the case where the x_i are independently $\mathrm{N}(\mu, \varphi)$ where μ is known but φ is unknown, then S is not merely sufficient but minimal sufficient.

Poisson case. In the case where the x_i are independently $\mathrm{P}(\lambda)$, then T is not merely sufficient but minimal sufficient.

Cauchy distribution. We say that x has a Cauchy distribution with location parameter θ and scale parameter 1, denoted $x \sim \mathrm{C}(\theta, 1)$ if it has density

$$p(x \mid \theta) = \pi^{-1}\{1 + (x - \theta)^2\}^{-1}.$$

It is hard to find examples of real data which follow a Cauchy distribution, but the distribution often turns up in counter-examples in theoretical statistics [for example, a mean of n variables with a $\mathrm{C}(\theta, 1)$ distribution has itself a $\mathrm{C}(\theta, 1)$ distribution and does not tend to normality as n tends to infinity in apparent contradiction of the Central Limit Theorem]. Suppose that $x_1, x_2, \ldots, x_n$ are independently $\mathrm{C}(\theta, 1)$. Then if $l(\theta \mid \boldsymbol{x}') \propto l(\theta \mid \boldsymbol{x})$ we must have

$$\prod \{1 + (x_k' - \theta)^2\} \propto \prod \{1 + (x_k - \theta)^2\}.$$

By comparison of the coefficients of θ^{2n} the constant of proportionality must be 1 and by comparison of the zeroes of both sides considered as polynomials in θ, viz. $x_k' \pm i$ and $x_k \pm i$ respectively, we see that the x_k' must be a permutation of the x_k and hence the order statistics $\boldsymbol{x}'_{(\mathrm{O})}$ and $\boldsymbol{x}_{(\mathrm{O})}$ of $\boldsymbol{x}'$ and $\boldsymbol{x}$ are equal. It follows that the order statistic $\boldsymbol{x}_{(\mathrm{O})}$ is a minimal sufficient statistic, and in particular there is no one-dimensional sufficient statistic. This sort of situation is unusual with the commoner statistical distributions, but we should be aware that it can arise.

A useful reference for advanced workers in this area is Huzurbazar (1976).

2.10 Conjugate prior distributions

Definition and difficulties

When the normal variance was first mentioned, it was stated that it helps if the prior is of such that the posterior is of a "nice" form, and this led to the suggestion that if a reasonable approximation to your prior beliefs could be managed by using (a multiple of) an inverse chi-squared distribution it would be sensible to employ this distribution. It is this thought which leads to the notion of conjugate families. The usual definition adopted is as follows.

Let l be a likelihood function $l(\theta\,|\,\boldsymbol{x})$. A class Π of prior distributions is said to form a *conjugate family* if the posterior density

$$p(\theta\,|\,\boldsymbol{x}) \propto p(\theta)l(\theta\,|\,\boldsymbol{x})$$

is in the class Π for all $\boldsymbol{x}$ whenever the prior density is in Π.

There is actually a difficulty with this definition, as was pointed out by Diaconis and Ylvisaker (1979, 1985). If Π is a conjugate family and $q(\theta)$ is *any* fixed function, then the family Ψ of densities proportional to $q(\theta)p(\theta)$ for $p\in\Pi$ is also a conjugate family. While this is a logical difficulty, we are in practice only interested in "natural" families of distributions which are at least simply related to the standard families that are tabulated. In fact there is a more precise definition available when we restrict ourselves to the exponential family (discussed in the next section), and there are not many cases discussed in this book that are not covered by that definition. Nevertheless, the usual definition gives the idea well enough.

Examples

Normal mean. In the case of several normal observations of known variance with a normal prior for the mean (discussed in Section 2.3), where

$$l(\theta\,|\,\boldsymbol{x}) \propto \exp\{-\tfrac{1}{2}(\theta-\bar{x})^2/(\varphi/n)\}$$

we showed that if the prior $p(\theta)$ is $\mathrm{N}(\theta_0, \varphi_0)$ then the posterior $p(\theta\,|\,\boldsymbol{x})$ is $\mathrm{N}(\theta_1, \varphi_1)$ for suitable θ_1 and φ_1. Consequently, if Π is the class of all normal distributions, then the posterior is in Π for all $\boldsymbol{x}$ whenever the prior is in Π. Note, however, that it would not do to let Π be the class of all normal distributions with any mean but fixed variance φ (at least unless we regard the sample size as fixed once and for all); Π must in some sense be "large enough".

Normal variance. In the case of the normal variance, where

$$l(\varphi | \boldsymbol{x}) \propto \varphi^{-n/2} \exp(-\tfrac{1}{2}S/\varphi)$$

we showed that if the prior $p(\varphi)$ is $S_0\chi_\nu^{-2}$ then the posterior $p(\varphi | \boldsymbol{x})$ is $(S_0+S)\chi_{\nu+n}^{-2}$. Consequently if Π is the class of distributions of constant multiples of inverse chi-squareds, then the posterior is in Π whenever the prior is. Again, it is necessary to take Π as a *two*-parameter rather than a one-parameter family.

Poisson distribution. Suppose $\boldsymbol{x} = (x_1, x_2, \ldots, x_n)$ is a sample from the Poisson distribution of mean λ, that is, $x_i \sim \mathrm{P}(\lambda)$. Then as we noted in the last section

$$l(\lambda | \boldsymbol{x}) \propto \lambda^T \exp(-n\lambda)$$

where $T = \sum x_i$. If λ has a prior distribution of the form

$$p(\lambda) \propto \lambda^{\nu/2-1} \exp(-\tfrac{1}{2}S_0\lambda)$$

that is, $\lambda \sim S_0^{-1}\chi_\nu^2$ so that λ is a multiple of a chi-squared random variable, then the posterior is

$$\begin{aligned} p(\lambda | \boldsymbol{x}) &\propto p(\lambda)\, l(\lambda | \boldsymbol{x}) \\ &\propto \lambda^{(\nu+2T)/2-1} \exp\{-\tfrac{1}{2}(S_0+2n)\lambda\}. \end{aligned}$$

Consequently if Π is the class distributions of constant multiples of chi-squared random variables, then the posterior is in Π whenever the prior is. There are three points to be noted. First, this family is closely related to, but different from, the conjugate family in the previous example. Second, the conjugate family consists of a family of continuous distributions although the observations are discrete; the point is that this discrete distribution depends on a continuous parameter. Third, the conjugate family in this case is usually referred to in terms of the gamma distribution, but the chi-squared distribution is preferred here in order to minimize the number of distributions we need to know about and because the reader is likely to refer to tables of chi-squared in any case; the two descriptions are of course equivalent.

Binomial distribution. Suppose that k has a binomial distribution of index n and parameter π. Then

$$l(\pi | k) \propto \pi^k(1-\pi)^{n-k}.$$

We say that π has a beta distribution with parameters α and β, denoted

$\mathrm{Be}(\alpha, \beta)$ if its density is of the form

$$p(\pi) \propto \pi^{\alpha-1}(1-\pi)^{\beta-1}.$$

(The fact that $\alpha - 1$ and $\beta - 1$ appear in the indices rather than α and β is for technical reasons). The beta distribution is described in more detail in the Appendix. If, then, π has a beta prior density, it is clear that it has a beta posterior density, so that the family of beta densities forms a conjugate family. It is a simple extension that this family is still conjugate if we have a sample of size k rather than just one observation from a binomial distribution.

Mixtures of conjugate densities

Suppose we have a likelihood $l(\theta \mid \boldsymbol{x})$ and $p_1(\theta)$ and $p_2(\theta)$ are both densities in a conjugate family Π which give rise to posteriors $p_1(\theta \mid \boldsymbol{x})$ and $p_2(\theta \mid \boldsymbol{x})$ respectively. We let α and β be any non-negative real numbers summing to unity, and write

$$p(\theta) = \alpha p_1(\theta) + \beta p_2(\theta).$$

Then it is very easy to see that the posterior corresponding to the prior $p(\theta)$ is

$$p(\theta \mid \boldsymbol{x}) = \alpha' p_1(\theta \mid \boldsymbol{x}) + \beta' p_2(\theta \mid \boldsymbol{x})$$

where

$$\frac{\alpha'/\beta'}{\alpha/\beta} = \frac{\int p_1(\theta)\, p(\theta \mid \boldsymbol{x})\, \mathrm{d}\boldsymbol{x}}{\int p_2(\theta)\, p(\theta \mid \boldsymbol{x})\, \mathrm{d}\boldsymbol{x}}.$$

More generally, it is clearly possible to take any convex combination of more than two priors in Π and obtain a corresponding convex combination of the respective posteriors. Strictly in accordance with the definition given, this would allow us to extend the definition of Π to include all such convex combinations, but this would not retain the "naturalness" of families such as the normal or the inverse chi-squared. However, the idea can be useful if, for example, you have a bimodal prior distribution. An example quoted by Diaconis and Ylvisaker (1985) is as follows. They observe that there is a big difference between spinning a coin on a table and tossing it in the air. While tossing often leads to about an even proportion of "heads" and "tails", spinning often leads to proportions like $\frac{1}{3}$ or $\frac{2}{3}$. They say that the reasons for this bias are not hard to infer, since the shape of the edge will be a strong determining factor—indeed magicians have coins that are slightly shaved; the eye cannot detect the shaving but the spun coin *always* comes up "heads". Assuming that they were not dealing with one of the said magician's coins, they thought that a fifty–fifty mixture of two beta densities, namely, $\mathrm{Be}(10, 20)$ and $\mathrm{Be}(20, 10)$, would seem a reasonable prior for the proportion π of heads (actually, they consider other possibilities as well). This is a bimodal distribution, which of course no beta density is, having modes, that is, maxima of the density, close to the modes $\pi = 0.32$ and $\pi = 0.68$ of the components. They then spun a coin 10 times, getting "heads" three times. This gives a posterior which is a B(13, 27)/B(10, 20) to B(23, 17)/B(20, 10), that is a $\Gamma(13)\Gamma(27)$ to $\Gamma(23)\Gamma(17)$ and so a 115 to 14, mixture of Be(13, 27) and Be(23, 17). We can deduce some properties of this posterior from those of the component betas. For example, the probability that π is greater than 0.5 is the sum of 115/129 times the probability that a Be(13, 27) is greater than 0.5 and 14/129 times the probability that a Be(27, 13) is greater than 0.5; and similarly for the mean.

These ideas are worth bearing in mind if you have a complicated prior which is not fully dominated by the data, and yet want to obtain a posterior about which at least something can be said without complicated numerical integration.

Is your prior really conjugate?

The answer to this question is, almost certainly, "No". Nevertheless, it is often the case that the family of conjugate priors is large enough that there is one that is sufficiently close to your real prior beliefs that the resulting posterior is barely distinguishable from the posterior that comes from using your real prior. When this is so, there are clear advantages in using a conjugate prior because of the greater simplicity of the computations. You should, however, be aware that cases can arise when no member of the conjugate family is, in the above sense, close enough, and then you may well have to proceed using numerical integration if you want to investigate the properties of the posterior.

2.11 The exponential family

Definition

It turns out that many of the common statistical distributions have a similar form. This leads to the definition that a density is from the *one-parameter exponential family* if it can be put into the form

$$p(x \mid \theta) = g(x)h(\theta)\exp\{t(x)\psi(\theta)\}$$

or equivalently if the likelihood of n independent observations $\boldsymbol{x} = (x_1, x_2, \ldots, x_n)$ from this distribution is

$$l(\theta \mid \boldsymbol{x}) \propto h(\theta)^n \exp\{\textstyle\sum t(x_i)\psi(\theta)\}.$$

It follows immediately from Neyman's Factorization Theorem that $\sum t(x_i)$ is sufficient for θ given $\boldsymbol{x}$.

Examples

Normal mean. If $x \sim \mathrm{N}(\theta, \varphi)$ with φ known then

$$p(x\,|\,\theta) = (2\pi\varphi)^{-\frac{1}{2}} \exp\{-\tfrac{1}{2}(x-\theta)^2/\varphi\}$$
$$= [(2\pi\varphi)^{-\frac{1}{2}} \exp(-\tfrac{1}{2}x^2/\varphi)] \exp(-\tfrac{1}{2}\theta^2/\varphi) \exp(x\theta/\varphi)$$

which is clearly of the above form.

Normal variance. If $x \sim \mathrm{N}(\theta, \varphi)$ with θ known then we can express the density in the appropriate form by writing

$$p(x\,|\,\varphi) = \{(2\pi)^{-\frac{1}{2}}\}\, \varphi^{-\frac{1}{2}} \exp\{-\tfrac{1}{2}(x-\theta)^2/\varphi\}.$$

Poisson distribution. In the Poisson case we can write

$$p(x\,|\,\lambda) = (1/x!) \exp(-\lambda) \exp\{x(\log\lambda)\}.$$

Binomial distribution. In the binomial case we can write

$$p(x\,|\,\pi) = \binom{n}{x}(1-\pi)^n \exp[x \log\{\pi/(1-\pi)\}].$$

Conjugate densities

When a likelihood function comes from the exponential family, so that

$$l(\theta\,|\,\boldsymbol{x}) \propto h(\theta)^n \exp\{\textstyle\sum t(x_i)\psi(\theta)\}$$

there is an unambiguous definition of a conjugate family—it is defined to be the family Π of densities such that

$$p(\theta) \propto h(\theta)^{\nu} \exp\{\tau\psi(\theta)\}.$$

This definition does fit in with the particular cases we have discussed before. For example, if x has a normal distribution $\mathrm{N}(\theta, \varphi)$ with unknown mean but known variance, the conjugate family as defined here consists of densities such that

$$p(\theta) \propto \{\exp(-\tfrac{1}{2}\theta^2/\varphi)\}^{\nu} \exp(\tau\theta/\varphi).$$

If we set $\tau = \nu\theta_0$, we see that

$$p(\theta) \propto \exp\{-\tfrac{1}{2}\nu(\theta^2 - 2\theta\theta_0)/\varphi\}$$
$$\propto (2\pi\varphi/\nu)^{-\frac{1}{2}} \exp\{-\tfrac{1}{2}(\theta-\theta_0)^2/(\varphi/\nu)\}$$

which is a normal $\mathrm{N}(\theta_0, \varphi/\nu)$ density. Although the notation is slightly different, the end result is the same as that obtained earlier.

Two-parameter exponential family

The one-parameter exponential family, as its name implies, only includes densities with one unknown parameter (and not even all of those which we shall encounter). There are a few cases in which we have *two* unknown parameters, most notably when the mean and variance of a normal distribution are both unknown, which will be considered in detail in the next section. It is this situation which prompts us to consider a generalization. A density is from the two-parameter exponential family if it is of the form

$$p(x\,|\,\theta, \varphi) = g(x)h(\theta, \varphi)\exp\{t(x)\psi(\theta, \varphi) + u(x)\chi(\theta, \varphi)\}$$

or equivalently if the likelihood of n independent observations $\boldsymbol{x} = (x_1, x_2, \ldots, x_n)$ takes the form

$$l(\theta, \varphi\,|\,\boldsymbol{x}) \propto h(\theta, \varphi)^n \exp\{\textstyle\sum t(x_i)\psi(\theta, \varphi) + \sum u(x_i)\chi(\theta, \varphi)\}.$$

Evidently the two-dimensional vector $(\sum t(x_i), \sum u(x_i))$ is sufficient for the two-dimensional vector (θ, φ) of parameters given $\boldsymbol{x}$. The family of densities conjugate to such a likelihood takes the form

$$p(\theta, \varphi) \propto h(\theta, \varphi)^{\nu} \exp\{\tau\psi(\theta, \varphi) + \upsilon\chi(\theta, \varphi)\}.$$

While the case of the normal distribution with both parameters unknown is of considerable theoretical and practical importance, we shall not encounter many other two-parameter families. The idea of the exponential family can easily be extended to a k-parameter exponential family in an obvious way, but there will be no need for more than two parameters in this book.

2.12 Normal mean and variance both unknown

Formulation of the problem

It is much more realistic to suppose that both parameters of a normal distribution are unknown rather than just one. So we consider the case where we have a set of observations $\boldsymbol{x} = (x_1, x_2, \ldots, x_n)$ which are $\mathrm{N}(\theta, \varphi)$ with θ and φ both unknown. Clearly

$$\begin{aligned} p(x\,|\,\theta, \varphi) &= (2\pi\varphi)^{-\frac{1}{2}} \exp\{-\tfrac{1}{2}(x-\theta)^2/\varphi\} \\ &= \{(2\pi)^{-\frac{1}{2}}\}\{\varphi^{-\frac{1}{2}}\exp(-\tfrac{1}{2}\theta^2/\varphi)\}\exp(x\theta/\varphi - \tfrac{1}{2}x^2/\varphi) \end{aligned}$$

from which it follows that the density is in the two-parameter exponential family as defined above. Further

$$
\begin{aligned}
l(\theta, \varphi \mid \boldsymbol{x}) &\propto p(\boldsymbol{x} \mid \theta, \varphi) \\
&= (2\pi\varphi)^{-\frac{1}{2}} \exp\{-\tfrac{1}{2}(x_1-\theta)^2/\varphi\} \\
&\quad \times \ldots \times (2\pi\varphi)^{-\frac{1}{2}} \exp\{-\tfrac{1}{2}(x_n-\theta)^2/\varphi\} \\
&\propto \varphi^{-n/2} \exp[-\tfrac{1}{2}\textstyle\sum(x_i-\theta)^2/\varphi] \\
&= \varphi^{-n/2} \exp[-\tfrac{1}{2}\{\textstyle\sum(x_i-\bar{x})^2 + n(\bar{x}-\theta)^2\}/\varphi] \\
&= \varphi^{-n/2} \exp[-\tfrac{1}{2}\{S + n(\bar{x}-\theta)^2\}/\varphi]
\end{aligned}
$$

where we define

$$S = \sum (x_i - \bar{x})^2$$

(rather than as $\sum(x_i-\mu)^2$ as in the case where the mean is known to be equal to μ). It is also convenient to define

$$s^2 = S/(n-1)$$

(rather than $s^2 = S/n$ as in the case where the mean is known).

It is worth noting that the two-dimensional vector $(\bar{x}, S)$ or equivalently $(\bar{x}, s^2)$ is clearly sufficient for (θ, φ) given $\boldsymbol{x}$.

Because this case can become quite complicated, we shall first consider the case of an indifference or "reference" prior. It is usual to take

$$p(\boldsymbol{x} \mid \theta, \varphi) \propto 1/\varphi$$

which is the product of the reference prior $p(\theta) \propto 1$ for θ and the reference prior $p(\varphi) \propto 1/\varphi$ for φ. The justification for this is that it seems unlikely that if you knew very little about either the mean or the variance, then being given information about the one would affect your judgements about the other. (Other possible priors will be discussed later). If we do take this reference prior, then

$$p(\theta, \varphi \mid \boldsymbol{x}) \propto \varphi^{-n/2-1} \exp[-\tfrac{1}{2}\{S + n(\bar{x}-\theta)^2\}/\varphi].$$

For reasons which will appear later it is convenient to set

$$\nu = n - 1$$

in the power of φ, but not in the exponential, so that

$$p(\theta, \varphi \mid \boldsymbol{x}) \propto \varphi^{-(\nu+1)/2-1} \exp[-\tfrac{1}{2}\{S + n(\bar{x}-\theta)^2\}/\varphi].$$

Marginal distribution of the mean

Now in many real problems what interests us is the mean θ, and φ is what is referred to as a *nuisance parameter*. In classical (sampling theory) statistics, nuisance parameters can be a real nuisance, but there is (at least in principle) no problem from a Bayesian viewpoint. All we need to do is to find the *marginal* (posterior) distribution of θ, and we recall from Section 1.4 on "Several random variables" that

$$\begin{aligned} p(\theta \mid \boldsymbol{x}) &= \int p(\theta, \varphi \mid \boldsymbol{x}) \, \mathrm{d}\varphi \\ &= \int_0^\infty \varphi^{-(\nu+1)/2-1} \exp\left[-\tfrac{1}{2}\{S+n(\bar{x}-\theta)^2\}/\varphi\right] \mathrm{d}\varphi. \end{aligned}$$

This integral is not too difficult—all we need to do is to substitute

$$x = \tfrac{1}{2}A/\varphi$$

where

$$A = \{S + n(\bar{x}-\theta)^2\}$$

and it reduces to a standard gamma function integral

$$\left(\int_0^\infty x^{(\nu+1)/2-1} \exp(-x) \, \mathrm{d}x\right) \Big/ A^{(\nu+1)/2}.$$

It follows that

$$p(\theta \mid \boldsymbol{x}) \propto \{S + n(\bar{x}-\theta)^2\}^{-(\nu+1)/2}$$

which is the required posterior distribution of θ. However, this is not the most convenient way to express the result. It is usual to define

$$t = \frac{\theta - \bar{x}}{s/\sqrt{n}}$$

where (as defined earlier) $s^2 = S/(n-1) = S/\nu$. Because the Jacobian $|\mathrm{d}\theta/\mathrm{d}t|$ of the transformation from θ to t is a constant, the posterior density of t is given by

$$\begin{aligned} p(t \mid \boldsymbol{x}) &\propto \{\nu s^2 + (st)^2\}^{-(\nu+1)/2} \\ &\propto \{1 + t^2/\nu\}^{-(\nu+1)/2}. \end{aligned}$$

A glance at the Appendix will show that this is the density of a random variable with Student's t distribution on ν degrees of freedom, so that we can write $t \sim \mathrm{t}_\nu$. The fact that the distribution of t depends on the single parameter ν makes it sensible to express the result in terms of this distribution rather than that of θ itself, which depends on $\bar{x}$ and S as well as on ν, and is consequently more complicated to tabulate. Note that as $\nu \rightarrow \infty$ the standard exponential limit shows that the density of t is ultimately proportional to $\exp(-\frac{1}{2}t^2)$, which is the standard normal form. On the other hand, if $\nu = 1$ we see that t has a standard Cauchy distribution $\mathrm{C}(0, 1)$, or equivalently that $\theta \sim \mathrm{C}(\bar{x}, s^2/n)$.

Because the density of Student's t is symmetrical about the origin, an HDR is also symmetrical about the origin, and so can be found simply from a table of percentage points.

Example of the posterior density for the mean

Consider the data on uterine weight of rats introduced earlier in Section 2.8 on "HDRs for the normal variance". With those data, $n = 20$, $\sum x_i = 420$, and $\sum x_i^2 = 9484$, so that $\bar{x} = 21.0$ and

$$S = \sum(x_i - \bar{x})^2 = \sum x_i^2 - (\sum x_i)^2/n = 664,$$
$$s/\sqrt{n} = \sqrt{(664/19 \times 20)} = 1.32.$$

We can deduce that that the posterior distribution of the true mean θ is given by

$$\frac{\theta - 21}{1.32} \sim t_{19}.$$

In principle, this tells us all we can deduce from the data if we have no very definite prior knowledge. It can help to understand what this means by looking for highest density regions. From tables of the t distribution the value exceeded by t_{19} with probability 0.025 is $t_{19,0.025} = 2.093$. It follows that a 95% HDR for θ is $21 \pm 1.32 \times 2.093$, i.e. the inverval (18, 24).

Marginal distribution of the variance

If we require knowledge about φ rather than θ, we use

$$p(\varphi|\boldsymbol{x}) = \int p(\theta, \varphi|\boldsymbol{x})\,d\theta$$
$$= \int_{-\infty}^{\infty} \varphi^{-(\nu+1)/2-1} \exp[-\tfrac{1}{2}\{S + n(\theta - \bar{x})^2\}/\varphi)\,d\theta$$
$$\propto \varphi^{-\nu/2-1} \exp(-\tfrac{1}{2}S/\varphi) \int_{-\infty}^{\infty} (2\pi\varphi/n)^{-\frac{1}{2}} \exp\{-\tfrac{1}{2}n(\theta - \bar{x})^2/\varphi\}\,d\theta$$
$$= \varphi^{-\nu/2-1} \exp(-\tfrac{1}{2}S/\varphi)$$

as the last integral is that of a normal density. It follows that the posterior density of the variance is $S\chi_\nu^{-2}$. Except that n is replaced by $\nu = n - 1$ the conclusion is the same as in the case where the mean is known. Similar consierations to those which arose then make it preferable to use HDRs based on log chi-squared, though with a different number of degrees of freedom.

Example of the posterior density of the variance

With the same data as before, if the mean is not known (which in real life it almost certainly would not be), the posterior density for the variance φ is $664\,\chi_{19}^{-2}$. Some idea of the meaning of this can be obtained from looking for a 95% HDR. Because values of χ_{19}^2 corresponding to an HDR for $\log\chi_{19}^2$ are found from the tables in the Appendix to be 9.267 and 33.921, a 95% HDR lies between 664/33.921 and 664/9.267, that is, the interval (20, 72). It may be worth noting that this does not differ very much from the interval (19, 67) which we found on the assumption that the mean was known.

Conditional density of the mean for given variance

We will find it useful in the next section to write the posterior in the form

$$p(\theta, \varphi\,|\,\boldsymbol{x}) = p(\varphi\,|\,\boldsymbol{x})p(\theta\,|\,\varphi, \boldsymbol{x})$$

Since

$$p(\theta, \varphi\,|\,\boldsymbol{x}) \propto \varphi^{-(\nu+1)/2-1} \exp\left[-\tfrac{1}{2}\{S+n(\bar{x}-\theta)^2\}/\varphi\right],$$
$$p(\varphi\,|\,\boldsymbol{x}) = \varphi^{-\nu/2-1} \exp\left(-\tfrac{1}{2}S/\varphi\right)$$

this implies that

$$p(\theta\,|\,\varphi, \boldsymbol{x}) \propto \varphi^{-\frac{1}{2}} \exp\left\{-\tfrac{1}{2}n(\theta-\bar{x})^2/\varphi\right\}$$

which as the density integrates to unity implies that

$$p(\theta\,|\,\varphi, \boldsymbol{x}) = (2\pi\varphi/n)^{-\frac{1}{2}} \exp\{-\tfrac{1}{2}n(\theta-\bar{x})^2/\varphi\}$$

that is, for given φ and $\boldsymbol{x}$, the distribution of the mean θ is $\mathrm{N}(\bar{x}, \varphi/n)$. This is the result we might have expected from our investigations of the case where the variance is known, although this time we have arrived at the result from conditioning on the variance in the case where neither parameter is truly known.

A distribution for the two-dimensional vector (θ, φ) of this form, in which φ has (a multiple of an) inverse chi-squared distribution and, for given φ, θ has a normal distribution, will be referred to as a *normal/chi-squared distribution*, although it is more commonly referred to as normal gamma or normal inverse gamma. (The chi-squared distribution is used to avoid unnecessary complications.)

It is possible to try to look at the *joint* posterior density of θ and φ, but two-dimensional distributions can be hard to visualize in the absence of independence, although numerical techniques can help. Some idea of an approach to this can be obtained from Box and Tiao (1973, Section 2.4).

2.13 Conjugate joint prior for the normal distribution

The form of the conjugate prior

In the last section, we considered a reference for a normal distribution with both parameters unknown, whereas in this section we shall consider a conjugate prior for this situation. It is in fact rather difficult to determine *which* member of the conjugate family to use when substantial prior information is available, and hence in practice the reference prior is often used in the hope that the likelihood dominates the prior. It is also the case that the manipulations necessary to deal with the conjugate prior are rather involved, although the end results are, of course, similar to those when we use a reference prior, with some of the parameters altered slightly. Part of the problem is the unavoidable notational complexity. Further, the notation is not agreed among the different writers on the subject. A new notation is introduced below.

We first recall that the likelihood is

$$l(\theta, \varphi \,|\, \boldsymbol{x}) \propto p(\boldsymbol{x} \,|\, \theta, \varphi) \propto \varphi^{-n/2} \exp\,[-\tfrac{1}{2}\{S + n(\theta - \bar{x})^2\}/\varphi].$$

Now suppose that your prior distribution of φ is (a multiple of) an inverse chi-squared on ν_0 degrees of freedom. It may be convenient to think of ν_0 as $l_0 - 1$, so that your prior knowledge about φ is in some sense worth l_0 observations. Thus

$$p(\varphi) \propto \varphi^{-\nu_0/2-1} \exp\,(-\tfrac{1}{2}S_0/\varphi).$$

Suppose that, conditional on φ, your prior distribution for θ is normal of mean θ_0 and variance φ/n_0, so that your prior knowledge is worth n_0 observations of variance φ or their equivalent. It is not necessarily the case that $n_0 = l_0$. Then

$$p(\theta \,|\, \varphi) = (2\pi\varphi/n_0)^{-\frac{1}{2}} \exp\,\{-\tfrac{1}{2}(\theta - \theta_0)^2/(\varphi/n_0)\}.$$

Thus the joint prior is a case of a *normal/chi-squared* distribution, which was referred to briefly at the end of the last section. Its joint density is

$$\begin{aligned} p(\theta, \varphi) &= p(\varphi)p(\theta \,|\, \varphi) \\ &\propto \varphi^{-(\nu_0+1)/2-1} \exp\,[-\tfrac{1}{2}\{S_0 + n_0(\theta - \theta_0)^2\}/\varphi] \\ &= \varphi^{-(\nu_0+1)/2-1} \exp\,\{-\tfrac{1}{2}Q_0(\theta)/\varphi\} \end{aligned}$$

where $Q_0(\theta)$ is the quadratic

$$Q_0(\theta) = n_0\theta^2 - 2(n_0\theta_0)\theta + (n_0\theta_0{}^2 + S_0).$$

It should be clear that by suitable choice of the parameters n_0, θ_0 and S_0, the quadratic $Q_0(\theta)$ can be made into any non-negative definite quadratic form.

By taking

$$\nu_0 = -1 \text{ (i.e. } l_0 = 0), \qquad n_0 = 0, \quad \text{and} \quad S_0 = 0$$

(so that the quadratic vanishes identically, that is, $Q_0(\theta) \equiv 0$) we obtain the reference prior

$$p(\theta, \varphi) \propto 1/\varphi.$$

It should be noted that if $n_0 \neq 0$ then $p(\theta, \varphi)$ is *not* a product of a function of θ and φ, so that θ and φ are *not* independent *a priori*. This does not mean that it is impossible to use a prior other than the reference prior in which the mean and the variance are independent *a priori*, but that such a prior will not be in the conjugate family, so that the posterior distribution will be complicated and it may need a lot of numerical investigation to find its properties.

Derivation of the posterior

If the prior is of a normal/chi-squared form, then the posterior is

$$\begin{aligned} p(\theta, \varphi \mid \boldsymbol{x}) &\propto p(\theta, \varphi)\, l(\theta, \varphi \mid \boldsymbol{x}) \\ &\propto \varphi^{-(\nu_0+n+1)/2-1} \\ &\quad \times \exp\left[-\tfrac{1}{2}\{(S_0+S)+n_0(\theta-\theta_0)^2+n(\theta-\bar{x})^2\}/\varphi\right] \\ &= \varphi^{-(\nu_1+1)/2-1} \exp\{-\tfrac{1}{2}Q_1(\theta)/\varphi\} \end{aligned}$$

where

$$\nu_1 = \nu_0 + n$$

and $Q_1(\theta)$ is another quadratic in θ, namely,

$$\begin{aligned} Q_1(\theta) &= (S_0+S)+n_0(\theta-\theta_0)^2+n(\theta-\bar{x})^2 \\ &= (n_0+n)\theta^2 - 2(n_0\theta_0+n\bar{x})\theta + (n_0\theta_0{}^2+n\bar{x}^2+S_0+S) \end{aligned}$$

which is in the form in which the prior was expressed, that is,

$$\begin{aligned} Q_1(\theta) &= S_1 + n_1(\theta-\theta_1)^2 \\ &= n_1\theta^2 - 2(n_1\theta_1)\theta + (n_1\theta_1{}^2 + S_1) \end{aligned}$$

if we define

$$\begin{aligned} n_1 &= n_0 + n \\ \theta_1 &= (n_0\theta_0 + n\bar{x})/n_1 \\ S_1 &= S_0 + n_0\theta_0{}^2 + n\bar{x}^2 - n_1\theta_1{}^2 \\ &= S_0 + S + (n_0{}^{-1} + n^{-1})^{-1}(\theta_0 - \bar{x})^2. \end{aligned}$$

(The second formula for S_1 follows from the first after a little manipulation—its importance is that it is less subject to rounding errors). This result has finally vindicated the claim that if the prior normal/chi-squared, then so is the posterior, so that the normal/chi-squared family is conjugate to the normal likelihood with both mean and variance unknown. Thus the posterior for φ is

$$\varphi \sim S_1 \chi_{\nu_1}^{-2}$$

and that for θ given φ is

$$\theta \mid \varphi \sim \mathrm{N}(\theta_1, \varphi/n).$$

Clearly we can adapt the argument used when we considered the reference prior to find the marginal distribution for θ. Thus the posterior distribution of

$$t = \frac{\theta - \theta_1}{s_1/\sqrt{n_1}}$$

where

$$s_1^{\,2} = S_1/\nu_1$$

is a Student's t distribution on ν_1 degrees of freedom, that is, $t \sim \mathrm{t}_{\nu_1}$.

It follows that if you use a conjugate prior, then your inferences should proceed as with the reference prior *except* that you have to replace ν by ν_1, S by S_1, n by n_1 and $\bar{x}$ by θ_1.

Example

An experimental station has had experience of growing wheat which leads it to believe that the yield per plot is more or less normally distributed with mean 100 and standard deviation 10. The station then wished to investigate the effect of a growth hormone on the yield per plot. In the absence of any other information, the prior distribution for the variance on the plots might be taken to have mean 300 and standard deviation 160. As for the mean, it is expected to be about 110, and this information is thought to be worth about 15 observations. To fit a prior of a normal/chi-squared form we first equate the mean and variance of (a multiple of) an inverse chi-squared distribution to 300 and 160^2, so that

$$S_0/(\nu_0 - 2) = 300, \qquad \text{so} \qquad 2S_0^{\,2}/(\nu_0 - 2)^2 = 180{,}000,$$

$$\frac{2S_0^{\,2}}{(\nu_0 - 2)^2(\nu_0 - 4)} = 25{,}600$$

from which $\nu_0 - 4 = 7$ and hence $\nu_0 = 11$ and $S_0 = 2700$. The other information gives $\theta_0 = 110$ and $n_0 = 15$.

Twelve plots treated with the hormone gave the following yields:

141, 102, 73, 171, 137, 91, 81, 157, 146, 69, 121, 134,

so that $n = 12$, $\sum x_i = 1423$, $\sum x_i^2 = 181{,}789$, and so $\bar{x} = 119$,

$$S = \sum(x_i - \bar{x})^2 = \sum x_i^2 - (\sum x_i)^2/n = 13{,}045.$$

The parameters of the posterior are

$$\begin{aligned}
\nu_1 &= \nu_0 + n = 23, \\
n_1 &= n_0 + n = 27, \\
\theta_1 &= (n_0\theta_0 + n\bar{x})/n_1 = 114, \\
S_1 &= S_0 + S + (n_0^{-1} + n^{-1})^{-1}(\theta_0 - \bar{x})^2 = 16{,}285, \\
s/\surd n_1 &= \surd 16{,}285/23 \times 27) = 5.1.
\end{aligned}$$

It follows that *a posteriori*

$$\frac{\theta - \theta_1}{s/\surd n_1} \sim \mathrm{t}_{\nu_1}$$

$$\varphi \sim S_1 \chi_{\nu_1}^{-2}.$$

In particular, φ is somewhere near $S_1/\nu_1 = 708$ [actually the exact mean of φ is $S_1/(\nu_1 - 2) = 775$]. Using tables of t on 23 degrees of freedom, a 95% HDR for θ is $\theta_1 \pm 2.069\ s/\surd n_1$, that is, the interval (103, 125). The chi-squared distribution can also be approximated by a normal distribution (see the Appendix).

Concluding remarks

While Bayesian techniques are *in principle* just as applicable when there are two or even more unknown parameters as when there is only one unknown parameter, the practical problems are considerably increased. The computational problems can be quite severe if the prior is not from the conjugate family, but even more importantly it is difficult to convince oneself that the prior has been specified satisfactorily. In the case of the normal distribution, the fact that if the prior is taken from the conjugate family the mean and variance are not usually independent makes it quite difficult to understand the nature of the assumption being made. Of course, the more data you have, the less the prior matters and hence some of the difficulties become less important.

Exercises on Chapter 2

1. Suppose we are given the 12 observations from a normal distribution:

15.644, 16.437, 17.287, 14.448, 15.308, 15.169,
18.123, 17.635, 17.259, 16.311, 15.390, 17.252

and we are told that the variance $\varphi = 1$. Find a 90% HDR for the posterior distribution of the mean assuming the usual reference prior.

2. A random sample of size n is to be taken from an $N(\theta, \varphi)$ distribution where φ is known. How large must n be to reduce the posterior variance of φ to the fraction φ/k of its original value (where $k > 1$)?

3. Our prior beliefs about a quantity θ are such that

$$p(\theta) \propto \begin{cases} 1 & (\theta \geqslant 0) \\ 0 & (\theta < 0) \end{cases}.$$

A random sample of size 25 is taken from an $N(\theta, 1)$ distribution and the mean of the observations is observed to be 0.33. Find a 95% HDR for θ.

4. Suppose that you have prior beliefs about an unknown quantity θ which can be approximated by an $N(\lambda, \varphi)$ distribution, while my beliefs can be approximated by an $N(\mu, \psi)$ distribution. Suppose further that the reasons that have led us to these conclusions do not overlap with one another. What distribution should represent our beliefs about θ when we take into account all the information available to both of us?

5. Suppose that you are interested in investigating how variable the performance of schoolchildren on a new mathematics test, and that you begin by trying this test out on children in 12 similar schools. It turns out that the average standard deviation is about 10 marks. You then want to try the test on a thirteenth school, which is fairly similar to those you have already investigated, and you reckon that the data on the other schools gives you a prior for the variance in this new school which has a mean of 100 and is worth 8 direct observations on the school. What is the posterior distribution for the variance if you then observe a sample of size 30 from the school of which the standard deviation is 13.2? Give an interval in which the variance lies with 90% posterior probability.

6. The following are the dried weights (in grammes) of a number of plants from a batch of seeds:

$$4.17,\ 5.58,\ 5.18,\ 6.11,\ 4.50,\ 4.61,\ 5.17,\ 4.53,\ 5.33,\ 5.14.$$

Give 90% HDRs for the mean and variance of the population from which they come.

7. Find a sufficient statistic for μ given an n-sample $\boldsymbol{x} = (x_1, x_2, \ldots, x_n)$ from the exponential distribution

$$p(\mu) = \mu^{-1} \exp(-x/\mu) \qquad (0 < x < \infty)$$

where the parameter μ can take any value in $0 < \mu < \infty$.

8. Find a (two-dimensional) sufficient statistic for (α, β) given an n-sample $\boldsymbol{x} = (x_1, x_2, \ldots, x_n)$ from the two-parameter gamma distribution

$$p(x \mid \alpha, \beta) = \{\beta^{\alpha}\Gamma(\alpha)\}^{-1} x^{\alpha-1} \exp(-x/\beta) \qquad (0 < x < \infty)$$

where the parameters α and β can take any values in $0 < \alpha < \infty$, $0 < \beta < \infty$.

9. Find a family of conjugate priors for the likelihood $l(x \mid \beta) = p(x \mid \alpha, \beta)$ where $p(x \mid \alpha, \beta)$ is as in question 8 above, but α is known.

10. Suppose that the vector $\boldsymbol{x} = (x, y, z)$ has a trinomial distribution depending on the index n and the parameter $\boldsymbol{\pi} = (\pi, \rho, \sigma)$ where $\pi + \rho + \sigma = 1$, that is,

$$p(\boldsymbol{x} \mid \boldsymbol{\pi}) = \frac{n!}{x!\,y!\,z!}\, \pi,^{x} \rho^{y} \sigma^{z} \qquad (x + y + z = n).$$

Show that this distribution is in the two-parameter exponential family.

11. Suppose that the results of a certain test are known, on the basis of general theory, to be normally distributed about the same mean μ with the same variance φ, neither of which is known. Suppose further that your prior beliefs about (μ, φ) can be represented by a normal/chi-squared distribution with

$$\nu_0 = 4, \qquad S_0 = 350, \qquad n_0 = 1, \qquad \theta_0 = 85.$$

Now suppose that 100 observations are obtained from the population with mean 89 and sample variance $s^2 = 30$. Find the posterior distribution of (μ, φ). Compare 50% prior and posterior HDRs for μ.

12. Establish the formula

$$(n_0{}^{-1} + n^{-1})^{-1}(\bar{x} - \theta_0)^2 = n\bar{x}^2 + n_0\theta_0{}^2 - n_1\theta_1{}^2$$

where $n_1 = n_0 + n$ and $\theta_1 = (n_0\theta_0 + n\bar{x})/n$, which was quoted in Section 2.13 as providing a formula for the parameter S_1 of the posterior distribution, in the case where both mean and variance are unknown, which is less susceptible to rounding errors.

3

Some Other Common Distributions

3.1 The binomial distribution

Conjugate prior

In this section, the parameter of interest is the probability π of success in a number of trials which can result in success (S) or failure (F), the trials being independent of one another and having the same probability of success. We suppose that there is a fixed number of n trials, so that we have an observation x (the number of successes) such that

$$x \sim \mathrm{B}(n, \pi)$$

from a binomial distribution of index n and parameter π, and so

$$\begin{aligned} p(x \mid \pi) &= \tbinom{n}{x} \pi^x (1-\pi)^{n-x} \qquad (x = 0, 1, \ldots, n) \\ &\propto \pi^x (1-\pi)^{n-x}. \end{aligned}$$

The binomial distribution was introduced in Section 1.3 on "Random variables" and its properties are of course summarized in the Appendix.

If your prior for π has the form

$$p(\pi) \propto \pi^{\alpha-1}(1-\pi)^{\beta-1} \qquad (0 \leqslant \pi \leqslant 1)$$

that is, if

$$\pi \sim \mathrm{Be}(\alpha, \beta)$$

has a beta distribution (which is also described in the Appendix), then the posterior evidently has the form

$$p(\pi \mid x) \propto \pi^{\alpha+x-1}(1-\pi)^{\beta+n-x-1}$$

that is

$$\pi \mid x \sim \mathrm{Be}(\alpha + x, \beta + n - x).$$

It is immediately clear that the family of beta distributions is conjugate to a binomial likelihood.

The family of beta distributions is illustrated in Fig. 3.1. Basically, any reasonably smooth unimodal distribution on [0, 1] is likely to be reasonably well approximated by some beta distribution, so that it is very often possible to approximate your prior beliefs by a member of the conjugate family, with all the simplifications that this implies. In identifying an appropriate member of the family, it is often useful to equate the mean

$$\mathsf{E}\pi = \alpha/(\alpha+\beta)$$

of $\mathrm{Be}(\alpha, \beta)$ to a value which represents your belief and $\alpha+\beta$ to a number which in some sense represents the number of observations which you reckon your prior information to be worth. (It is arguable that it should be $\alpha+\beta+1$ or $\alpha+\beta+2$ that should equal this number, but in practice this will make no real difference). Alternatively, you could equate the mean to a value which represents your beliefs about the location of π and the variance

$$\mathcal{V}\pi = \frac{\alpha\beta}{(\alpha+\beta)^2(\alpha+\beta+1)}$$

of $\mathrm{Be}(\alpha, \beta)$ to a value which represents how spread out your beliefs are.

Odds and log-odds

We sometimes find it convenient to work in terms of the odds on success against failure, defined by

$$\lambda = \pi/(1-\pi)$$

so that $\pi = \lambda/(1+\lambda)$. One reason for this is the relationship mentioned in the Appendix that if $\pi \sim \mathrm{Be}(\alpha, \beta)$ then

$$\frac{\beta}{\alpha}\lambda = \frac{\beta\pi}{\alpha(1-\pi)} \sim \mathrm{F}_{2\alpha,2\beta}$$

has Snedecor's F distribution. Moreover, the log-odds

$$\Lambda = \log \lambda = \log \{\pi/(1-\pi)\}$$

is close to having Fisher's z distribution; more precisely

$$\tfrac{1}{2}\Lambda + \tfrac{1}{2}\log(\beta/\alpha) \sim \mathrm{z}_{2\alpha,2\beta}.$$

It is then easy to deduce from the properties of the z distribution given in the Appendix that

$$\mathsf{E}\Lambda \cong \log\{(\alpha-\tfrac{1}{2})/(\beta-\tfrac{1}{2})\}$$
$$\mathcal{V}\Lambda \cong \alpha^{-1}+\beta^{-1}.$$

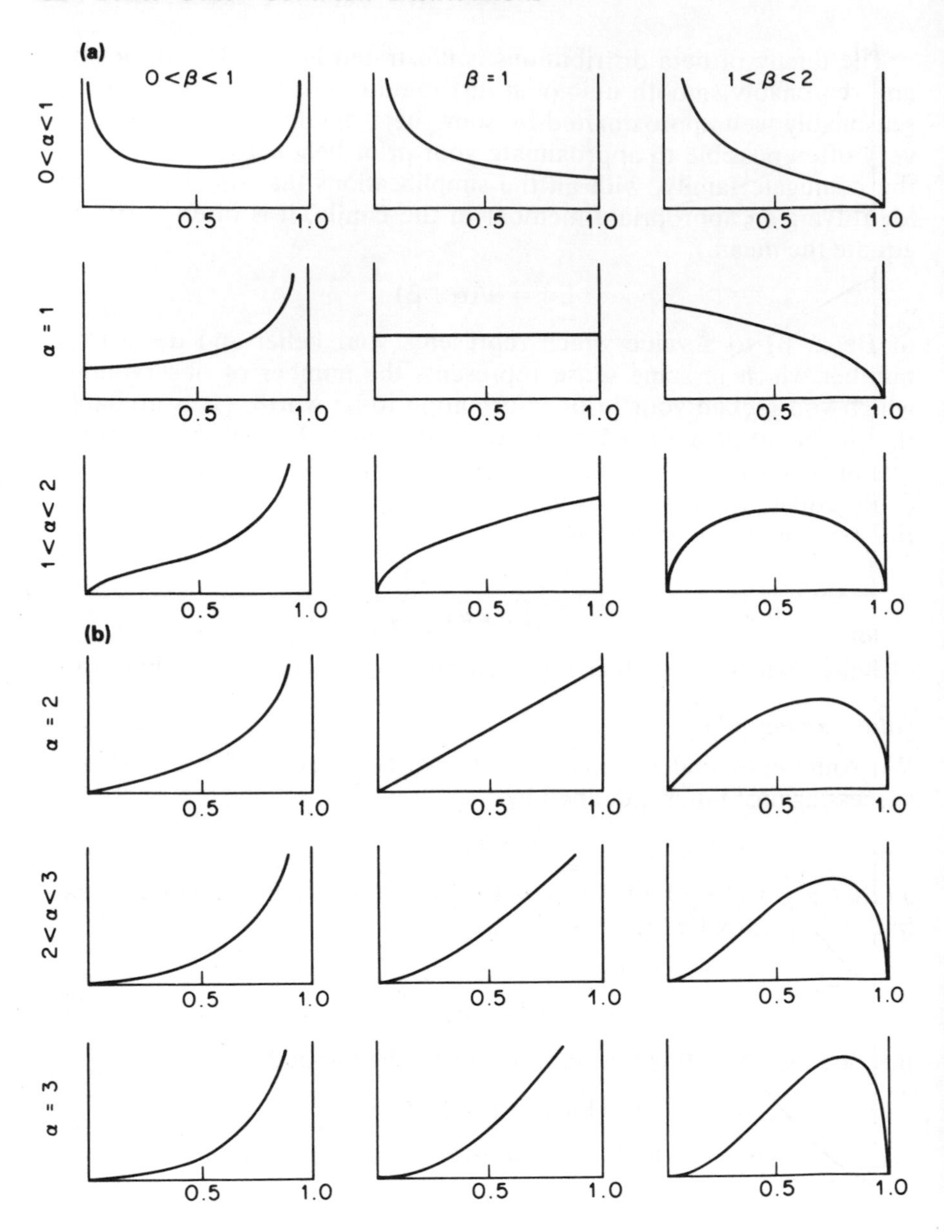

Fig. 3.1 Examples of beta densities

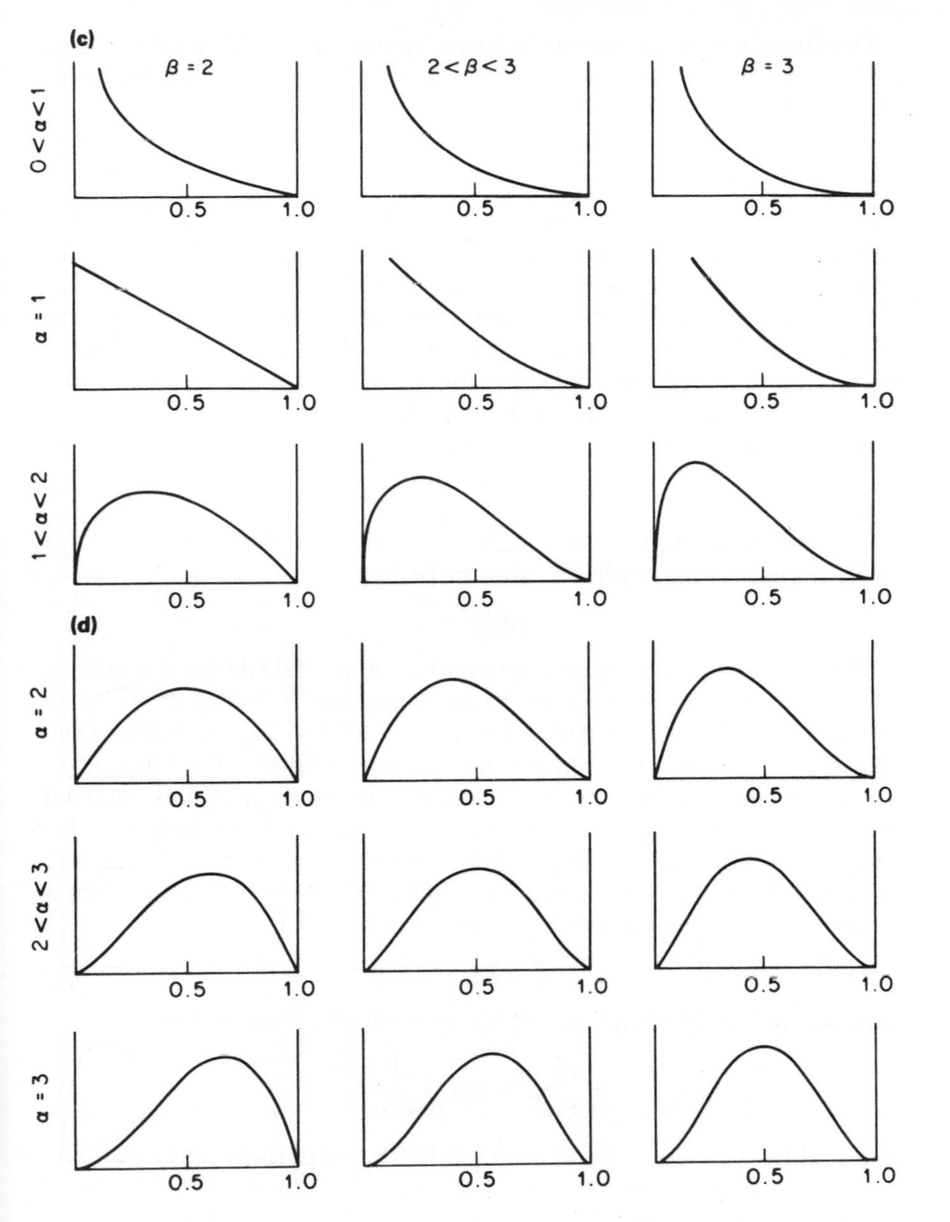
(c)
β = 2
2<β<3
β = 3
0<α<1
α = 1
1<α<2
(d)
α = 2
2<α<3
α = 3
0.5
1.0

One reason why it is useful to consider the odds and log-odds is that tables of the F and z distributions are more readily available than tables of the beta distribution.

Highest density regions

Tables of HDRs of the beta distribution are available [see, for example, Novick and Jackson (1974, Table A.15) or Isaacs *et al.* (1974, Table 43)], but it is not necessary or particularly desirable to use them. (The reason is related to the reason for not using HDRs for the inverse chi-squared distribution as such.) In the next section we shall discuss the choice of a reference prior for the unknown parameter π. It turns out that there are several possible candidates for this honour, but there is at least a reasonably strong case for using a prior

$$p(\pi) \propto \pi^{-1}(1-\pi)^{-1}.$$

Using the usual change-of-variable rule $p(\Lambda) \propto p(\pi)\ |d\pi/d\Lambda|$, it is easily seen that this implies a uniform prior

$$p(\Lambda) \propto 1$$

in the log-odds Λ. As argued in Section 2.8 on "HDRs for the normal variance", this would seem to be an argument in favour of using an interval in which the *posterior distribution of* $\Lambda = \log \lambda$ is higher than anywhere outside. The Appendix includes tables of values of F corresponding to HDRs for log F, and the distribution of Λ as deduced above as clearly very nearly that of log F. Hence in seeking for, for example, a 90% interval for π when $\pi \sim \text{Be}(\alpha, \beta)$, we should first look up values $\underline{F}$ and $\bar{F}$ corresponding to a 90% HDR for $\log F_{2\alpha,2\beta}$. Then a suitable interval for values of the odds λ is given by

$$\underline{F} \leqslant \beta\lambda/\alpha \leqslant \bar{F}$$

from which it follows that a suitable interval of values of π is

$$\frac{\alpha\underline{F}}{\beta+\alpha\underline{F}} \leqslant \pi \leqslant \frac{\alpha\bar{F}}{\beta+\alpha\bar{F}}.$$

If the tables were going to be used solely for this purpose, they could be better arranged to avoid some of the arithmetic involved at this stage, but as they are used for other purposes and do take a lot of space, the minimal extra arithmetic is justifiable.

Although this is not the reason for using these tables, a helpful thing about them is that we need not tabulate values of $\underline{F}$ and $\bar{F}$ for $\beta > \alpha$. This is because if F has an $F_{\alpha,\beta}$ distribution then F^{-1} has an $F_{\beta,\alpha}$

distribution. It follows that if an HDR for $\log F$ is $(\log \underline{F}, \log \bar{F})$ then an HDR for $\log F^{-1}$ is $(-\log \bar{F}, -\log \underline{F})$, and so if (α, β) is replaced by (β, α) then the interval $(\underline{F}, \bar{F})$ is simply replaced by $(1/\bar{F}, 1/\underline{F})$. There is no such simple relationship in tables of HDRs for F itself or in tables of HDRs for the beta distribution.

Example

It is my guess that about 20% of the best known (printable) limericks have the same word at the end of the last line as at the end of the first. However, I am not very sure about this, so I would say that my prior information was only "worth" some nine observations. If I seek a conjugate prior to represent my beliefs, I need to take

$$\alpha/(\alpha+\beta) = 0.20$$
$$\alpha+\beta = 9.$$

These equations imply that $\alpha = 1.8$ and $\beta = 7.2$. There is no particular reason to restrict α and β to integer values, but on the other hand prior information is rarely very precise, so it seems simpler to take $\alpha = 2$ and $\beta = 7$. Having made these conjectures, I then looked at one of my favourite books of light verse, Silcock (1952), and found that it included 12 limericks, of which two (both by Lear) have the same word at the ends of the first and last lines. This leads me to a posterior which is Be(4, 17). I can obtain some idea of what this distribution is like by looking for a 90% HDR. From interpolation in the tables in the Appendix, values of F corresponding to a 90% HDR for $\log F_{34,8}$ are $\underline{F} = 0.42$ and $\bar{F} = 2.85$. It follows that an appropriate interval of values of $F_{8,34}$ is $(1/\bar{F}, 1/\underline{F})$, that is (0.35, 2.38), so that an appropriate interval for π is

$$\frac{4\times 0.35}{17+4\times 0.35} \leqslant \pi \leqslant \frac{4\times 2.38}{17+4\times 2.38}$$

that is (0.08, 0.36).

If for some reason we want HDRs for π itself, instead of for $\Lambda = \log \lambda$, then we can use the tables quoted above [namely Novick and Jackson (1974, Table A.15) or Isaacs *et al.* (1974, Table 43)]. Alternatively, Novick and Jackson (1974, Section 5.5), point out that a reasonable approximation can be obtained by finding the median of the posterior distribution and looking for a 90% interval such that the probability of being between the lower bound and the median is 45% and the probability of being between the median and the upper bound is 45%. The usefulness of this procedure lies in the ease with which it can be followed using tables of the percentage points of the beta distribution alone should tables of HDRs be unavailable. It can even be used in connection with the nomogram which constitutes Table 17 of Pearson and Hartley (ed.) (1954, 1958, 1966), although the accuracy resulting leaves something to be desired. On the whole, the use of the tables of values of F corresponding to HDRs for log F, as described above, seems preferable.

Predictive distribution

The posterior distribution is clearly of the form $\text{Be}(\alpha, \beta)$ for some α and β (which, of course, include x and $n-x$ respectively), so that the predictive distribution of the next observation $y \sim \text{B}(m, \pi)$ after we have the single observation x on top of our previous background information is

$$\begin{aligned} p(y\,|\,x) &= \int p(y\,|\,\pi)p(\pi\,|\,x)\mathrm{d}\pi \\ &= \int \binom{m}{y}\pi^{y}(1-x)^{m-y}\text{B}(\alpha, \beta)^{-1}\pi^{\alpha-1}(1-x)^{\beta-1}\mathrm{d}\pi \\ &= \binom{m}{y}\text{B}(\alpha,\beta)^{-1}\text{B}(\alpha+y, \beta+m-y) \end{aligned}$$

This distribution is known as the *beta-binomial distribution*, or sometimes as the *Pólya distribution* [see, for example, Calvin (1984)]. I shall not have a great deal of use for it in this book. It is considered, for example, in Raiffa and Schlaifer (1961, Section 7.11). We shall encounter a related distribution, the *beta-Pascal distribution* in Section 7.3 when we consider informative stopping rules.

3.2 Reference prior for the binomial likelihood

Bayes' postulate

The Rev. Thomas Bayes himself in Bayes (1763) put forward arguments in favour of a uniform prior

$$p(\pi) = \begin{cases} 1 & (0 \leqslant \pi \leqslant 1) \\ 0 & (\text{otherwise}) \end{cases}$$

(which, unlike the choice of a prior uniform over $-\infty < \theta < \infty$, is a proper density in that it integrates to unity) as the appropriate one to use when we are "completely ignorant". This choice of prior has long been known as *Bayes' postulate*, as distinct from his theorem. The same prior was used by Laplace (1774). It is a member of the conjugate family, to wit $\text{Be}(1, 1)$.

Bayes' arguments are quite intricate, and still repay study. Nevertheless, he seems to have had some doubts about the validity of the postulate, and these doubts appear to have been partly responsible for the fact that his paper was not published in his lifetime, but rather communicated posthumously by his friend Richard Price.

The postulate seems intuitively reasonable, in that it seems to treat all values on a level and thus reflect the fact that there does not seem to be any reason for preferring any one value to any other. However, we should not be too hasty in endorsing it because ignorance about the value of π presumably implies ignorance about the value of any function of π, and yet when the change of variable rule is used a uniform prior for π will not usually imply a uniform prior for any function of π.

One possible argument for it is as follows. A "natural" estimator for the parameter π of a binomial distribution of index n is the observed proportion x/n of successes, and it might seem a sensible estimator to use when we have no prior information. It is in fact the *maximum likelihood* estimator, that is, the value of π for which the likelihood

$$l(\pi \mid x) \propto \pi^x(1-\pi)^{n-x}$$

is a maximum. In classical or sampling theory statistics it is also commended for various reasons which do not usually carry much weight with Bayesians, for example, that it is *unbiased*, that is,

$$\mathsf{E}(x/n) = \pi$$

(the expectation being taken over repeated sampling) whatever the value of π is. Indeed, it is not hard to show that it is a *minimum variance unbiased estimator (MVUE)*.

Now if we have a $\mathrm{Be}(\alpha, \beta)$ prior and so obtain a posterior which is $\mathrm{Be}(\alpha+x, \beta+n-x)$, it might seem natural to say that a good estimator for π would be obtained by finding that value at which the posterior density is a maximum, that is, the posterior mode. This procedure is clearly related to the idea of maximum likelihood. Since the posterior mode occurs at

$$(\alpha+x-1)/(\alpha+\beta+n-2)$$

as is easily checked by differentiation, this posterior mode coincides with x/n if and only if $\alpha = \beta = 1$, that is, the prior is uniform.

Jeffreys (1939, 1948, 1961, Section 3.1) argued that "if we take the Bayes–Laplace rule right up to the extremes we are led to results that do not correspond to anybody's way of thinking". His argument was that, in some cases, even when we know nothing the possibility that π is very close to 0 or 1 should not be ruled out, and he cited Mendelian genetics as an area of study where such possibilities could occur.

Haldane's prior

Another suggestion, due to Haldane (1931), is to use a Be(0, 0) prior, which has density

$$p(\pi) \propto \pi^{-1}(1-\pi)^{-1}$$

which is an improper density and is equivalent (by the usual change-of-variable argument) to a prior uniform in the log-odds

$$\Lambda = \log\{\pi/(1-\pi)\}.$$

An argument for this prior based on the "naturalness" of the estimator x/n when $x \sim \mathrm{B}(n, \pi)$ is that the mean of the posterior distribution for π, namely, $\mathrm{Be}(\alpha+x, \beta+n-x)$, is

$$(\alpha+x)/(\alpha+\beta+n),$$

which coincides with x/n if and only if $\alpha = \beta = 0$. (There is a connection here with the classical notion of the unbiasedness of x/n).

Another argument that has been used for this prior is that since any observation always increases either α or β, it corresponds to the greatest possible ignorance to take α and β as small as possible. For a beta density to be proper (that is, to have a finite integral and so be normalizable so that its integral is unity) it is necessary and sufficient that α and β should both be strictly greater than 0. This can then be taken as an indication that the right reference prior is Be(0, 0).

A point against this choice of prior is that if we have one observation with probability of success π, then use of this prior results in a posterior which is Be(1, 0) if that observation is a success and Be(0, 1) if it is a failure. However, a Be(1, 0) distribution gives infinitely more weight to values near 1 than to values away from 1, and so it would seem that a sample with just one success in it would lead us to conclude that all future observations will result in successes, which seems unreasonable on the basis of so small an amount of evidence.

The arc-sine distribution

A possible compromise between Be(1, 1) and Be(0, 0) is $\mathrm{Be}(\frac{1}{2}, \frac{1}{2})$, that is, the (proper) density

$$p(\pi) \propto \pi^{-\frac{1}{2}}(1-\pi)^{-\frac{1}{2}}.$$

This distribution is sometimes called the *arc-sine distribution* (cf. Feller, 1968, Vol. 1, III.4). In the next section, we will see that a general principle known as *Jeffreys' rule* suggests that this is the correct reference prior to use. However, Jeffreys' rule is a guideline which cannot be followed blindly, so that in itself does not settle the matter.

The $\mathrm{Be}(\frac{1}{2}, \frac{1}{2})$ prior can easily be shown (by the usual change-of-variable rule $p(\psi) = p(\pi)\,|\mathrm{d}\pi/\mathrm{d}\psi|$) to imply a uniform prior for

$$\psi = \sin^{-1}\sqrt{\pi}.$$

This transformation is related to the transformation of the *data* when

$$x \sim \mathrm{B}(n, \pi)$$

in which z is defined by

$$z = \sin^{-1}\sqrt{(x/n)}.$$

This transformation was first introduced in Section 1.5 on "Means and variances", where we saw that it results in the approximations

$$\mathsf{E}z \cong \sin^{-1}\sqrt{\pi} = \psi,$$
$$\mathcal{V}z \cong 1/4n.$$

Indeed, it turns out that

$$z \approx \mathrm{N}(\psi, 1/4n)$$

where the symbol $\approx$ means "is approximately distributed as" (see Section 3.10 on "Approximations based on the likelihood"). To the extent that this is so, it follows that the transformation $\psi = \psi(\pi)$ puts the likelihood in data translated form, and hence that a uniform prior in ψ, that is, a $\mathrm{Be}(\frac{1}{2}, \frac{1}{2})$ prior for π, is an appropriate reference prior.

Conclusion

The three possibilities above are not the only ones that have been suggested. For example, Zellner (1977) suggested the use of a prior

$$p(\pi) \propto \pi^{\pi}(1-\pi)^{1-\pi}$$

[see also the references in Berger (1985, Section 3.3.4)]. However, this is difficult to work with because it is not in the conjugate family.

In fact, the three suggested conjugate priors $\mathrm{Be}(0, 0)$, $\mathrm{Be}(\frac{1}{2}, \frac{1}{2})$ and $\mathrm{Be}(1, 1)$ (and for that matter Zellner's prior) do not differ enough to make much difference with even a fairly small amount of data, and the discussion above on the problem of a suitable reference prior may be too lengthy, except for the fact that it does underline the difficulty in giving a precise meaning to the notion of a prior distribution that represents "knowing nothing". The reader may find it worth while to try a few examples to see how little difference there is between the possible priors in particular cases.

In practice, the use of Be(0, 0) is favoured here, although it must be admitted that one reason for this is that it ties in with the use of HDRs found from tables of values of F based on HDRs for log F and hence obviates the need for a separate set of tables for the beta distribution. But in any case we could use the method based on these tables and the results would not be very different from those based on any other appropriate tables.

3.3 Jeffreys' rule

Fisher's information

In Section 2.1 on the "Nature of Bayesian inference", the log-likelihood function was defined as

$$L(\theta\,|\,x) = \log l(\theta\,|\,x).$$

In this section we shall sometimes use l for $l(\theta\,|\,x)$, L for $L(\theta\,|\,x)$ and p for the probability density function $p(x\,|\,\theta)$. The fact that the likelihood can be multiplied by any constant implies that the log-likelihood contains an arbitrary additive constant.

An important concept in classical statistics which arises, for example, in connection with the Cramèr–Rao bound for the variance of an unbiased estimator, is that of the information provided by an experiment which was defined in a paper by Fisher (1925a) as

$$I(\theta\,|\,x) = -\,\mathsf{E}\partial^2(\log p)/\partial\theta^2$$

the expectation being taken over possible values of x for fixed θ. It is important to note that the information depends on the *distribution* of the data rather than on any particular value, so that if we carry out an experiment and observe, for example, that $\tilde{x} = 3$, then the information is no different from the information if $\tilde{x} = 5$; basically it is to do with what can be expected from an experiment *before* rather than after it has been performed. It may be helpful to note that strictly speaking it should be denoted

$$I(\theta\,|\,\tilde{x}).$$

Because the log-likelihood differs from $\log p(x\,|\,\theta)$ by a constant, all their derivatives are equal, and we can equally well define the information by

$$I(\theta\,|\,x) = -\mathsf{E}\partial^2 L/\partial\theta^2.$$

It is useful to prove two lemmas.

Lemma 1. $\mathsf{E}\partial L/\partial\theta = 0.$

Proof.
$$\begin{aligned}\mathsf{E}\partial L/\partial\theta &= \int\{\partial(\log l)/\partial\theta\}p\,\mathrm{d}x = \int\{\partial(\log p)/\partial\theta\}p\,\mathrm{d}x\\ &= \int\{(\partial p/\partial\theta)/p\}p\,\mathrm{d}x = \int(\partial p/\partial\theta)\,\mathrm{d}x\\ &= \frac{\mathrm{d}}{\mathrm{d}\theta}\int p\,\mathrm{d}x = \frac{\mathrm{d}}{\mathrm{d}\theta}1 = 0\end{aligned}$$

since in any reasonable case it makes no difference whether differentiation with respect to θ is carried out inside or outside the integral with respect to x.

Lemma 2. $I(\theta\,|\,x) = \mathsf{E}(\partial L/\partial\theta)^2$.

Proof.
$$\begin{aligned}I(\theta\,|\,x) &= -\mathsf{E}\partial^2(\log l)/\partial\theta^2 = -\int\{\partial^2(\log p)/\partial\theta^2\}\,p\,\mathrm{d}x\\ &= -\int\frac{\partial}{\partial\theta}\left(\frac{\partial p/\partial\theta}{p}\right)p\,\mathrm{d}x\\ &= -\int\left(\frac{\partial^2 p/\partial\theta^2}{p}\right)p\,\mathrm{d}x + \int\left(\frac{(\partial p/\partial\theta)^2}{p^2}\right)p\,\mathrm{d}x\\ &= -\int(\partial^2 p/\partial\theta^2)\,\mathrm{d}x + \int\{\partial(\log p)/\partial\theta\}^2\,p\,\mathrm{d}x\\ &= -\frac{\mathrm{d}^2}{\mathrm{d}\theta^2}1 + \int(\partial L/\partial\theta)^2\,p\,\mathrm{d}x\\ &= \mathsf{E}(\partial L/\partial\theta)^2.\end{aligned}$$

The information from several observations

If we have n independent observations $\boldsymbol{x} = (x_1, x_2, \ldots, x_n)$, then the probability densities multiply, so the log-likelihoods add. Consequently, if we define

$$I(\theta\,|\,\boldsymbol{x}) = -\mathsf{E}\partial^2 L(\theta\,|\,\boldsymbol{x})/\partial\theta^2.$$

then by linearity of expectation

$$I(\theta\,|\,\boldsymbol{x}) = n\,I(\theta\,|\,x)$$

where x is any one of the x_i. This accords with the intuitive idea that n times as many observations should give us n times as much information about the value of an unknown parameter.

Jeffreys' prior

In a Bayesian context, the important thing to note is that if we transform the unknown parameter θ to $\psi = \psi(\theta)$ then

$$\frac{\partial\{\log l(\psi\,|\,x)\}}{\partial\psi} = \frac{\partial\{\log l(\theta\,|\,x)\}}{\partial\theta}\,\frac{\mathrm{d}\theta}{\mathrm{d}\psi}.$$

Squaring and taking expectations over values of x (and noting that $\mathrm{d}\theta/\mathrm{d}\psi$ does not depend on x), it follows that

$$I(\psi\,|\,x) = I(\theta\,|\,x)\,(\mathrm{d}\theta/\mathrm{d}\psi)^2.$$

It follows from this that if a prior density

$$p(\theta) \propto \sqrt{I(\theta\,|\,x)}$$

is used, then by the usual change-of-variable rule

$$p(\psi) \propto \sqrt{I(\psi\,|\,x)}.$$

It is because of this property that Jeffreys (1939, 1948, 1961, Section 3.10) suggested that the density

$$p(\theta) \propto \sqrt{I(\theta\,|\,x)}$$

provided a suitable reference prior (the use of this prior is sometimes called *Jeffreys' rule*). This rule has the valuable property that the prior is *invariant* in that, whatever scale we choose to measure the unknown parameter in, the same prior results when the scale is transformed to any particular scale. This seems a highly desirable property of a reference prior. In Jeffrey's words, "any arbitrariness in the choice of parameters could make no difference to the results".

Examples

Normal mean. For the normal mean with known variance, the log-likelihood is

$$L(\theta\,|\,x) = -\tfrac{1}{2}(x-\theta)^2/\varphi + \text{constant}$$

so that

$$\partial^2 L/\partial\theta^2 = -1/\varphi$$

which does not depend on x, so that

$$I(\theta\,|\,x) = 1/\varphi$$

implying that we should take a prior

$$p(\theta) \propto 1/\sqrt{\varphi} = \text{constant}$$

which is the rule suggested earlier for a reference prior.

Normal variance. In the case of the normal variance

$$L(\varphi\,|\,x) = -\tfrac{1}{2}\log\varphi - \tfrac{1}{2}(x-\theta)^2/\varphi + \text{constant}$$

so that

$$\partial^2 L/\partial\varphi^2 = \tfrac{1}{2}\varphi^{-2} - (x-\theta)^2/\varphi^3.$$

Because $\mathsf{E}(x-\theta)^2 = \mathcal{V}x = \varphi$, it follows that

$$I(\varphi\,|\,x) = -\tfrac{1}{2}\varphi^{-2} + \varphi/\varphi^3 = \tfrac{1}{2}\varphi^{-2}$$

implying that we should take a prior

$$p(\varphi) \propto 1/\varphi$$

which again is the rule suggested earlier for a reference prior.

Binomial parameter. In this case

$$L(\pi\,|\,x) = x \log \pi + (n-x) \log (1-\pi) + \text{constant}$$

so that

$$\partial^2 L/\partial\pi^2 = -x/\pi^2 - (n-x)/(1-\pi)^2.$$

Because $\mathsf{E}x = n\pi$ it follows that

$$I(\pi\,|\,x) = n\pi/\pi^2 + (n-n\pi)/(1-\pi)^2 = n\pi^{-1}(1-\pi)^{-1}$$

implying that we should take a prior

$$p(\pi) \propto \pi^{-\frac{1}{2}}(1-\pi)^{-\frac{1}{2}}$$

that is, $\pi \sim \mathrm{Be}(\tfrac{1}{2}, \tfrac{1}{2})$, or π has an arc-sine distribution, which is one of the rules suggested earlier as possible choices for the reference prior in this case.

Warning

While Jeffreys' rule is suggestive, it cannot be applied blindly. Apart from anything else, the integral defining the information can diverge; it is easily seen to do so for the Cauchy distribution $\mathrm{C}(\theta, 1)$, for example. It should be thought of as a guideline that is well worth considering, particularly if there is no other obvious way of finding a prior distribution. Generally speaking, it is less useful if there are more unknown parameters than one, although an outline of the generalization to that case is given below for reference

Several unknown parameters

If there are several unknown parameters $\boldsymbol{\theta} = (\theta_1, \theta_2, \ldots, \theta_k)$, the information $I(\boldsymbol{\theta}\,|\,x)$ provided by a single observation is defined as a *matrix*, the element in row i, column j, of which is

$$(I(\boldsymbol{\theta}\,|\,x))_{i,j} = -\mathsf{E}(\partial^2 L/\partial\theta_i\partial\theta_j).$$

As in the one parameter case, if there are several observations $\boldsymbol{x} = (x_1, x_2, \ldots, x_n)$ we obtain

$$I(\boldsymbol{\theta}\,|\,\boldsymbol{x}) = n\,I(\boldsymbol{\theta}\,|\,x).$$

If we then transform to new parameters $\boldsymbol{\psi} = (\psi_1, \psi_2, \ldots, \psi_k)$ where $\boldsymbol{\psi} = \boldsymbol{\psi}(\boldsymbol{\theta})$, then if J is the matrix the element in row i, column j of which is

$$J_{i,j} = \partial\theta_i/\partial\psi_j$$

then it is quite easy to see that

$$I(\boldsymbol{\psi}\,|\,\boldsymbol{x}) = JI(\boldsymbol{\theta}\,|\,\boldsymbol{x})J^{\mathrm{T}}$$

where J^{T} is the transpose of J, and hence that the determinant $\det I$ of the information matrix satisfies

$$\det I(\boldsymbol{\psi}\,|\,\boldsymbol{x}) = \{\det I(\boldsymbol{\theta}\,|\,\boldsymbol{x})\}\,(\det J)^2.$$

Because $\det J$ is the Jacobian determinant, it follows that

$$p(\boldsymbol{\theta}) \propto \surd\{\det I(\boldsymbol{\theta}\,|\,\boldsymbol{x})\}$$

provides an invariant prior for the multi-parameter case.

Example

Normal mean and variance both unknown. In this case the log-likelihood is

$$L(\theta,\varphi\,|\,x) = -\tfrac{1}{2}\log\varphi - \tfrac{1}{2}(x-\theta)^2/\varphi + \text{constant}$$

so that $\partial L/\partial\theta = (x-\theta)/\varphi$ and $\partial L/\partial\varphi = -\tfrac{1}{2}\varphi^{-1} + \tfrac{1}{2}(x-\theta)^2/\varphi^2$ and hence

$$\partial L/\partial\theta^2 = -1/\varphi;$$
$$\partial^2 L/\partial\theta\partial\varphi = -(x-\theta)/\varphi^2;$$
$$\partial^2 L/\partial\varphi^2 = \tfrac{1}{2}\varphi^{-2} - (x-\theta)^2/\varphi^3.$$

Because $\mathsf{E}x = \theta$ and $\mathsf{E}(x-\theta)^2 = \mathcal{V}x = \varphi$, it follows that

$$I(\theta,\varphi\,|\,x) = \begin{pmatrix} \varphi^{-1} & 0 \\ 0 & \tfrac{1}{2}\varphi^{-2} \end{pmatrix}$$

and so that

$$\det I(\theta,\varphi\,|\,x) = \tfrac{1}{2}\varphi^{-3}.$$

This implies that we should use the reference prior

$$p(\theta, \psi) \propto \varphi^{-3/2}.$$

It should be noted that this is *not* the same as the reference prior recommended earlier for use in this case, viz.

$$p(\theta, \psi) \propto \varphi^{-1}.$$

However, I would still prefer to use the prior recommended earlier. The invariance argument does not take into account the fact that in most such problems our judgement about the mean would not be affected by anything we were told about the variance or vice versa, and on those grounds it seems reasonable to take a prior which is the product of the reference priors for the mean and the variance separately.

The example underlines the fact that we have to be rather careful about the choice of a prior in multi-parameter cases. It is also worth mentioning that it is very often the case that when there are parameters which can be thought of as representing "location" and "scale" respectively, then it would usually be reasonable to think of these parameters as being independent *a priori*, just as suggested above in the normal case.

3.4 The Poisson distribution

Conjugage prior

A discrete random variable x is said to have a Poisson distribution of mean λ if it has the density

$$p(x \mid \lambda) = \frac{\lambda^x}{x!} \exp(-\lambda).$$

This distribution often occurs as a limiting case of the binomial distribution as the index $n \to \infty$ and the parameter $\pi \to 0$ but their product $n\pi \to \lambda$. It is thus a useful model for rare events, such as the number of radioactive decays in a fixed time interval, when we can split the interval into an arbitrarily large number of sub-intervals in any of which a particle might decay, although the probability of a decay in any particular sub-interval is small (though constant).

Suppose that we have n observations $\boldsymbol{x} = (x_1, x_2, \ldots, x_n)$ from such a distribution, so that the likelihood is

$$l(\lambda \mid \boldsymbol{x}) \propto \lambda^T \exp(-n\lambda)$$

where T is the sufficient statistic

$$T = \sum x_i.$$

We have already seen in Section 2.10 on "Conjugate prior distributions" that the appropriate conjugate density is

$$p(\lambda) \propto \lambda^{\nu/2-1} \exp(-\tfrac{1}{2}S_0\lambda)$$

that is, $\lambda \sim S_0^{-1}\chi_\nu^2$ so that λ is a multiple of a chi-squared random variable. Then the posterior density is

$$p(\lambda \mid \boldsymbol{x}) \propto \lambda^{(\nu+2T)/2-1} \exp\{-\tfrac{1}{2}(S_0 + 2n)\lambda\}$$

that is

$$\lambda \mid \boldsymbol{x} \sim S_1^{-1}\chi_{\nu'}^2$$

where

$$S_1 = S_0 + 2n, \quad \nu' = \nu + 2T.$$

Reference prior

This is a case where we can try to use Jeffreys' rule. The log-likelihood resulting from a single observation x is

$$L(\lambda \mid x) = x \log \lambda - \lambda + \text{constant}$$

so that $\partial^2 L/\partial\lambda^2 = -x/\lambda^2$ and hence

$$I(\lambda \mid x) = \lambda/\lambda^2 = 1/\lambda.$$

Consequently Jeffreys' rule suggests the prior

$$p(\lambda) \propto \lambda^{-\frac{1}{2}}$$

which corresponds to $\nu = 1$, $S_0 = 0$ in the conjugate family, and is easily seen to be equivalent to a prior uniform in $\psi = \sqrt{\lambda}$. It may be noted that there is a sense in which this is intermediate between a prior uniform in $\log \lambda$ and one uniform in λ itself, since as $k \to 0$

$$\frac{\lambda^k - 1}{k} = \frac{\exp(k \log \lambda) - 1}{k} \to \log \lambda$$

so that there is a sense in which the transformation from λ to $\log \lambda$ can be regarded as a "zeroth power" transformation (cf. Box and Cox, 1964).

On the other hand, it could be argued that λ is a scale parameter between 0 and $+\infty$ that the right reference prior should therefore be

$$p(\lambda) \propto 1/\lambda$$

which is uniform in $\log \lambda$ and corresponds to $\nu = 0$, $S_0 = 0$ in the conjugate family. However, the difference this would make in practice would almost always be negligible.

Example

The numbers of misprints spotted on the first few pages of an early draft of this book were

$$3, \quad 4, \quad 2, \quad 1, \quad 2, \quad 3.$$

It seems reasonable that these numbers should constitute a sample from a Poisson distribution of unknown mean λ. If you had no knowledge of my skill as a typist, you might adopt the reference prior uniform in $\sqrt{\lambda}$ for which $\nu = 1$, $S_0 = 0$. Since

$$n = 6, \qquad T = \sum x_i = 15$$

your posterior for λ would then be $S_1^{-1}\chi_{\nu'}^{\ 2}$, that is, $12^{-1}\chi_{31}^{\ 2}$. This distribution has mean and variance

$$\nu'/S_1 = 2.6, \qquad 2\nu'/S_1^{\ 2} = 0.43.$$

Of course, I have some experience of my own skill as a typist, so if I considered these figures, I would have used a prior with a mean of about 3 and variance about 4. (As a matter of fact, subsequent re-readings have caused me to adjust my prior beliefs about λ in an upwards direction!) If then I seek a prior in the conjugate family, I need

$$\nu/S_0 = 3, \qquad 2\nu/S_0^{\ 2} = 4.$$

which implies $\nu = 4.5$ and $S_0 = 1.5$. This means that my posterior has $\nu' = 34.5$, $S_1 = 13.5$ and so has mean and variance

$$\nu'/S_1 = 2.6, \qquad 2\nu'/S_1^{\ 2} = 0.38.$$

The difference between the two posteriors is not great and of course would become less and less as more data were included in the analysis. It would be easy enough to give HDRs. According to arguments presented in other cases, it would seem to be appropriate to use HDRs for the chi (rather than the chi-squared distribution), but it really would not make much difference if the regions were based on HDRs for chi-squared or on values of chi-squared corresponding to HDRs for log chi-squared.

Predictive distribution

Once we know that λ has a posterior distribution

$$p(\lambda) \propto \lambda^{\nu'/2-1} \exp(-\tfrac{1}{2}S_1\lambda)$$

then since

$$p(x\,|\,\lambda) \propto (\lambda^x/x!) \exp(-\lambda)$$

it follows that the predictive distribution

$$\begin{aligned} p(x) &= \textstyle\int p(x\,|\,\lambda)\,p(\lambda)\,\mathrm{d}\lambda \\ &\propto \int_0^\infty (\lambda^x/x!) \exp(-\lambda)\,\lambda^{\nu'/2-1} \exp(-\tfrac{1}{2}S_1\lambda)\,\mathrm{d}\lambda \\ &\propto \{(S_1+1)^{-\frac{1}{2}\nu'-x-1}/x!\} \int_0^\infty z^{\frac{1}{2}\nu'+x-1} \exp(-z)\,\mathrm{d}z \\ &\propto \{(S_1+1)^{-x}/x!\}\,\Gamma(x+\tfrac{1}{2}\nu') \end{aligned}$$

(dropping a factor which depends on ν' alone). Setting $\pi = 1-(S_1+1)^{-1}$ we can find the constant by reference to the Appendix. In fact, at least when $\frac{1}{2}\nu'$ is an integer, the predictive distribution is *negative binomial*, that is

$$x \sim \mathrm{NB}(\tfrac{1}{2}\nu', \pi).$$

Further, although this point is not very important, it is not difficult to see that the negative binomial distribution can be generalized to the case where $\frac{1}{2}\nu'$ is not an integer. All we need to do is to replace some factorials by corresponding gamma functions and note that (using the functional equation for the gamma function)

$$\Gamma(\tfrac{1}{2}\nu'+x)/\Gamma(\tfrac{1}{2}\nu') = (\tfrac{1}{2}\nu'+x-1)(\tfrac{1}{2}\nu'+x-2)\ldots(\tfrac{1}{2}\nu'+1)\tfrac{1}{2}\nu'$$

so that we can write the general binomial coefficient as

$$\binom{\frac{1}{2}\nu'+x-1}{x} = \frac{\Gamma(\frac{1}{2}\nu'+x)}{x!\,\Gamma(\frac{1}{2}\nu')}.$$

The negative binomial distribution is usually defined in terms of a sequence of independent trials each of which results in success or failure with the same probabilities π and $1-\pi$ (such trials are often called *Bernoulli trials*) and considering the number x of failures before the nth success. There will not be much more use for this distribution in this book, but it is interesting to see it turning up here in a rather different context.

3.5 The uniform distribution

Preliminary definitions

The *support* of a density $p(x|\theta)$ is defined as the set of values of x for which it is non-zero. A simple example of a family of densities in which the support depends on the unknown parameter is the family of uniform distributions (defined below). While problems involving the uniform distribution do not arise very often in practice, it is worthwhile seeing what complications can arise in cases where the support does depend on the unknown parameter.

It is useful to begin with a few definitions. The *indicator function* of any set A is defined by

$$I_A(x) = \begin{cases} 1 & (x \in A) \\ 0 & (x \notin A) \end{cases}.$$

This is sometimes called the *characteristic function* of the set A in some other branches of mathematics, but not in probability and statistics (where the term characteristic function is applied to the Fourier–Stieltjes transform of the distribution function).

We say that y has a *Pareto* distribution with parameters ξ and γ and write

$$y \sim \mathrm{Pa}(\xi, \gamma)$$

if it has density

$$\begin{aligned} p(y|\xi, \gamma) &= \begin{cases} \gamma\xi^{\gamma} y^{-\gamma-1} & (y > \xi) \\ 0 & (y \leqslant \xi) \end{cases} \\ &\propto y^{-\gamma-1} I_{(\xi, \infty)}(y). \end{aligned}$$

This distribution is often used as a model for distributions of income. A survey of its properties and applications can be found in Arnold (1983).

We say that x has a *uniform distribution* (or a *rectangular distribution*) on (α, β) and write

$$x \sim \mathrm{U}(\alpha, \beta)$$

if it has density

$$p(x \mid \alpha, \beta) = \begin{cases} (\beta-\alpha)^{-1} & (\alpha<x<\beta) \\ 0 & (\text{otherwise}) \end{cases}$$
$$\propto (\beta-\alpha)^{-1} I_{(\alpha,\beta)}(x)$$

so that all values in the interval (α, β) are equally likely.

Uniform distribution with a fixed lower endpoint

Now we suppose we have n independent observations $\boldsymbol{x} = (x_1, x_2, \ldots, x_n)$ such that

$$x_i \sim \mathrm{U}(0, \theta)$$

for each i, where θ is a single unknown parameter. Then

$$p(\boldsymbol{x} \mid \theta) = \begin{cases} \theta^{-n} & (\theta<x_i<\theta \quad \text{for all } i) \\ 0 & (\text{otherwise}) \end{cases}$$

It is now easy to see that we can write the likelihood as

$$l(\theta \mid \boldsymbol{x}) \propto \begin{cases} \theta^{-n} & (\theta>x_i \quad \text{for all } i) \\ 0 & (\text{otherwise}) \end{cases}$$

Defining

$$M = \max\{x_1, x_2, \ldots, x_n\}$$

it is clear that

$$l(\theta \mid \boldsymbol{x}) \propto \begin{cases} \theta^{-n} & (\theta>M) \\ 0 & (\text{otherwise}) \end{cases}$$
$$\propto \theta^{-n} I_{(M,\infty)}(\theta).$$

Because the likelihood depends on the data through M alone, it follows that M is sufficient for θ given $\boldsymbol{x}$.

It is now possible to see that the Pareto distribution provides the conjugate prior for the above likelihood. For if θ has prior

$$p(\theta) \propto \theta^{-\gamma-1} I_{(\xi,\infty)}(\theta)$$

then the posterior is

$$\begin{aligned} p(\theta \mid \boldsymbol{x}) &\propto p(\theta)\, l(\theta \mid \boldsymbol{x}) \\ &\propto \theta^{-\gamma-1} I_{(\xi,\infty)}(\theta)\, \theta^{-n} I_{(M,\infty)}(\theta) \\ &\propto \theta^{-(\gamma+n)-1} I_{(\xi,\infty)}(\theta)\, I_{(M,\infty)}(\theta). \end{aligned}$$

If now we write

$$\gamma' = \gamma + n$$
$$\xi' = \max\{\xi, M\}$$

so that $\theta > \xi'$ if and only if $\theta > \xi$ and $\theta > M$ and hence

$$I_{(\xi',\infty)}(\theta) = I_{(\xi,\infty)}(\theta)\, I_{(M,\infty)}(\theta)$$

we see that

$$p(\theta \mid \boldsymbol{x}) \propto \theta^{-\gamma'-1} I_{(\xi',\infty)}(\theta).$$

It follows that if the prior is $\text{Pa}(\xi, \gamma)$ then the posterior is $\text{Pa}(\xi', \gamma')$ and hence that the Pareto distribution does indeed provide the conjugate family. We should note that neither the uniform nor the Pareto distribution falls into the exponential family, so that we are not here employing the unambiguous definition of conjugacy given in Section 2.11 on "The exponential family". Although this means that the cautionary remarks of Diaconis and Ylvisaker (1979, 1985) (quoted in Section 2.10 on "Conjugate prior distributions") apply, there is no doubt of the "naturalness" of the Pareto distribution in this context.

The general uniform distribution

The case where both parameters of a uniform distribution are unknown is less important, but it can be dealt with similarly. In this case, it turns out that an appropriate family of conjugate prior distributions is given by the *bilateral bivariate Pareto distribution.* We say that the ordered pair (y, z) has such a distribution and write

$$(y, z) \sim \text{Pabb}(\xi, \eta, \gamma)$$

if the joint density is

$$p(y, z \mid \xi, \eta, \gamma) = \begin{cases} \gamma(\gamma+1)(\xi-\eta)^{\gamma}(z-y)^{-\gamma-2} & (y<\eta<\xi<z) \\ 0 & (\text{otherwise}) \end{cases}$$
$$\propto (z-y)^{-\gamma-2} I_{(\xi',\infty)}(z)\, I_{(-\infty,\eta)}(y).$$

Now suppose we have n independent observations $\boldsymbol{x} = (x_1, x_2, \ldots, x_n)$ such that

$$x_i \sim \text{U}(\alpha, \beta)$$

where α and β are unknown. Then

$$p(\boldsymbol{x} \mid \alpha, \beta) = \begin{cases} (\beta-\alpha)^{-n} & (\alpha<x_i<\beta \quad \text{for all } i) \\ 0 & (\text{otherwise}) \end{cases}$$

Defining

$$M = \max\{x_1, x_2, \ldots, x_n\}$$
$$m = \min\{x_1, x_2, \ldots, x_n\}$$

it is clear that the likelihood $l(\alpha, \beta \mid \boldsymbol{x})$ can be written as

$$l(\alpha, \beta \mid \boldsymbol{x}) = (\beta - \alpha)^{-n} I_{(M,\infty)}(\beta)\, I_{(-\infty, m)}(\alpha).$$

Because the likelihood depends on the data through m and M alone, it follows that (m, M) is sufficient for (α, β) given $\boldsymbol{x}$.

It is now possible to see that the bilateral bivariate Pareto distribution provides the conjugate prior for the above likelihood. For if (α, β) has prior

$$p(\alpha, \beta) \propto (\beta - \alpha)^{-\gamma-2} I_{(\xi,\infty)}(\beta)\, I_{(-\infty,\eta)}(\alpha)$$

then the posterior is

$$\begin{aligned} p(\alpha, \beta \mid \boldsymbol{x}) &\propto p(\alpha, \beta)\, l(\alpha, \beta \mid \boldsymbol{x}) \\ &\propto (\beta - \alpha)^{-(\gamma+n)-2} I_{(\xi,\infty)}(\beta)\; I_{(M,\infty)}(\beta)\, I_{(-\infty,\eta)}(\alpha)\, I_{(-\infty,m)}(\alpha). \end{aligned}$$

If now we write

$$\gamma = \gamma + n$$
$$\xi' = \max\{\xi, M\}$$
$$\eta' = \min\{\eta, m\}$$

we see that

$$p(\alpha, \beta \mid \boldsymbol{x}) \propto (\beta - \alpha)^{-\gamma'-2} I_{(\xi',\infty)}(\beta) I_{(-\infty,\eta')}(\alpha).$$

It follows that if the prior is $\text{Pabb}(\xi, \eta, \gamma)$ then the posterior is $\text{Pabb}(\xi', \eta', \gamma')$ and hence that the bilateral bivariate Pareto distribution does indeed provide the conjugate prior.

The properties of this and of the ordinary Pareto distribution are, as usual, described in the Appendix.

Examples

I realize that the case of the uniform distribution, and in particular the case of a uniform distribution on $(0, \theta)$, must be of considerable importance, since it is considered in virtually all the standard textbooks on statistics. Strangely, however, none of the standard references seems to be able to find any reasonably plausible practical case in which it arises [with apologies to DeGroot (1970, Section 9.7) if his case really does arise]. In the circumstances, consideration of examples is deferred until the next section, and even then the example considered will be artificial.

3.6 Reference prior for the uniform distribution

Lower limit of the interval fixed

If $\boldsymbol{x} = (x_1, x_2, \ldots, x_n)$ consists of independent random variables with $U(0, \theta)$ distributions, then

$$p(\boldsymbol{x}|\theta) = \begin{cases} \theta^{-n} & (\theta > M) \\ 0 & (\text{otherwise}) \end{cases}$$

where $M = \max\{x_1, x_2, \ldots, x_n\}$, so that the likelihood can be written in the form

$$l(\theta|\boldsymbol{x}) = (M/\theta)^n I_{(M,\infty)}(\theta)$$

after multiplying by a constant (as far as θ is concerned). Hence

$$l(\theta|\boldsymbol{x}) = g(\psi(\theta) - t(\boldsymbol{x}))$$

with

$$\begin{aligned} g(y) &= \exp(-ny) I_{(0,\infty)}(y), \\ t(\boldsymbol{x}) &= \log M, \\ \psi(\theta) &= \log \theta. \end{aligned}$$

It follows that the likelihood is *data-translated*, and the general argument about data-translated likelihoods in Section 2.4 now suggests that we take a prior which is at least locally uniform in $\psi = \log\theta$, that is, $p(\psi) \propto 1$. In terms of the parameter θ, the usual change-of-variable rule shows that this means

$$p(\theta) \propto 1/\theta$$

which is the same prior that is conventionally used for variances. This is not a coincidence, but represents the fact that both are measures of spread (strictly, the standard deviation is more closely analogous to θ in this case, but a prior for the variance proportional to the reciprocal of the variance corresponds to a prior for the standard deviation proportional to the reciprocal of the standard deviation). As the density of $\text{Pa}(\xi, \gamma)$ is proportional to $y^{-\gamma-1} I_{(\xi,\infty)}(y)$, the density proportional to $1/\theta$ can be regarded as the limit $\text{Pa}(0, 0)$ of $\text{Pa}(\xi, \gamma)$ as $\xi \to 0$ and $\gamma \to 0$. Certainly if the likelihood is $\theta^{-n} I_{(M,\infty)}(\theta)$, then the posterior is $\text{Pa}(M, n)$, which is what would be expected when the general rule is applied to the particular case of a $\text{Pa}(0, 0)$ prior.

Example

A very artificial example can be obtained by taking groups of random digits from Neave (1978, Table 7.1) ignoring all values greater than some value θ. A sample of 10 such values is:

0.49487; 0.52802; 0.28667; 0.62058; 0.14704; 0.18519; 0.17889; 0.14554; 0.29480; 0.46317.

This sample was constructed using the value $\theta = 0.75$, but we want to investigate how far this method succeeds in giving information about θ, so we note that the posterior is Pa(0.62058, 10). Since the density function of a Pareto distribution $\mathrm{Pa}(\xi, \gamma)$ decreases monotonically beyond ξ, an HDR must be of the form (ξ, x) for some x, and since the distribution function is (see the Appendix)

$$F(x) = [1 - (\xi/x)^{\gamma}]\, I_{(\xi,\, \infty)}(x)$$

a 90% HDR for θ is $(0.62058, x)$ where x is such that

$$0.90 = [1 - (0.62058/x)^{10}]$$

and so is 0.78126. Thus a 90% HDR for θ is the interval (0.62, 0.78). We can see that the true value of θ in this artificial example does turn out to lie in the 90% HDR.

Both limits unknown

In the two parameter case, when $x \sim \mathrm{U}(\alpha, \beta)$ where $\alpha < \beta$ are both unknown and x is any one of the observations, we note that it is easily shown that

$$\theta = \mathsf{E}x = \tfrac{1}{2}(\alpha + \beta)$$
$$\varphi = \mathcal{V}x = (\beta - \alpha)^2/12.$$

Very similar arguments to those used in the case of the normal distribution with mean and variance both unknown in Section 2.12 can now be deployed to suggest independent priors uniform in θ in $\log \varphi$ so that

$$p(\theta, \varphi) \propto 1/\varphi.$$

But

$$\frac{\partial(\theta, \varphi)}{\partial(\alpha, \beta)} = \begin{vmatrix} \tfrac{1}{2} & \tfrac{1}{2} \\ -(\beta - \alpha)/6 & (\beta - \alpha)/6 \end{vmatrix} \propto (\beta - \alpha)$$

so that this corresponds to

$$\begin{aligned} p(\alpha, \beta) = p(\theta, \varphi)\frac{\partial(\theta, \varphi)}{\partial(\alpha, \beta)} &\propto \frac{12}{(\beta - \alpha)^2}(\beta - \alpha) \\ &\propto 1/(\beta - \alpha). \end{aligned}$$

It may be noted that the density of Pabb(ξ, η, γ) is proportional to

$$(z-y)^{-\gamma-2}I_{(\xi,\infty)}(z)I_{(-\infty,\eta)}(y)$$

so that in some sense a density

$$\begin{aligned} p(\alpha,\beta) &\propto (\beta-\alpha)^{-2}I_{(\xi,\infty)}(\beta)I_{(-\infty,\xi')}(\alpha) \\ &\propto (\beta-\alpha)^{-2}I_{(\alpha,\beta)}(\xi) \end{aligned}$$

might be regarded as a limit of Pabb(ξ,η,γ) as $\eta\to\xi$ and $\gamma\to 0$. Integrating over a uniform prior for ξ, which might well seem reasonable, gives

$$\begin{aligned} p(\alpha,\beta) &\propto \int(\beta-\alpha)^{-2}I_{(\alpha,\beta)}(\xi)\,\mathrm{d}\xi = \int_{\alpha}^{\beta}(\beta-\alpha)^{-2}\mathrm{d}\xi \\ &\propto 1/(\beta-\alpha). \end{aligned}$$

If the likelihood takes the form

$$l(\alpha,\beta\,|\,\boldsymbol{x}) \propto (\beta-\alpha)^{-n}I_{(M,\infty)}(\beta)I_{(-\infty,m)}(\alpha)$$

then the posterior from this prior is Pabb$(M,m,n-1)$. Thus our reference prior could be regarded as a Pabb$(-\infty,\infty,-1)$ distribution, and if we think of it as such the same formulae as before can be used.

The rule $p(\alpha,\beta)\propto(\beta-\alpha)^{-2}$ or Pabb$(-\infty,\infty,0)$ corresponds to $p(\theta,\varphi)\propto\varphi^{-3/2}$ which is the prior Jeffreys' rule gave us in the normal case with both parameters unknown (see Section 3.3 on Jeffreys' rule).

3.7 The tramcar problem

The discrete uniform distribution

Occasionally we encounter problems to do with the discrete uniform distribution. We say that x has a discrete uniform distribution on $[\alpha,\beta]$ and write

$$x\sim \mathrm{UD}(\alpha,\beta)$$

if

$$p(x)=(\beta-\alpha+1)^{-1} \qquad (x=\alpha,\mathrm{a}+1,\alpha+2,\ldots,\beta).$$

One context in which it arises was cited by Jeffreys (1939, 1948, 1961, Section 4.8). He says, "The following problem was suggested to me several years ago by Professor M. H. A. Newman. A man travelling in a foreign country has to change trains at a junction and goes into the town, of the existence of which he has only just heard. He has no idea of its size. The first thing that he sees is a tramcar numbered 100. What can he infer about the number of tramcars in the town? It may be assumed for the purpose that they are numbered consecutively from 1 upwards."

Clearly if there are ν tramcars in the town and we are equally likely to see any one of the tramcars, then the number n of the car observed has a discrete uniform distribution $\mathrm{UD}(1, \nu)$. Jeffreys suggests that (assuming ν is not too small) we can deal with this problem by analogy with problems involving a continuous distribution $\mathrm{U}(0, \nu)$. In the absence of a prior information, the arguments of the previous section suggest a reference prior $p(\nu) \propto 1/\nu$ in the latter case, so his suggestion is that the prior for ν in a problem involving a discrete uniform distribution $\mathrm{UD}(1, \nu)$ should be, at least approximately, proportional to $1/\nu$. But if

$$p(\nu) \propto 1/\mathrm{v} \qquad (\nu = 1, 2, 3, \ldots),$$
$$p(n \mid \nu) = 1/\nu \qquad (n = 1, 2, \ldots, \nu)$$

then by Bayes' Theorem

$$p(\nu \mid n) \propto \nu^{-2} \qquad (\nu = n, n+1, n+2, \ldots).$$

It follows that

$$p(\nu \mid n) = \nu^{-2} \Big/ \left(\sum_{\mu \geqslant n} \mu^{-2} \right).$$

In particular, the posterior probability that $\nu \geqslant \lambda$ is approximately

$$\left(\sum_{\mu \geqslant \lambda} \mu^{-2} \right) \Big/ \left(\sum_{\mu \geqslant n} \mu^{-2} \right).$$

Approximating the sums by integrals and noting that $\int \mu^{-2} \mathrm{d}\mu = \mu^{-1}$, this is approximately n/λ. Consequently the posterior median is $2n$, and so 200 if we observed tramcar number 100.

The argument seems rather unconvincing, because it puts quite a lot of weight on the prior as opposed to the likelihood and yet the arguments for the prior are not very strong, but we may agree with Jeffreys that it may be "worth recording". It is hard to take the reference prior suggested terribly seriously, although if we had a lot more data, then it would not matter what prior was used.

3.8 The first digit problem; invariant priors

A prior in search of an explanation

The problem considered in this section is not really one of statistical inference as such. What is introduced here is another argument that can sometimes be taken into account in deriving a prior distribution—that of invariance. To introduce the notion, we consider a population which appears to be invariant in a particular sense.

The problem

The problem we are going to consider in this section has a long history going back to Newcomb (1881). Recent references include Knuth (1969, Section 4.2.4B), Raimi (1976) and Turner (1987).

Newcomb's basic observation, in the days where large tables of logarithms were in frequent use, was that the early pages of such tables tended to look dirtier and more worn than the later ones. This appears to suggest that numbers whose logarithms we need to find are more likely to have 1 as their first digit than 9. If we then look up a few tables of physical constants, we can gain some idea as to whether this is borne out. For example, *Whitaker's Almanack* (1988, p. 202) quotes the areas of 40 European countries (in square kilometres) as

28,778; 453; 83,849; 30,513; 110,912; 9251; 127,869; 43,069; 1399; 337,032; 547,026; 108,178; 248,577; 6; 131,944; 93,030; 103,000; 70,283; 301,225; 157; 2586; 316; 1; 40,844; 324,219; 312,677; 92,082; 237,500; 61; 504,782; 449,964; 41,293; 23,623; 130,439; 20,768; 78,772; 14,121; 5,571,000; 0.44; 255,804.

The first significant digits of these are distributed as follows:

Digit	1	2	3	4	5	6	7	8	9	Total
Frequency	10	7	6	6	3	2	2	1	3	40
Percentage	25	17.5	15	15	7.5	5	5	2.5	7.5	100

We will see that there are grounds for thinking that the distribution should be approximately as follows:

Digit	1	2	3	4	5	6	7	8	9	Total
Percentage	30	18	12	10	8	7	6	5	5	100

A solution

The argument for this distribution runs as follows. The quantities we measure are generally measured in an arbitrary scale, and we would expect that if we measured them in another scale (thus in the case of the example above, we might measure areas in square miles instead of square kilometres), then the *population* of values (or at least of their first significant figures) would look much the same, although individual values would of course change. This implies that if θ is a randomly chosen constant, then for any fixed c the transformation

$$\theta \rightarrow \psi(\theta) = c\theta$$

should leave the probability distribution of values of constants alone. This means that if the functional form of the density of values of θ is

$$p(\theta) = f(\theta)$$

then the corresponding density of values of ψ will be

$$p(\psi) = f(\psi).$$

Using the usual change-of-variable rule, we know that $p(\psi) = p(\theta) \, |d\theta/d\psi|$, so that we are entitled to deduce that

$$f(c\theta) = f(\psi) = f(\theta)c^{-1}.$$

But if $f(\theta)$ is any function such that $f(\theta) = cf(c\theta)$ for all c and θ, then we may take $c = 1/\theta$ to see that $f(\theta) = (1/\theta)f(1)$ so that $f(\theta) \propto 1/\theta$. It seems therefore that the distribution of constants that are likely to arise in a scientific context should, at least approximately, satisfy

$$p(\theta) \propto 1/\theta.$$

Naturally, the reservations expressed in Section 2.5 on "Locally uniform priors" about the use of improper priors as representing genuine prior beliefs over a whole infinite range still apply. But it *is* possible to regard the prior $p(\theta) \propto 1/\theta$ for such constants as valid over any interval (a, b) where $0 < a < b < \infty$ which is not too large. So we consider those constants between

$$10^k = a \quad \text{and} \quad 10^{k+1} = b.$$

Because

$$\int_a^b d\theta/\theta = \log_e(b/a) = \log_e 10$$

the prior density for constants θ between a and b is

$$1/(\theta \log_e 10)$$

and so the probability that such a constant has first digit d, that is, that it lies between da and $(d+1)a$, is

$$\int_{da}^{(d+1)a} d\theta/(\theta \log_e 10) = \log_e(1+d^{-1})/\log_e 10 = \log_{10}(1+d^{-1}).$$

Since this is true for *all* values of k, and any constant lies between 10^k

and 10^{k+1} for *some* k, it seems reasonable to conclude that the probability that a physical constant has first digit d is approximately

$$\log_{10}(1 + d^{-1})$$

which is the density tabulated earlier. This is sometimes known as *Benford's Law* because of the work of Benford (1938) on this problem.

This sub-section was headed "A solution" rather than "The solution" because a number of other reasons for this density have been adduced. Nevertheless, it is quite an interesting solution. It also leads us into the whole notion of invariant priors.

Haar priors

It is sometimes the case that your prior beliefs about a parameter θ are in some sense symmetrical. Now when a mathematician hears of symmetry, he or she tends immediately to think of groups, and the notions above generalize very easily to general symmetry groups. If the parameter values θ can be thought of as members of an abstract group Θ, then the fact that your prior beliefs about θ are not altered when the values of θ are all multiplied by the same value c can be expressed by saying that the transformation

$$\theta \rightarrow \psi(\theta) = c\theta$$

should leave the probability distribution of values of the parameter alone. A density which is unaltered by this operation for arbitrary values of c is known as a *Haar measure* or, in this context, as a *Haar prior* or an *invariant prior*. Such priors are, in general, unique (at least up to multiplicative constants about which there is an arbitrariness if the priors are improper). This is just the condition used above to deduce Benford's Law, except that $c\theta$ is now to be interpreted in terms of the multiplicative operation of the symmetry group, which will not, in general, be ordinary multiplication.

This gives another argument for a uniform prior for the mean θ of a normal distribution $N(\theta, \varphi)$ of known variance, since it might well seem that adding the same constant to all possible values of the mean would leave your prior beliefs unaltered—there seems to be a symmetry under additive operations. If this is so, then the transformation

$$\theta \rightarrow \psi(\theta) = c + \theta$$

should leave the functional form of the prior density for θ unchanged, and it is easy to see that this is the case if and only if $p(\theta)$ is constant. A similar argument about the multiplicative group might be used about an unknown variance when the mean is known to produce the usual reference prior $p(\psi) \propto 1/\psi$. A good discussion of this approach and some references can be found in Berger (1985, Section 3.3.2).

3.9 The circular normal distribution

Distributions on the circle

In this section, the variable is an angle running from 0° to 360°, that is, from 0 to 2π radians. Such variables occur in a number of contexts, for example in connection with the homing ability of birds and in various problems in astronomy and crystallography. Useful references for such problems are Mardia (1972) and Batschelet (1981).

The only distribution for such angles which will be considered is the so-called *circular normal* or *von Mises* distribution. An angle η is said to have such a distribution with mean direction μ and concentration parameter κ if

$$p(\eta \mid \mu, \kappa) = \{2\pi I_0(\kappa)\}^{-1} \exp\{\kappa \cos(\eta - \mu) \quad (0 \leqslant \eta < 2\pi)$$

and when this is so we write

$$\eta \sim \mathrm{M}(\mu, \kappa).$$

The function $I_0(\kappa)$ is the modified Bessel function of the first kind and order zero, but as far as we are concerned it may as well be regarded as defined by

$$I_0(\kappa) = (2\pi)^{-1} \int_0^{2\pi} \exp\{\kappa \cos(\eta - \mu\}\, d\eta.$$

It is tabulated in many standard tables, for example, British Association (1937) or Abramowitz and Stegun (1964, Section 9.7.1). It can be shown that

$$I_0(\kappa) = \sum_{r=0}^{\infty} (\tfrac{1}{2}\kappa)^{2r}/(r!)^2.$$

The circular normal distribution was originally introduced by von Mises (1918). It plays a prominent role in statistical inference on the circle and its importance there is almost the same as that of the normal distribution on the line. There is a relationship with the normal distribution, since as $\kappa \to \infty$ the distribution of

$$\sqrt{\kappa}(\eta - \mu)$$

approaches the standard normal form N(0, 1) and hence $\mathrm{M}(\mu, \kappa)$ is approximately $\mathrm{N}(\mu, 1/\kappa)$. It follows that the concentration parameter is analogous to the precision of a normal distribution. This is related to the fact that asymptotically for large z

$$I_0(\kappa) \sim (2\pi z)^{-\frac{1}{2}} \exp(z).$$

However, the equivalent of the central limit theorem does *not* result in convergence to the circular normal distribution. Further, the circular normal distribution is not in the exponential family. It should not be confused with the so-called wrapped normal distribution.

The likelihood of n observations $\boldsymbol{\eta} = (\eta_1, \eta_2, \ldots, \eta_n)$ from an $\mathrm{M}(\mu, \kappa)$ distribution is

$$\begin{aligned} l(\mu, \kappa \mid \boldsymbol{\eta}) &\propto \{I_0(\kappa)\}^{-n} \exp\{\textstyle\sum \kappa \cos(\eta_i - \mu)\} \\ &= \{I_0(\kappa)\}^{-n} \exp\{\kappa \cos\mu \textstyle\sum \cos\eta_i + \kappa \sin\mu \textstyle\sum \sin\eta_i\} \end{aligned}$$

so that if we define

$$\begin{aligned} c &= n^{-1} \textstyle\sum \cos\eta_i \\ s &= n^{-1} \textstyle\sum \sin\eta_i \end{aligned}$$

then (c, s) is sufficient for (μ, κ) given $\boldsymbol{\eta}$, and indeed

$$l(\mu, \kappa \mid \boldsymbol{\eta}) \propto \{I_0(\kappa)\}^{-n} \exp\{n\kappa c \cos\mu + n\kappa s \sin\mu\}.$$

If we define

$$\begin{aligned} \rho &= \sqrt{(c^2 + s^2)} \\ \hat{\mu} &= \tan^{-1}(s/c) \end{aligned}$$

then we obtain

$$c = \rho \cos\hat{\mu} \quad \text{and} \quad s = \rho \sin\hat{\mu}$$

and hence

$$l(\mu, \kappa \mid \boldsymbol{\eta}) \propto \{I_0(\kappa)\}^{-n} \exp\{n\kappa\rho \cos(\mu - \hat{\mu})\}.$$

(It may be worth noting that it can be shown by differentiating ρ^2 with respect to the η_i that ρ is a maximum when all the observations are equal and that it then equals unity). It is easy enough now to construct a family of conjugate priors, but for simplicity let us consider a reference prior

$$p(\mu, \kappa) \propto 1.$$

It seems reasonable enough to take a uniform prior in μ and to take independent priors for μ and κ, but it is not so clear that a uniform prior in κ is sensible. Schmitt (1969, Section 10.2) argues that a uniform prior in κ is a sensible compromise and notes that there are difficulties in using a prior proportional to $1/\kappa$ since, unlike the precision of a normal variable, the concentration parameter of a circular normal distribution can actually equal zero. If this is taken as the prior, then of course

$$p(\mu, \kappa \,|\, \boldsymbol{\eta}) \propto \{I_0(\kappa)\}^{-n} \exp\{n\kappa\rho \cos(\mu - \hat{\mu})\}.$$

Example

Batschelet (1981, Example 4.3.1) quotes data on the time of day of major traffic accidents in a major city. In an obvious sense, the time of day can be regarded as a circular measure, and it is meaningful to ask what is the mean time of day at which accidents occur and how tightly clustered about this time these times are. Writing

$$\eta = 360\,\{h + (m/60)\}/24$$

the $n = 21$ observations are as follows:

hr	min	η	$\cos\eta$	$\sin\eta$
00	56	14°	0.9703	0.2419
03	08	47°	0.6820	0.7314
04	52	73°	0.2923	0.9563
07	16	109°	−0.2250	0.9455
08	08	122°	−0.5299	0.8480
10	00	150°	−0.8660	0.5000
11	24	171°	−0.9877	0.1564
12	08	182°	−0.9994	−0.0349
13	28	202°	−0.9272	−0.3746
14	16	214°	−0.8290	−0.5592
16	20	245°	−0.4226	−0.9063
16	44	251°	−0.3256	−0.9455
17	04	256°	−0.2419	−0.9703
17	20	260°	−0.1736	−0.9848
17	24	261°	−0.1564	−0.9877
18	08	272°	0.0349	−0.9994
18	16	274°	0.0698	−0.9976
18	56	284°	0.2419	−0.9703
19	32	293°	0.3907	−0.9205
20	52	313°	0.6820	−0.7314
22	08	332°	0.8829	−0.4695
Total			−2.5381	−6.4723
			$c = -0.1209$	$s = -0.3082$

This results in $\rho = 0.3311$ and $\tan\hat{\mu} = 2.5492$ and so (allowing for the signs of c and s) $\hat{\mu} = 248°\,34'$ (or in terms of a time-scale 16 hr 34 min) and so the posterior density takes the form

$$p(\mu, \kappa \mid \eta) \propto \exp\{-n\log\{I_0(\kappa)\} + n\kappa\rho\cos(\mu - \hat{\mu})\}$$

where ρ and $\hat{\mu}$ take these values. It is, however, difficult to understand what this means without experience of this distribution, and yet there is no simple way of finding HDRs. This, indeed, is one reason why a consideration of the circular normal distribution has been included, since it serves to emphasize that there are cases where it is difficult if not impossible to avoid numerical integration.

Construction of an HDR by numerical integration

By writing $\lambda = (\frac{1}{2}\kappa)^2$ and taking the first few terms in the power series quoted above for $I_0(\kappa)$ we see that for $0 \leqslant \kappa \leqslant 2.0$

$$I_0(\kappa) = 1 + \lambda + \lambda^2/4 + \lambda^3/36$$

to within 0.002. We can thus deduce some values for $I_0(\kappa)$ and $\kappa\rho$, viz.:

κ	0.0	0.5	1.0	1.5	2.0
$\lambda = (\frac{1}{2}\kappa)^2$	0.0	0.062	0.250	0.562	1.000
$I_0(\kappa)$	1.0	1.063	1.266	1.647	2.278
$\log\{I_0(\kappa)\}$	0.0	0.061	0.236	0.499	0.823
$\kappa\rho$	0.0	0.166	0.331	0.497	0.662

As $n = 21$, this implies that (ignoring the constant) the posterior density is

$\mu \backslash \kappa$	0.0	0.5	1.0	1.5	2.0
158°	1.000	0.278	0.007	0.000	0.000
203°	1.000	3.267	0.960	0.045	0.000
248°	1.000	9.070	7.352	0.959	0.034
293°	1.000	3.267	0.960	0.045	0.000
338°	1.000	0.278	0.007	0.000	0.000

In order to say anything about the marginal density of μ, we need to integrate out κ. In order to do this, we can use Simpson's Rule. The integral of a function between a and b can thus be approximated by the sum

$$\int_a^b f(x)\,dx \propto f(x_0) + 4f(x_1) + 2f(x_2) + 4f(x_3) + f(x_4)$$

where the x_i are equally spaced with $x_0 = a$ and $x_4 = b$. Applying it to the figures above, we can say that very roughly the density of μ is proportional to the following values:

μ	158	203	248	293	338
Density	2.13	16.17	55.84	16.17	2.13

Integrating over intervals of values of μ using the (even more crude) approximation

$$\int_{a-45}^{a+45} f(x)\,\mathrm{d}x \propto f(a-45)+4f(a)+f(a+45)$$

(and taking the densities below 158 and above 338 to be negligible) the probabilities that μ lies in intervals centred on various values are proportional to the values stated:

Centre	158	203	248	293	338	Total
Value	25	123	256	123	25	552

It follows that the probability that μ lies in the range (203, 293) is about $256/552 = 0.46$, and thus this interval is close to being a 45% HDR.

Remarks

The main purpose of this section is to show in some detail, albeit with very crude numerical methods, how a Bayesian approach can deal with a problem which does not lead to a neat posterior distribution, the values of which are tabulated and readily available. In practice, if we need to approach such a problem, we would have to have recourse to numerical integration techniques on a computer, but the basic ideas would be much the same.

3.10 Approximations based on the likelihood

Maximum likelihood

Suppose, as usual, that we have independent observations $\boldsymbol{x} = (x_1, x_2, \ldots, x_n)$ whose distribution depends on an unknown parameter θ about which we want to make inferences. Sometimes it is useful to quote the posterior mode, that is, that value of θ at which the posterior density is a maximum, as a single number giving some idea of the location of the posterior distribution of θ; it could be regarded as the ultimate limit of the idea of an HDR. However, some Bayesians are opposed to the use of *any* single number in this way [see, for example, Box and Tiao (1973, Section A5.6)].

If the likelihood dominates the prior, the posterior mode will occur very close to the point $\hat{\theta}$ at which the likelihood is a maximum. Use of $\hat{\theta}$ is known as the *method of maximum likelihood* and is originally due to Fisher (1922). One notable point about maximum likelihood estimators is that if $\psi(\theta)$ is any function of θ then it is easily seen that

$$\hat{\psi} = \psi(\hat{\theta})$$

because the point at which $p(x \mid \theta)$ is a maximum is not affected by how it is labelled. This invariance is not true of the exact position of the maximum of the posterior, nor indeed of HDRs, because these are affected by the factor $\mathrm{d}\psi/\mathrm{d}\theta$.

We should note that the maximum likelihood estimator is often found by the Newton–Raphson method. We suppose that the likelihood is $l(\theta \mid \boldsymbol{x})$ and that its logarithm (in terms of which it is often easier to work) is $L(\theta \mid \boldsymbol{x})$. In order to simplify the notation, we may sometimes omit explicit reference to the data and write $L(\theta)$ for $L(\theta \mid \boldsymbol{x})$. We seek $\hat{\theta}$ such that

$$\partial l(\theta \mid \boldsymbol{x})/\partial\theta \,|_{\theta=\hat{\theta}} = 0$$

or equivalently that it satisfies the so-called likelihood equation

$$L'(\hat{\theta}) = 0.$$

Iterative methods

If θ_k is an approximation to $\hat{\theta}$ then using Taylor's Theorem

$$0 = L'(\hat{\theta}) = L'(\theta_k) + (\hat{\theta} - \theta_k) L''(\theta^*)$$

where θ^* is between $\hat{\theta}$ and $\hat{\theta}_k$. In most cases $L''(\theta^*)$ will not differ much from $L''(\hat{\theta})$ and neither will differ much from its expectation over $\boldsymbol{x}$. However

$$\mathsf{E}L''(\theta) = -I(\theta \mid \boldsymbol{x})$$

where $I(\theta \mid \boldsymbol{x})$ is Fisher's information which was introduced earlier in Section 3.3 in connection with Jeffreys' rule. We note that, although $L''(\theta)$ does depend on the value $\boldsymbol{x}$ observed, the information $I(\theta \mid \boldsymbol{x})$ depends on the *distribution* of the random variable $\tilde{\boldsymbol{x}}$ rather than on the value $\boldsymbol{x}$ observed on this particular occasion, and to this extent the notation, good though it is for other purposes, is misleading. However, the value of $I(\hat{\theta} \mid \boldsymbol{x})$ does depend on $\boldsymbol{x}$, because $\hat{\theta}$ does.

It follows that as $k \to \theta$ the value of $L''(\theta^*)$ tends to $-I(\theta \mid \boldsymbol{x})$, so that a better approximation than θ_k will usually be provided either by

$$\theta_{k+1} = \theta_k - L'(\theta_k)/L''(\theta_k),$$

the Newton–Raphson method, or by

$$\theta_{k+1} = \theta_k + L'(\theta_k)/I(\theta_k | \boldsymbol{x}),$$

the *method of scoring for parameters*. The latter method was first published in a paper by Fisher (1925a).

It has been shown by Kale (1961) that the method of scoring will usually be the quicker process for large n unless high accuracy is ultimately required. In perverse cases both methods can fail to converge or can converge to a root which does not give the absolute maximum.

Approximation to the posterior density

We can also observe that in the neighbourhood of $\hat{\theta}$

$$L(\theta) = L(\hat{\theta}) + \tfrac{1}{2}(\theta - \hat{\theta})^2 L''(\hat{\theta}) + \ldots$$

so that approximately

$$\begin{aligned} l(\theta | \boldsymbol{x}) &= \exp\{L(\hat{\theta}) + \tfrac{1}{2}(\theta - \hat{\theta})^2 L''(\hat{\theta})\} \\ &\propto \exp\{\tfrac{1}{2}(\theta - \hat{\theta})^2 L''(\hat{\theta})\} \end{aligned}$$

Hence the likelihood is approximately proportional to an $\mathrm{N}(\hat{\theta}, -1/L''(\hat{\theta}))$ density, and so approximately to an $\mathrm{N}(\hat{\theta}, 1/I(\hat{\theta} | \boldsymbol{x}))$ density. We can thus construct approximate HDRs by using this approximation to the likelihood and assuming that the likelihood dominates the prior.

Examples

Normal variance. For the normal variance (with known mean θ)

$$L(\varphi) = \log l(\varphi | \boldsymbol{x}) = -\tfrac{1}{2}n \log \varphi - \tfrac{1}{2}S/\varphi + \text{constant}$$

where $S = \sum (X_i - \theta)^2$, so that

$$L'(\varphi) = -\tfrac{1}{2}n/\varphi + \tfrac{1}{2}S/\varphi^2.$$

In this case the likelihood equation is solved without recourse to iteration to give

$$\hat{\varphi} = S/n$$

Further

$$\begin{aligned} L''(\varphi) &= \tfrac{1}{2}n/\varphi^2 - S/\varphi^3, \\ -1/L''(\hat{\varphi}) &= 2S^2/n^3. \end{aligned}$$

Alternatively

$$I(\varphi\,|\,\boldsymbol{x}) = -\mathsf{E}L''(\varphi) = -\tfrac{1}{2}n/\varphi^2 + \mathsf{E}S/\varphi^3$$

and as $S \sim \varphi\chi_n^{\ 2}$, so that $\mathsf{E}S = n\varphi$, we have

$$I(\varphi\,|\,\boldsymbol{x}) = \tfrac{1}{2}n/\varphi^2,$$
$$1/I(\hat{\varphi}\,|\,\boldsymbol{x}) = 2S^2/n^3.$$

Of course there is no need to use an iterative method to find $\hat{\varphi}$ in this case, but the difference between the formulae for $L''(\varphi)$ and $I(\varphi\,|\,\boldsymbol{x})$ is illustrative of the extent to which the Newton–Raphson method and the method of scoring differ from one another. The results suggest that we approximate the posterior distribution of φ [which we found to be $(S_0+S)\chi_{\nu+n}^{\ -2}$ if we took a conjugate prior] by

$$\varphi \sim \mathrm{N}(S/n,\ 2S^2/n^3).$$

With the data we considered in Section 2.8 on HDRs for the normal variance we had $n = 20$ and $S = 664$, so that $2S^2/n^3 = 110.224$. The approximation would suggest a 95% HDR between 664/20 $\pm 1.96\sqrt{110.224}$, that is, the interval (13, 54) as opposed to the interval (19, 67) which was found in Section 2.8.

This example is deceptively simple—the method is of greatest use when analytic solutions are difficult or impossible. Further, the accuracy is greater when sample sizes are larger.

Poisson distribution. We can find another deceptively simple example by supposing that $\boldsymbol{x} = (x_1, x_2, \ldots, x_n)$ is an n-sample from $\mathrm{P}(\lambda)$ and that $T = \sum x_i$ so that (as shown in Section 3.4)

$$l(\lambda\,|\,\boldsymbol{x}) \propto (\lambda^T/T!)\exp(-n\lambda)$$
$$L(\lambda) = T\log\lambda - n\lambda + \text{constant}$$
$$L'(\lambda) = (T/\lambda) - n$$

and the likelihood equation is again solved without iteration, this time giving $\hat{\lambda} = T/n = \bar{x}$. Further

$$L''(\lambda) = -T/\lambda^2$$
$$I(\lambda\,|\,\boldsymbol{x}) = n\lambda/\lambda^2 = n/\lambda$$

and $I(\hat{\lambda}\,|\,\boldsymbol{x}) = L''(\hat{\lambda}) = n^2/T$. This suggests that we can approximate the posterior of λ (which we found to be $(S_0+2n)\chi_{\nu+2T}^{\ 2}$ if we took a conjugate prior) by

$$\lambda \sim \mathrm{N}(T/n,\ T/n^2).$$

Cauchy distribution. Suppose $\boldsymbol{x} = (x_1, x_2, \ldots, x_n)$ is an n-sample from $\mathrm{C}(\theta, 1)$ so that

$$p(\boldsymbol{x}|\theta) = \prod \pi^{-1}\{1+(x_i-\theta)^2\}^{-1} \quad (-\infty<x<\infty)$$
$$L(\theta) = \text{constant} - \sum \log\{1+(x_i-\theta)^2\}$$
$$L'(\theta) = 2\sum (x_i-\theta)/\{1+(x_i-\theta)^2\}$$
$$L''(\theta) = 2\sum\{(x_i-\theta)^2-1\}/\{1+(x_i-\theta)^2\}^2$$
$$L'(\theta)/L''(\theta) = \frac{\sum (x_i-\theta)/\{1+(x_i-\theta)^2\}}{\sum\{(x_i-\theta)^2-1\}/\{1+(x_i-\theta)^2\}^2}.$$

It is easily seen that

$$I(\theta|\boldsymbol{x}) = -\mathsf{E}L''(\theta) = (4n/\pi)\int_0^\infty (1-x^2)/(1+x^2)^3\,dx.$$

On substituting $x = \tan\psi$ and using standard reduction formulae, it follows that

$$I(\theta|\boldsymbol{x}) = n/2$$

from which it can be seen that successive approximations to $\hat{\theta}$ can be found using the method of scoring by setting

$$\theta_{k+1} = \theta_k + (4/n)\sum (x_i-\theta)/\{1+(x_i-\theta)^2\}.$$

The iteration could, for example, be started from the sample median, that is, that one of the observations which is in the middle when they are arranged in increasing order. For small n the iteration may not converge, or may converge to the wrong answer (see Barnett, 1966), but the process usually behaves satisfactorily.

Real-life data from a Cauchy distribution are rarely encountered, but the following values are simulated from a $\mathrm{C}(\theta, 1)$ distribution (the value of θ being, in fact, 0):

$$-0.774;\ 0.597;\ 7.575;\ 0.397;\ -0.865;\ -0.318;\ -0.125;\ 0.961;\ 1.039$$

The sample median of the $n = 9$ values is 0.397. If we take this as our first approximation θ_0 to $\hat{\theta}$, then

$$\theta_1 = 0.107;\ \theta_2 = 0.201;\ \theta_3 = 0.173;\ \theta_4 = 0.181;\ \theta_5 = 0.179$$

and all subsequent θ_k equal 0.179, which is in fact the correct value of $\hat{\theta}$. Since $I(\theta|\boldsymbol{x}) = n/2 = 9/2$, an approximate 95% HDR for θ is $0.179 \pm 1.96\sqrt{(2/9)}$, that is, the interval $(-0.74, 1.10)$. This does include the true value, which we happen to know is 0, but of course the value of n has been chosen unrealistically small in order to illustrate the method without too much calculation.

It would also be possible in this case to carry out an iteration based on the Newton–Raphson method

$$\theta_{k+1} = \theta_k - L'(\theta_k)/L''(\theta_k)$$

using the formula given above for $L'(\theta)/L''(\theta)$, but as explained above it is in general better to use the method of scoring.

Extension to more than one parameter

If we have two parameters, say θ and φ, which are both unknown, a similar argument shows that the maximum likelihood occurs at $(\hat{\theta}, \hat{\varphi})$ where

$$\partial L/\partial\theta = \partial L/\partial\varphi = 0.$$

Similarly, if (θ_k, φ_k) is an approximation, a better one is $(\theta_{k+1}, \varphi_{k+1})$ where

$$\begin{pmatrix}\theta_{k+1}\\ \varphi_{k+1}\end{pmatrix} = \begin{pmatrix}\theta_k\\ \varphi_k\end{pmatrix} - \begin{pmatrix}\partial^2 L/\partial\theta^2 & \partial^2 L/\partial\theta\partial\varphi\\ \partial^2 L/\partial\theta\partial\varphi & \partial^2 L/\partial\varphi^2\end{pmatrix}^{-1} \begin{pmatrix}\partial L/\partial\theta\\ \partial L/\partial\varphi\end{pmatrix}$$

where the derivatives are evaluated at (θ_k, φ_k) and the matrix of second derivatives can be replaced by its expectation, which is minus the information matrix as defined in Section 3.3 on Jeffreys' rule.

Further, the likelihood and hence the posterior can be approximated by a bivariate normal distribution of mean $(\hat{\theta}, \hat{\varphi})$ and variance–covariance matrix whose inverse is equal to minus the matrix of second derivatives (or the information matrix) evaluated at $(\hat{\theta}, \hat{\varphi})$.

All of this extends in an obvious way to the case of more than two unknown parameters.

Example

We shall consider only one, very simple, case, that of a normal distribution of unknown mean and variance. In this case

$$\begin{aligned} L(\theta, \varphi) &= -\tfrac{1}{2}n \log \varphi - \tfrac{1}{2}\{S + n(\bar{x} - \theta)^2\}/\varphi \\ \partial L/\partial\theta &= n(\bar{x} - \theta)/\varphi \\ \partial L/\partial\varphi &= -\tfrac{1}{2}n/\varphi + \tfrac{1}{2}\{S + n(\bar{x} - \theta)^2\}/\varphi^2 \end{aligned}$$

where $S = \sum (x_i - \bar{x})^2$, so that

$$\begin{aligned} \hat{\theta} &= \bar{x}, \\ \hat{\varphi} &= S/n = ns^2/(n-1). \end{aligned}$$

Further, it is easily seen that

$$\begin{pmatrix} \partial^2 L/\partial\theta^2 & \partial^2 L/\partial\theta\partial\varphi \\ \partial^2 L/\partial\theta\partial\varphi & \partial^2 L/\partial\varphi^2 \end{pmatrix} = \begin{pmatrix} -n/\varphi & -n(\bar{x}-\theta)/\varphi^2 \\ -n(\bar{x}-\theta)/\varphi^2 & \tfrac{1}{2}n/\varphi^2 - \{S + n(\bar{x}-\theta)^2\}/\varphi^3 \end{pmatrix}$$

which at $(\hat{\theta}, \hat{\varphi})$ reduces to

$$\begin{pmatrix} -n/\hat{\varphi} & 0 \\ 0 & -n/2\hat{\varphi}^2 \end{pmatrix}.$$

Because the off-diagonal elements vanish, the posteriors for θ and φ are approximately independent. Further, we see that approximately

$$\theta \sim \mathrm{N}(\hat{\theta}, \hat{\varphi}/n) \quad \text{and} \quad \varphi \sim \mathrm{N}(\hat{\varphi}, 2\hat{\varphi}^2/n).$$

In fact we found in Section 2.12 on normal mean and variance both unknown that with standard reference priors, the posterior for θ and φ is a normal/chi-squared distribution and the marginals are such that

$$(\bar{x}-\theta)/(s/\sqrt{n}) \sim \mathrm{t}_{n-1} \quad \text{and} \quad \varphi \sim S\chi_{n-1}^{-2}$$

which implies that the means and variances are

$$\begin{aligned} \mathsf{E}\theta &= \bar{x} = \hat{\theta} & \mathscr{V}\theta &= (n-1)s^2/n(n-3) \cong \hat{\varphi}/n, \\ \mathsf{E}\varphi &= S/(n-3) \cong \hat{\varphi}, & \mathscr{V}\varphi &= 2S^2/(n-3)^2(n-5) \cong 2\hat{\varphi}^2/n. \end{aligned}$$

This shows that for large n the approximation is indeed valid.

Exercises on Chapter 3

1. Laplace claimed that the probability that an event which has occurred n times, and has not hitherto failed, will occur again is $(n+1)/(n+2)$ [see Laplace (1774)]. Suggest grounds for this assertion.

2. Find a suitable interval of 90% posterior probability to quote in a case when your posterior distribution for an unknown parameter π is Be(20, 12), and compare this interval with similar intervals for the cases of Be(20.5, 12.5) and Be(21, 13) posteriors. Comment on the relevance of the results to the choice of a reference prior for the binomial distribution.

3. Suppose that you have a prior distribution for the probability π of success in a certain kind of gambling game which has mean 0.4, and

that you regard your prior information as equivalent to 12 trials. You then play the game 25 times and win 12 times. What is your posterior distribution for π?

4. Show that if $g(x) = \sinh^{-1}\surd(x/n)$ then

$$g'(x) = \tfrac{1}{2}n^{-1}[(x/n)\{1+x/n\}]^{-\frac{1}{2}}.$$

Deduce that if $x \sim \mathrm{NB}(n, \pi)$ has a negative binomial distribution of index n and parameter π and $z = g(x)$ then $\mathsf{E}z \cong \sinh^{-1}\surd(x/n)$ and $\mathcal{V}z \cong 1/4n$. What does this suggest as a reference prior for π?

5. The following data were collected by von Bortkiewicz (1898) on the number of men killed by a horse in certain Prussian army corps in 20 years, the unit being one army corps for one year:

Number of deaths:	0	1	2	3	4	5 and more
Number of units:	144	91	32	11	2	0.

Give an interval in which the mean number λ of such deaths in a particular army corps in a particular year lies with 95% probability.

6. Recalculate the answer to the previous question assuming that you had a prior distribution for λ of mean 0.66 and standard deviation 0.115.

7. Suppose that x has a Pareto distribution $\mathrm{Pa}(\xi, \gamma)$ where ξ is known but γ is unknown, that is,

$$p(x\,|\,\gamma) = \gamma\xi^{\gamma}x^{-\gamma-1}I_{(\xi,\infty)}(x).$$

Use Jeffreys' rule to find a suitable reference prior for γ.

8. Consider a uniform distribution on the interval (α, β), where the values of α and β are unknown, and suppose that the joint distribution of α and β is a bilateral bivariate Pareto distribution with $\gamma = 2$. How large a random sample must be taken from the uniform distribution in order that the coefficient of variation (that is, the standard deviation divided by the mean) of the length $\beta - \alpha$ of the interval should be reduced to 0.01 or less?

9. What could you conclude if you observed *two* tramcars numbered, say, 71 and 100?

10. In Section 3.8 we discussed Newcomb's observation that the front

pages of a well-used table of logarithms tend to get dirtier than the back pages. What if we had an *antilogarithm* table, that is, a table giving the value of x when $\log_{10} x$ is given? Which pages of such a table would be the dirtiest?

11. Suppose that the prior distribution $p(\mu, \sigma)$ for the parameters μ and σ of a Cauchy distribution

$$p(x \mid \mu, \sigma) = \frac{1}{\pi} \frac{\sigma}{\sigma^2 + (x - \mu)^2}$$

is uniform in μ and σ, and that two observations $x_1 = 2$ and $x_2 = 6$ are available from this distribution. Calculate the value of the posterior density $p(\mu, \sigma \mid \boldsymbol{x})$ (ignoring the factor $1/\pi^2$) to two decimal places for $\mu = 0, 2, 4, 6, 8$ and $\sigma = 1, 2, 3, 4, 5$. Use Simpson's rule to approximate the posterior marginal density of μ, and hence go on to find an approximation to the posterior probability that $3 < \mu < 5$.

12. Show that if the log-likelihood $L(\theta \mid x)$ is a concave function of θ for each scalar x, then the log-likelihood function $L(\theta \mid \boldsymbol{x})$ for θ given an n-sample $\boldsymbol{x} = (x_1, x_2, \ldots, x_n)$ has a unique maximum. Prove that this is the case if the observations x_i come from a logistic density

$$p(x \mid \theta) = \exp(\theta - x) / \{1 + \exp(\theta - x)\}^2 \quad (-\infty < x < \infty)$$

where θ is an unknown real parameter. Fill in the details of the Newton–Raphson method and the method of scoring for finding the position of the maximum, and suggest a suitable starting point for the algorithms.

4

Hypothesis testing

4.1 Hypothesis testing

Introduction

If preferred, the reader may begin with the example at the end of this section, then return to the general theory at the beginning.

Classical hypothesis testing

Most simple problems in which tests of hypotheses arise are of the following general form. There is one unknown parameter θ which is known to be from a set Θ, and we want to know whether $\theta \in \Theta_0$ or $\theta \in \Theta_1$ where

$$\Theta_0 \cup \Theta_1 = \Theta \qquad \text{and} \qquad \Theta_0 \cap \Theta_1 = \varnothing.$$

Usually we are able to make use of a set of observations $x_1, x_2, \ldots, x_n$ whose density $p(x \mid \theta)$ depends on θ. It is convenient to denote the set of all possible observations $\boldsymbol{x} = (x_1, x_2, \ldots, x_n)$ by $\mathscr{X}$.

In the language of classical statistics, it is usual to refer to

$$\mathrm{H}_0\colon \theta \in \Theta_0 \qquad \text{as the} \qquad \textit{null hypothesis}$$

and to

$$\mathrm{H}_1\colon \theta \in \Theta_1 \qquad \text{as the} \qquad \textit{alternative hypothesis}$$

and to say that if we decide to reject H_0 when it is true then we have made a *Type I error* while if we decide to accept H_0 when it is false then we have made a *Type II error*.

A test is decided by a *rejection region* R where

$$R = \{\boldsymbol{x};\ \text{observing } \boldsymbol{x} \text{ would lead to the rejection of } \mathrm{H}_0\}.$$

Classical statisticians then say that decisions between tests should be based on the probabilities of Type I errors, that is,

$$\mathrm{P}(R \mid \theta) \qquad \text{for} \qquad \theta \in \Theta_0$$

and of Type II errors, that is,

$$1-\mathrm{P}(R\,|\,\theta) \qquad \text{for} \qquad \theta\in\Theta_1.$$

In general, the smaller the probability of Type I error, the larger the probability of Type II error and vice versa. Consequently, classical statisticians recommend a choice of R which in some sense represents an optimal balance between the two types of error. Very often R is chosen so that the probability of a Type II error is as small as possible subject to the requirement that the probability of a Type I error is always less than or equal to some fixed value α known as the *size* of the test. This theory is to be found in most books on statistical inference and is to be found in its fullest form in Lehmann (1959, 1986).

Difficulties with the classical approach

Other points will be made later about the comparison between the classical and the Bayesian approaches, but one thing to note at the outset is that in the classical approach we consider the probability (for various values of θ) of a set R to which the vector $\boldsymbol{x}$ of observations does, or does not, belong. Consequently, we are concerned not merely with the single vector of observations we actually made, but also with others we *might* have made but *did not*. Thus, classically, if we suppose that $x\sim\mathrm{N}(\theta, 1)$ and we wish to test whether H_0: $\theta=0$ or H_1: $\theta>0$ is true (negative values being supposed impossible), then we reject H_0 on the basis of a single observation $x=3$ because the probability that an N(0, 1) random variable is 3 *or greater* is 0.001350, even though we certainly did not make an observation greater than 3. This aspect of the classical approach led Jeffreys (1939, 1948, 1961, Section 7.2) to remark that:

"What the use of P implies, therefore, is that a hypothesis that may be true may be rejected because it has not predicted observable results that have not occurred."

Note, however, that the form of the model, in this case the assumption of normally distributed observations of unit variance, does depend on an assumption about the whole distribution of all possible observations.

The Bayesian approach

The Bayesian approach is in many ways more straightforward. All we need to do is to calculate the posterior probabilities

$$p_0=\mathrm{P}(\theta\in\Theta_0\,|\,\boldsymbol{x}), \qquad p_1=\mathrm{P}(\theta\in\Theta_1\,|\,\boldsymbol{x})$$

and decide between H_0 and H_1 accordingly. (We note that $p_0+p_1=1$ as $\Theta = \Theta_0 \cup \Theta_1$ and $\varnothing = \Theta_0 \cap \Theta_1$.)

Although posterior probabilities of hypotheses are our ultimate goal we also need prior probabilities

$$\pi_0 = \mathrm{P}(\theta \in \Theta_0), \qquad \pi_1 = \mathrm{P}(\theta \in \Theta_1)$$

to find them. (We note that $\pi_0 + \pi_1 = 1$ just as $p_0 + p_1 = 1$.) It is also useful to consider the *prior odds* on H_0 against H_1, namely,

$$\pi_0/\pi_1,$$

and the *posterior odds* on H_0 against H_1, namely,

$$p_0/p_1.$$

(The notion of odds was originally introduced in the very first section of this book). Observe that if your prior odds are close to 1 then you regard H_0 as more or less as likely as H_1 *a priori*, while if the ratio is large you regard H_0 as relatively likely and when it is small you regard it as relatively unlikely. Similarly remarks apply to the interpretation of the posterior odds.

It is also useful to define the *Bayes factor* B in favour of H_0 against H_1 as

$$B = \frac{(p_0/p_1)}{(\pi_0/\pi_1)} = \frac{p_0\pi_1}{p_1\pi_0}.$$

The interest in the Bayes factor is that it can sometimes be interpreted as the "odds in favour of H_0 against H_1 that are *given by the data*". It is worth noting that because $p_0/p_1 = B(\pi_0/\pi_1)$ and $p_1 = 1 - p_0$ we can find the posterior probability p_0 of H_0 from its prior probability and the Bayes factor by

$$p_0 = \frac{1}{[1+(\pi_1/\pi_0)B^{-1}]} = \frac{1}{[1+\{(1-\pi_0)/\pi_0\}B^{-1}]}.$$

The above interpretation is clearly valid when the hypotheses are *simple*, that is,

$$\Theta_0 = \{\theta_0\} \qquad \text{and} \qquad \Theta_1 = \{\theta_1\}$$

for some θ_0 and θ_1. For if so, then $p_0 \propto \pi_0 \, p(\boldsymbol{x}|\theta_0)$ and $p_1 \propto \pi_1 \, p(\boldsymbol{x}|\theta_1)$ so that

$$\frac{p_0}{p_1} = \frac{\pi_0}{\pi_1} \frac{p(\boldsymbol{x}|\theta_0)}{p(\boldsymbol{x}|\theta_1)}$$

and hence the Bayes factor is

$$B = \frac{p(\boldsymbol{x}\,|\,\theta_0)}{p(\boldsymbol{x}\,|\,\theta_1)}.$$

It follows that B is the *likelihood ratio* of H_0 against H_1 which most statisticians (whether Bayesian or not) view as the odds in favour of H_0 against H_1 that are given by the data.

However, the interpretation is not quite as simple when H_0 and H_1 are *composite*, that is, contain more than one member. In such a case it is convenient to write

$$\rho_0(\theta) = p(\theta)/\pi_0 \qquad \text{for } \theta \in \Theta_0$$

and

$$\rho_1(\theta) = p(\theta)/\pi_1 \qquad \text{for } \theta \in \Theta_1$$

where $p(\theta)$ is the prior density of θ, so that $\rho_0(\theta)$ is the restriction of $p(\theta)$ to Θ_0 renormalized to give a probability density over Θ_0, and similarly for $\rho_1(\theta)$. We then have

$$\begin{aligned} p_0 &= \mathrm{P}(\theta \in \Theta_0\,|\,\boldsymbol{x}) \\ &= \int_{\theta \in \Theta_0} p(\theta\,|\,\boldsymbol{x})\,\mathrm{d}\theta \\ &\propto \int_{\theta \in \Theta_0} p(\theta)p(\boldsymbol{x}\,|\,\theta)\,\mathrm{d}\theta \\ &\propto \pi_0 \int_{\theta \in \Theta_0} p(\boldsymbol{x}\,|\,\theta)\,\rho_0(\theta)\,\mathrm{d}\theta \end{aligned}$$

the constant of proportionality depending solely on $\boldsymbol{x}$. Similarly

$$p_1 \propto \pi_1 \int_{\theta \in \Theta_1} p(\boldsymbol{x}\,|\,\theta)\,\rho_1(\theta)\,\mathrm{d}\theta$$

and hence the Bayes factor is

$$B = \frac{(p_0/p_1)}{(\pi_0/\pi_1)} = \frac{\int_{\theta \in \Theta_0} p(\boldsymbol{x}\,|\,\theta)\,\rho_0(\theta)\,\mathrm{d}\theta}{\int_{\theta \in \Theta_1} p(\boldsymbol{x}\,|\,\theta)\rho_1(\theta)\,\mathrm{d}\theta}$$

which is the ratio of "weighted" (by ρ_0 and ρ_1) likelihoods of Θ_0 and Θ_1.

Because this expression for the Bayes factor involves ρ_0 and ρ_1 as well as the likelihood function $p(\boldsymbol{x}\,|\,\theta)$ itself, the Bayes factor cannot be regarded as a measure of the relative support for the hypotheses provided *solely* by the data. Sometimes, however, B will be relatively little affected within reasonable limits by the choice of ρ_0 and ρ_1, and then we *can* regard B as a measure of relative support for the hypotheses provided by the data. When this is so, the Bayes factor is reasonably objective and might, for example, be included in a scientific report so that different users of the data could determine their personal posterior odds by multiplying their personal prior odds by the factor.

It may be noted that the Bayes factor is referred to by a few authors simply as the factor. Jeffreys (1939, 1948, 1961) denoted it K, but did not give it a name. A number of authors, most notably Peirce (1878) and (independently) Good (1950, 1983 and elsewhere) refer to the logarithm of the Bayes factor as the *weight of evidence*. The point of taking the logarithm is, of course, that if we have several experiments about two simple hypotheses, then the Bayes factors multiply, and so the weight of evidence adds.

Example

According to Watkins (1986, Section 13.3) the electroweak theory predicted the existence of a new particle, the W particle, of a mass m of 82.4 ± 1.1 GeV. Experimental results showed that such a particle existed and had a mass of 82.1 ± 1.7 GeV. If we take the mass to have a normal prior and likelihood and assume that the values after the $\pm$ signs represent known standard deviations, and if we are prepared to take both the theory and the experiment into account, then we can conclude that the posterior for the mass is $N(\theta_1, \varphi_1)$ where

$$\varphi_1 = (1.1^{-2} + 1.7^{-2})^{-1} = 0.853 = 0.92^2$$
$$\theta_1 = 0.853(82.4/1.1^2 + 82.1/1.7^2) = 82.3$$

(following the procedure of Section 2.2 on "Normal prior and likelihood"). Suppose that for some reason it was important to know whether or not this mass was less than 83.0 GeV. Then since the prior distribution is $N(82.4, 1.1^2)$ the prior probability π_0 of this hypothesis is given by

$$\pi_0 = P(m \leqslant 83.0) = \Phi((83.0 - 82.4)/1.1) = \Phi(0.55)$$

where Φ is the distribution function of the standard normal distribution. From tables of the normal distribution it follows that $\pi_0 \cong 0.7088$ so that the prior odds are

$$\pi_0/(1 - \pi_0) \cong 2.43.$$

Similarly, the posterior probability of the hypothesis that $m \leqslant 83.0$ is

$p_0 = \Phi((83.0 - 82.3)/0.92) = \Phi(0.76) = 0.7764$, and hence the posterior odds are

$$p_0/(1-p_0) \cong 3.47.$$

The Bayes factor is thus

$$B = \frac{(p_0/p_1)}{(\pi_0/\pi_1)} = \frac{p_0\pi_1}{p_1\pi_0} = \frac{3.47}{2.43} = 1.43.$$

In this case the experiment has not much altered beliefs about the hypothesis under discussion, and this is represented by the nearness of B to 1.

Comment

A point about hypothesis tests well worth making is that they "are traditionally used as a method for testing between two *terminal acts* [but that] in *actual practice* [they] are far more commonly used [when we are] *given* the outcome of a sample [to decide whether] *any* final or terminal decision [should] be reached or should judgement be suspended until more sample evidence is available" (Schlaifer, 1961, Section 13.2).

4.2 One-sided hypothesis tests

Definition

A hypothesis testing situation of the type described in the previous section is said to be *one-sided* if the set Θ of possible values of the parameter θ is the set of real numbers or a subset of it and either

$$\theta_0 < \theta_1 \quad \text{whenever} \quad \theta_0 \in \Theta_0 \text{ and } \theta_1 \in \Theta_1$$

or

$$\theta_0 > \theta_1 \quad \text{whenever} \quad \theta_0 \in \Theta_0 \text{ and } \theta_1 \in \Theta_1.$$

From the Bayesian point of view there is nothing particularly special about this situation. The interesting point is that this is one of the few situations in which classical results, and in particular the use of P-values, has a Bayesian justification.

P-values

This is one of the places where it helps to use the "tilde" notation to emphasize which quantities are random. If $\tilde{x} \sim \mathrm{N}(\theta, \varphi)$ where φ is known and the reference prior $p(\theta) \propto 1$ is used, then the posterior distribution of θ given $\tilde{x} = x$ is $\mathrm{N}(x, \varphi)$. We consider now the situation in which we wish to test H_0: $\theta \leqslant \theta_0$ versus H_1: $\theta > \theta_0$. Then if we observe that $\tilde{x} = x$ we have a posterior probability

$$\begin{aligned} p_0 &= \mathrm{P}(\tilde{\theta} \leqslant \theta_0 \mid \tilde{x} = x) \\ &= \Phi((\theta_0 - x)/\sqrt{\varphi}). \end{aligned}$$

Now the classical *P-value* (sometimes called the *exact significance level*) against H_0 is defined as the probability, when $\theta = \theta_0$, of observing an $\tilde{x}$ "at least as extreme" as the actual data x and so is

$$\begin{aligned} P\text{-value} &= \mathrm{P}(\tilde{x} \geqslant x \mid \theta = \theta_0) \\ &= 1 - \Phi((x - \theta_0)/\sqrt{\varphi}) \\ &= \Phi((\theta_0 - x)/\sqrt{\varphi}) \\ &= p_0. \end{aligned}$$

For example, if we observe a value of x 1.5 standard deviations above θ_0 then a Bayesian using the reference prior would conclude that the posterior probability of the null hypothesis is $\Phi(-1.5) = 0.0668$, whereas a classical statistician would report a P-value of 0.0668. Of course $p_1 = 1 - p_0 = 1 - P$-value, so the posterior odds are

$$\frac{p_0}{p_1} = \frac{p_0}{1 - p_0} = \frac{P\text{-value}}{1 - P\text{-value}}\,.$$

In such a case, the prior distribution could perhaps be said to imply prior odds of 1 (but beware!—this comes from taking $\infty/\infty = 1$), and so we get a Bayes factor of

$$B = \frac{p_0}{1 - p_1} = \frac{P\text{-value}}{1 - P\text{-value}}$$

implying that

$$\begin{aligned} p_0 &= P\text{-value} = B/(1 + B) = (1 + B^{-1})^{-1} \\ p_1 &= 1/(1 + B). \end{aligned}$$

On the other hand, the classical probabilities of Type I and Type II errors do *not* have any close correspondence to the probabilities of hypotheses, and to that extent the increasing tendency of classical statisticians to quote P-values rather than just the probabilities of Type I and Type II errors is to be welcomed, even though a full Bayesian analysis would be better.

A *partial* interpretation of the traditional use of the probability of a Type I error (sometimes called a *significance level*) is as follows. A result is significant at level α if and only if the P-value is less than or equal to α, and hence if and only if the posterior probability

$$p_0 = \mathrm{P}(\tilde{\theta} \leqslant \theta_0 \,|\, \tilde{x} = x) \leqslant \alpha$$

or equivalently

$$p_1 = \mathrm{P}(\tilde{\theta} > \theta_0 \,|\, \tilde{x} = x) \geqslant 1 - \alpha.$$

4.3 Lindley's method

A compromise with classical statistics

The following method appears to have been first suggested by Lindley (1965, Section 5.6), and has since been advocated by a few other authors, for example, Zellner (1971, Section 10.2; 1974, Section 3.7).

Suppose, as is common in classical statistics, that you wish to conduct a test of a point (or sharp) null hypothesis

$$\mathrm{H}_0\colon \theta = \theta_0 \qquad \text{versus} \qquad \mathrm{H}_1\colon \theta \neq \theta_0.$$

Suppose further that your prior knowledge is *vague* or *diffuse*, so that you have no particular reason to believe that $\theta = \theta_0$ rather than that $\theta = \theta_1$ where θ_1 is any value in the neighbourhood of θ_0.

The suggested procedure depends on finding the posterior distribution of θ using a reference prior. To conduct a significance test at level α it is then suggested that you find a $100(1-\alpha)\%$ highest density region (HDR) from the posterior distribution and reject $\mathrm{H}_0\colon \theta = \theta_0$ if and only if θ_0 is *outside* this HDR.

Example

With the data in Section 2.8 on "HDRs for the normal variance" we found the posterior distribution of the variance φ to be

$$\varphi \sim 664\,\chi_{20}^{-2}$$

so that an interval corresponding to a 95% HDR for $\log\chi^2$ is (19, 67). Consequently, on the basis of these data we should reject a null hypothesis $\mathrm{H}_0\colon \varphi = 16$ at the 5% level, but on the other hand we should not reject a null hypothesis $\mathrm{H}_0\colon \varphi = 20$ at that level.

Discussion

This procedure is appropriate only when prior information is vague or diffuse and even then it is not often the best way of summarizing posterior beliefs; clearly, the significance level is a very incomplete expression of these beliefs. For many problems, including that considered in the example above, I think that this method should be seen as mainly of historical interest in that it gave a way of arriving at results related to those in classical statistics and thus helped to wean statisticians brought up on these methods towards the Bayesian approach as one which can give results like these as special cases, as well as having its own distinctive conclusions. It can, however, have a use in situations where there are several unknown parameters and the complete posterior is difficult to describe or take in. Thus, when we come to consider the analysis of variance in Sections 6.5 and 6.6, we shall use the significance level as described in this section to give some idea of the size of the treatment effect.

4.4 Point (or sharp) null hypotheses with prior information

When are point null hypotheses reasonable?

As was mentioned in the previous section, it is very common in classical statistics to conduct a test of a point (or sharp) null hypothesis

$$\mathrm{H}_0\colon \theta = \theta_0 \qquad \text{versus} \qquad \mathrm{H}_1\colon \theta \neq \theta_0.$$

In such a case the full-scale Bayesian approach (as opposed to the compromise described in the previous section) gives rise to conclusions which differ *radically* from the classical answers.

Before going on to the answers, a few basic comments on the whole problem are in order. First, tests of point null hypotheses are often performed in inappropriate circumstances. It will virtually never be the case that one seriously entertains the hypothesis that $\theta = \theta_0$ *exactly*, a point which classical statisticians fully admit (cf. Lehmann, 1959, 1986, Sections 4.5, 5.2). More reasonable would be the null hypothesis

$$\mathrm{H}_0\colon \theta \in \Theta_0 = (\theta_0 - \varepsilon,\ \theta_0 + \varepsilon)$$

where $\varepsilon > 0$ is so chosen that all $\theta \in \Theta_0$ can be considered "indistinguishable" from θ_0. An example in which this might arise would be an attempt to analyse a chemical by observing some aspect, described by a parameter θ, of its reaction with a known chemical. If it were desired to test whether or not the unknown chemical was a specific compound, with a reaction strength θ_0 known to an accuracy of ε, it would be reasonable to test

$$\mathrm{H}_0\colon \theta \in \Theta_0 = (\theta_0 - \varepsilon,\ \theta_0 + \varepsilon) \qquad \text{versus} \qquad \mathrm{H}_1\colon \theta \notin \Theta_0.$$

An example where ε might be extremely close to zero is a test for ESP (extrasensory perception) with θ_0 representing the hypothesis of *no* extra-sensory perception. (The only reason that ε would probably not be zero have is that an experiment designed to test for ESP probably would not lead to a perfectly well-defined θ_0). Of course, there are also many decision problems that would lead to a null hypothesis of the above form with a *large* ε, but such problems will rarely be well approximated by testing a point null hypothesis.

The question arises, if a realistic null hypothesis is H_0: $\theta \in \Theta_0 = (\theta_0 - \varepsilon, \theta_0 + \varepsilon)$, when is it reasonable to approximate it by H_0: $\theta = \theta_0$? From a Bayesian viewpoint, it will be reasonable if and only if the posterior probabilities are close. This will certainly happen if the likelihood function is approximately constant on Θ_0, but this is a very strong condition, and one can often manage with less.

A case of nearly constant likelihood

Suppose that $x_1, x_2, \ldots, x_n$ are independently $N(\theta, \varphi)$ where φ is known. Then we know from Section 2.3 on "Several normal observations with a normal prior" that the likelihood is proportional to an $N(\bar{x}, \varphi/n)$ density for θ. Now over the interval $\theta \in \Theta_0 = (\theta_0 - \varepsilon, \theta_0 + \varepsilon)$ this likelihood varies by a factor

$$\frac{(2\pi\varphi/n)^{-\frac{1}{2}} \exp[-\tfrac{1}{2}\{\bar{x} - (\theta_0 + \varepsilon)\}^2/(\varphi/n)]}{(2\pi\varphi/n)^{-\frac{1}{2}} \exp[-\tfrac{1}{2}\{\bar{x} - (\theta_0 - \varepsilon)\}^2/(\varphi/n)]} = \exp\left(\frac{2\varepsilon n}{\varphi}(\bar{x} - \theta_0)\right).$$

It follows that if we define z to be the statistic

$$z = \frac{|\bar{x} - \theta_0|}{\sqrt{(\varphi/n)}}$$

used in classical tests of significance, and

$$k \geqslant \frac{\varepsilon}{\sqrt{(\varphi/n)}} z$$

then the likelihood varies over Θ_0 by a factor which is at most $\exp(2k)$. Hence, provided that ε is reasonably small, there is a useful bound on the variation of the likelihood.

For example, if ε can be taken to be 0.0025 and

$$k = \frac{0.0025}{\sqrt{(\varphi/n)}} z$$

then the likelihood varies by at most $\exp(2k)$ over $\Theta_0 = (\theta_0 - \varepsilon, \theta_0 + \varepsilon)$. More specifically, if $z = 2$, $\varphi = 1$ and $n = 25$, then k becomes

$$k = (0.0025 \times \sqrt{25})(2) = 0.025$$

and $\exp(2k) = 1.05 = 1/0.95$. In summary, if all values within ± 0.0025 of θ_0 are regarded as indistinguishable from θ_0, then we can feel reassured that the likelihood function does not vary by more than 5% over this range of indistinguishable values, and if the interval can be made even smaller then the likelihood is even nearer to being constant.

Note that the bound depends on $|\bar{x} - \theta_0|$ as well as on φ/n.

The Bayesian method for point null hypotheses

We shall now develop a theory for testing point null hypotheses, which can then be compared with the classical theory. If there is doubt as to the adequacy of the point null hypothesis as a representation of the real null hypothesis, it is always possible to test an interval null hypothesis directly by Bayesian methods and compare the results (and this will generally be easier than checking the constancy of the likelihood function).

You cannot use a continuous prior density to conduct a test of H_0: $\theta = \theta_0$ because that would of necessity give $\theta = \theta_0$ a prior probability of zero and hence a posterior probability of zero. A reasonable way of proceeding is to give $\theta = \theta_0$ a prior probability of $\pi_0 > 0$ and to assign a probability *density* $\pi_1 \rho_1(\theta)$ to values $\theta \neq \theta_0$ where $\pi_1 = 1 - \pi_0$ and $\rho_1(\theta)$ integrates to unity. If you are thinking of the hypothesis $\theta = \theta_0$ as an approximation to a hypothesis $\theta \in (\theta_0 - \varepsilon, \theta_0 + \varepsilon)$ then π_0 is really your prior probability for the whole interval $\theta \in (\theta_0 - \varepsilon, \theta_0 + \varepsilon)$.

You can then derive the predictive density $p(\boldsymbol{x})$ of a vector $\boldsymbol{x} = (x_1, x_2, \ldots, x_n)$ of observations in the form

$$p(\boldsymbol{x}) = \pi_0 p(\boldsymbol{x} | \theta_0) + \pi_1 \int \rho_1(\theta) p(\boldsymbol{x} | \theta) \, d\theta.$$

Writing

$$p_1(\boldsymbol{x}) = \int \rho_1(\theta) p(\boldsymbol{x} | \theta) \, d\theta$$

for what might be called the predictive distribution under the alternative hypothesis we see that

$$p(\boldsymbol{x}) = \pi_0 p(\boldsymbol{x} | \theta_0) + \pi_1 p_1(\boldsymbol{x}).$$

It follows that the posterior probabilities are

$$p_0 = \frac{\pi_0 p(\boldsymbol{x} | \theta_0)}{\pi_0 p(\boldsymbol{x} | \theta_0) + \pi_1 p_1(\boldsymbol{x})} = \frac{\pi_0 p(\boldsymbol{x} | \theta_0)}{p(\boldsymbol{x})}$$

$$p_1 = \frac{\pi_1 p_1(\boldsymbol{x})}{\pi_0 p(\boldsymbol{x} | \theta_0) + \pi_1 p_1(\boldsymbol{x})} = \frac{\pi_1 p_1(\boldsymbol{x})}{p(\boldsymbol{x})}$$

and so the Bayes factor is

$$B = \frac{(p_0/p_1)}{(\pi_0/\pi_1)} = \frac{p(\boldsymbol{x}\,|\,\theta_0)}{p_1(\boldsymbol{x})}.$$

Of course it is possible to find the posterior probabilities p_0 and p_1 in terms of the Bayes factor B and the prior probability π_0 as noted in Section 4.1 when hypothesis testing in general was discussed.

Sufficient statistics

Sometimes we have a sufficient statistic $t = t(\boldsymbol{x})$ for $\boldsymbol{x}$ given θ, so that

$$p(\boldsymbol{x}\,|\,\theta) = p(t\,|\,\theta)p(\boldsymbol{x}\,|\,t)$$

where $p(\boldsymbol{x}\,|\,t)$ is *not* a function of θ. Clearly in such a case

$$\begin{aligned} p_1(\boldsymbol{x}) &= \int \rho_1(\theta)p(t\,|\,\theta)p(\boldsymbol{x}\,|\,t)\,\mathrm{d}\theta \\ &= \left(\int \rho_1(\theta)p(t\,|\,\theta)\,\mathrm{d}\theta\right)p(\boldsymbol{x}\,|\,t) \\ &= p_1(t)p(\boldsymbol{x}\,|\,t) \end{aligned}$$

so that we can cancel a common factor $p(\boldsymbol{x}\,|\,t)$ to obtain

$$p_0 = \frac{\pi_0 p(t\,|\,\theta_0)}{\pi_0 p(t\,|\,\theta_0) + \pi_1 p_1(t)} = \frac{\pi_0 p(t\,|\,\theta_0)}{p(t)}$$

$$p_1 = \frac{\pi_1 p_1(t)}{\pi_0 p(t\,|\,\theta_0) + \pi_1 p_1(t)} = \frac{\pi_1 p_1(t)}{p(t)}.$$

and the Bayes factor is

$$B = \frac{(p_0/p_1)}{(\pi_0/\pi_1)} = \frac{p(t\,|\,\theta_0)}{p_1(t)}.$$

In short, $\boldsymbol{x}$ can be replaced by t in the formulae for p_0, p_1 and the Bayes factor B.

Many of the ideas in this section should become clearer after a reading of the next section, in which the particular case of the normal mean is explored in detail.

4.5 Point null hypotheses for the normal distribution

Calculation of the Bayes factor

Suppose $\boldsymbol{x} = (x_1, x_2, \ldots, x_n)$ is a vector of independently $\mathrm{N}(\theta, \varphi)$ random variables, and that φ is known. Because of the remarks at the end of the last section, we can work entirely in terms of the sufficient statistic

$$\bar{x} \sim \mathrm{N}(\theta, \varphi/n).$$

We have to make some assumption about the density $\rho_1(\theta)$ of θ under the alternative hypothesis, and clearly one of the most natural things to do is to suppose that this density is normal, say $\mathrm{N}(\mu, \psi)$. Strictly, this should be regarded as a density on values of θ other than θ_0, but when probabilities are found by integration of this density, the odd point will make no difference. It will usually seem sensible to take $\mu = \theta_0$ as, presumably, values close to θ_0 are more likely than those far away, and this assumption will accordingly be made from now on. We note that the standard deviation $\sqrt{\psi}$ of the density of θ under the alternative hypothesis is supposed to be considerably greater than the width 2ε of the interval of values of θ considered "indistinguishable" from θ_0.

It is quite easy to find the predictive distribution $p_1(\bar{x})$ of $\bar{x}$ under the alternative, namely,

$$p_1(\bar{x}) = \int \rho_1(\theta) p(\bar{x} \mid \theta) \, \mathrm{d}\theta,$$

by writing

$$\bar{x} = (\bar{x} - \theta) + \theta$$

as in Section 2.2 on "Normal prior and likelihood". Then because, independently of one another, $\bar{x} - \theta \sim \mathrm{N}(0, \varphi/n)$ and $\theta \sim \mathrm{N}(\theta_0, \psi)$, the required density of $\bar{x}$ is $\mathrm{N}(\theta_0, \psi + \varphi/n)$.

It follows that the Bayes factor B is

$$B = \frac{p(\bar{x} \mid \theta_0)}{p_1(\bar{x})}$$

$$= \frac{\{2\pi\varphi/n\}^{-\frac{1}{2}} \exp\left[-\frac{1}{2}(\bar{x} - \theta_0)^2/(\varphi/n)\right]}{\{2\pi(\psi + \varphi/n)\}^{-\frac{1}{2}} \exp\left[-\frac{1}{2}(\bar{x} - \theta_0)^2/(\psi + \varphi/n)\right]}$$

$$= \{1 + n\psi/\varphi\}^{\frac{1}{2}} \exp\left[-\frac{1}{2}(\bar{x} - \theta_0)^2 \, \varphi^{-1} n \{1 + \varphi/n\psi\}^{-1}\right].$$

It is now useful to write

$$z = |\bar{x} - \theta_0| / \sqrt{(\varphi/n)}$$

for the statistic used in classical tests of significance. With this definition

$$B = \{1 + n\psi/\varphi\}^{\frac{1}{2}} \exp\left[-\frac{1}{2}z^2\{1 + \varphi/n\psi\}^{-1}\right].$$

The posterior probability p_0 can now be found in terms of the prior probability π_0 and the Bayes factor B by the usual formula

$$p_0 = \frac{1}{[1+(\pi_1/\pi_0)B^{-1}]} = \frac{1}{[1+\{(1-\pi_0)/\pi_0\}B^{-1}]}$$

derived in Section 4.1 when we first met hypothesis tests.

Numerical examples

For example, if $\pi_0 = \frac{1}{2}$ and $\psi = \varphi$, then the values

$$z = 1.96,\ n = 15$$

give rise to a Bayes factor

$$B = \{1+15\}^{\frac{1}{2}} \exp\left[-\tfrac{1}{2}(1.96)^2\{1+\tfrac{1}{15}\}^{-1}\right] = 0.66$$

and hence to a posterior probability

$$p_0 = [1+0.66^{-1}]^{-1} = 0.4.$$

This result is quite extraordinarily different from the conclusion that a classical statistician would arrive at with the same data. Such a person would say that, since z has a sampling distribution that is N(0, 1), a value of z that is, in modulus, 1.96 or greater would arise with probability only 5% (that is, the two-tailed P-value of $z = 1.96$ is 0.05), and consequently would reject the null hypothesis that $\theta = \theta_0$ at the 5% level. With the above assumptions about prior beliefs, we have, on the contrary, arrived at a posterior probability of 40% that the null hypothesis is true! Some further sample values are as follows (cf. Berger, 1985, Section 4.3):

P-value (2-tailed)	z \ n	1	5	10	20	50	100	1000
0.1	1.645	0.418	0.442	0.492	0.558	0.655	0.725	0.891
0.05	1.960	0.351	0.331	0.367	0.424	0.521	0.600	0.823
0.01	2.576	0.212	0.134	0.140	0.163	0.216	0.273	0.535
0.001	3.291	0.086	0.024	0.026	0.024	0.034	0.045	0.124

The results of classical and Bayesian analyses differ more and more as the sample size $n \to \infty$. For fixed z, it is easy to see that B is asymptotically

$$B \approx (n\psi/\varphi)^{\frac{1}{2}} \exp\left[-\tfrac{1}{2}z^2\right]$$

and hence $B \to \infty$. Consequently $1-p_0$ is of order. $1/\sqrt{n}$ and thus $p_0 \to 1$. So, with the specified prior, the result that $z=1.96$, which a classical statistician would regard as just sufficient to result in rejection of the null hypothesis at the 5% level irrespective of the value of n, can result in an arbitrarily high posterior probability p_0 of the null hypothesis. Despite this, beginners in statistical techniques often gain the impression that if some data are significant at the 5% level then in some sense the null hypothesis has a probability after the event of at most 5%.

A specific example of a problem with a large sample size arises in connection with Weldon's dice data, quoted by Fisher (1925b, Sections 18 and 23). It transpired that when 12 dice were thrown 26,306 times, the mean and variance of the number of dice showing more than 4 were 4.0524 and 2.6983, as compared with a theoretical mean of $12 \times \frac{1}{3} = 4$ for fair dice. Approximating the binomial distribution by a normal distribution leads to a z statistic of

$$z = (4.0524-4)/\sqrt{(2.6983/26{,}306)} = 5.17.$$

The corresponding two-tailed P-value is approximately $2\varphi(z)/z$ where φ is the density function of the standard normal distribution (cf. Abramowitz and Stegun, 1965, Section 26.2.12), so about 1 in 4,000,000. However, a Bayesian analysis (assuming $\psi = \varphi$ and $\pi_0 = \frac{1}{2}$ as usual) depends on a Bayes factor

$$\begin{aligned} B &= (1+26{,}306)^{\frac{1}{2}} \exp\left[-\tfrac{1}{2}(5.17)^2 \{1+(26{,}306)^{-1}\}^{-1}\right] \\ &= 0.00025 \end{aligned}$$

and so to a posterior probability of 1 in 4000 that the dice were fair. This is small, but nevertheless the conclusion is not as startling as that which the classical analysis leads to.

Lindley's paradox

This result is sometimes known as *Lindley's paradox* (cf. Lindley, 1957; Bartlett, 1957; Shafer, 1982) and sometimes as *Jeffreys' paradox*, because it was in essence known to Jeffreys [see Jeffreys (1939; 1948; 1961, Section 4.2)], although he did not refer to it as a paradox. A useful recent reference is Berger and Delampady (1987).

It does relate to something which has been noted by users of statistics. Lindley once pointed out (see Zellner (1974, Section 3.7)) that experienced workers often lament that for large sample sizes, say 5000, as encountered in survey data, use of the usual t-statistic and the 5% significance level shows that the values of parameters are usually different from zero and that many such workers sense that with such a large sample the 5% level is not the right thing to use, but do not know what else to use [see also Jeffreys (1939; 1948; 1961, Appendix B)]. On the other hand, in many scientific contexts it is unrealistic to use a very large sample because, for instance, systematic bias may vitiate it or the observer may tire; see Wilson (1952, Section 9.6) or Baird (1962, Section 2.8).

Since the result is so different from that found by so many statisticians, it is important to check that it does not depend very precisely on the nature of the prior distribution which led to it.

We assumed that the prior probability π_0 of the null hypothesis was $\frac{1}{2}$, and this assumption does seem "natural" and could be said to be "objective"; in any case a slight change in the value of π_0 would not make much difference to the qualitative feel of the results.

We also assumed that the prior density of θ under the alternative hypothesis was normal of mean θ_0 with some variance ψ. In fact the precise choice of $\rho_1(\theta)$ does not make a great deal of difference unless $|\bar{x}-\theta_0|$ is *large*. Lindley (1957) took $\rho_1(\theta)$ to be a uniform distribution

$$U(\theta_0-\tfrac{1}{2}\sqrt{\psi},\ \theta_0+\tfrac{1}{2}\sqrt{\psi})$$

over an interval centred on θ_0, while Jeffreys (1939; 1948; 1961, Section 5.2) argues that it should be a Cauchy distribution, that is,

$$\rho_1(\theta)=\frac{1}{\pi}\,\frac{\sqrt{\psi}}{\psi+(\theta-\theta_0)^2}$$

although his arguments are far from overwhelming and do not seem to have convinced anyone else. An examination of Lindley's and Jeffreys' work will show that in general terms they arrive at similar conclusions to those derived above.

There is also a scale parameter ψ in the distribution $\rho_1(\theta)$ to be decided on (and this is true whether this distribution is normal, uniform or Cauchy). Although it seems reasonable that ψ should be chosen proportional to φ, there does not seem to be any convincing argument for choosing this to have any particular value (although Jeffreys tries to give a rational argument for the Cauchy form in general, he seems to have no argument for the choice of ψ beyond saying that it should be proportional to φ). But it is easily seen that the *effect* of taking

$$\psi=k\varphi$$

on B and p_0 is just the same as taking $\psi=\varphi$ if *n is multiplied by a factor k*. It should be noted that it will *not* do to let $\psi\to\infty$ and thus to take

$\rho_1(\theta)$ as a uniform distribution on the whole real line, because this is equivalent to multiplying n by a factor which tends to ∞ and so leads to $B \to \infty$ and $p_0 \to 1$. It would clearly not be sensible to use a procedure which always gave the null hypothesis a posterior probability of unity. In any case, as Jeffreys points out (1939; 1948; 1961, Section 5.0), "the mere fact that it has been suggested that [θ] is zero corresponds to some presumption that it is fairly small".

A bound which does not depend on the prior distribution

In fact it is possible to give a bound on B which does not depend on any assumptions about $\rho_1(\theta)$. For

$$\begin{aligned} p_1(\bar{x}) &= \int \rho_1(\theta)\, p(\bar{x}\,|\,\theta)\,\mathrm{d}\theta \\ &\leqslant p(\bar{x}\,|\,\hat{\theta}) \end{aligned}$$

where $\hat{\theta}$ is the *maximum likelihood* estimator of θ, that is,

$$p(\bar{x}\,|\,\hat{\theta}) = \sup_{\theta} p(\bar{x}\,|\,\theta).$$

In the case being considered, $\bar{x}$ has a normal distribution of mean θ and hence $\hat{\theta} = \bar{x}$, so that

$$p_1(\bar{x}) \leqslant p(\bar{x}\,|\,\bar{x}) = (2\pi\varphi/n)^{-\frac{1}{2}}.$$

It follows that the Bayes factor satisfies

$$B = \frac{p(\bar{x}\,|\,\theta_0)}{p_1(\bar{x})} \geqslant \frac{\{2\pi\varphi/n\}^{-\frac{1}{2}} \exp\left[-\frac{1}{2}(\bar{x}-\theta_0)^2/(\varphi/n)\right]}{\{2\pi\varphi/n\}^{-\frac{1}{2}}}$$

so writing $z = |\bar{x}-\theta_0|/\sqrt{(\varphi/n)}$ as before, we see that

$$B \geqslant \exp\left(-\tfrac{1}{2}z^2\right)$$

implying a corresponding lower bound on p_0. Some sample values (assuming that $\pi_0 = \frac{1}{2}$) are as follows:

P-value (2-tailed)	z	Bound on B	Bound on p_0
0.1	1.645	0.258	0.205
0.05	1.960	0.146	0.128
0.01	2.576	0.036	0.035
0.001	3.291	0.004	0.004

[cf. Berger (1985, Section 4.3); Berger further claims that if $\pi_0 = \frac{1}{2}$ and $z > 1.68$ then $p_0 \geqslant (P\text{-value}) \times (1.25\, z)$].

As an example, if $z = 1.96$ then the Bayes factor B is *at least* 0.146 and hence the posterior probability of the null hypothesis is *at least*

0.128. Unlike the results derived earlier assuming a more precise form for $\rho_1(\theta)$, the bounds no longer depend on the sample size, but it should still be noted that the conclusion does not accord at all well with the classical result of significance at the 5% level.

The case of an unknown variance

In the case where φ is unknown, similar conclusions follow, although there are a few more complications. It will do no harm if the rest of this section is ignored at a first reading (or even at a second).

We first need to find the density $p(x\,|\,\theta_0)$. If φ is unknown, then as was shown in Section 2.12 on "Normal mean and variance both unknown"

$$p(\boldsymbol{x}\,|\,\theta_0,\varphi) \propto \varphi^{-n/2} \exp\left[-\tfrac{1}{2}\{S+n(\bar{x}-\theta_0)^2\}/\varphi\right]$$

where $S=\sum(x_i-\bar{x})^2$. Using a reference prior $p(\varphi)\propto 1/\varphi$ for φ, it is easy to integrate φ out much as was done there to obtain

$$\begin{aligned} p(\boldsymbol{x}\,|\,\theta_0) &= \int p(\boldsymbol{x},\varphi\,|\,\theta_0)\,\mathrm{d}\varphi = \int p(\varphi)p(\boldsymbol{x}\,|\,\theta_0,\varphi)\,\mathrm{d}\varphi \\ &\propto \int \varphi^{-n/2} \exp\left[-\tfrac{1}{2}\{S+n(\bar{x}-\theta_0)^2\}/\varphi\right]\mathrm{d}\varphi \\ &\propto \{1+t^2/\nu\}^{-(\nu+1)/2} \end{aligned}$$

where $\nu=n-1$ and

$$t=\frac{\theta-\bar{x}}{s/\sqrt{n}},$$

$$s^2=S/\nu.$$

It is now necessary to find the predictive density $p_1(\boldsymbol{x})$ under the alternative hypothesis. To do this, we first return to

$$p(\boldsymbol{x}\,|\,\theta,\varphi) \propto \varphi^{-n/2} \exp\left[-\tfrac{1}{2}\{S+n(\bar{x}-\theta)^2\}/\varphi\right].$$

Assuming a prior $\theta\sim\mathrm{N}(\theta_0,\psi)$ we can integrate θ out thus:

$$p_1(\boldsymbol{x}\,|\,\varphi) = \int p(\boldsymbol{x},\theta\,|\,\varphi)\,\mathrm{d}\theta = \int p(\theta)p(\boldsymbol{x}\,|\,\theta,\varphi)\,\mathrm{d}\theta$$

$$\propto \int \varphi^{-n/2}\psi^{-\frac{1}{2}} \exp\left[-\tfrac{1}{2}\left(\frac{S}{\varphi}+\frac{\mathrm{n}}{\varphi}(\bar{x}-\theta)^2+\frac{1}{\psi}(\theta-\theta_0)^2\right)\right]\mathrm{d}\theta$$

$$\propto \varphi^{-n/2}\psi^{-\frac{1}{2}} \exp\left[-\tfrac{1}{2}\left(\frac{S}{\varphi}+\frac{n\bar{x}^2}{\varphi}+\frac{\theta_0{}^2}{\psi}-\frac{(n\bar{x}/\varphi+\theta_0/\psi)^2}{n/\varphi+1/\psi}\right)\right]$$

$$\times \int \exp\left[-\tfrac{1}{2}\left(\frac{n}{\theta}+\frac{1}{\psi}\right)\left(\theta-\frac{n\bar{x}/\varphi+\theta_0/\psi}{n/\varphi+1/\psi}\right)^2\right]\mathrm{d}\theta.$$

The last integral is of course proportional to $(n/\varphi+1/\psi)^{-\frac{1}{2}}$ and so to $(1+n\psi/\varphi)^{-\frac{1}{2}}$, while a little manipulation shows that

$$\frac{n\bar{x}^2}{\psi}+\frac{\theta_0{}^2}{\psi}-\frac{(n\bar{x}/\varphi+\theta_0/\psi)^2}{n/\varphi+1/\psi}=\frac{(n/\varphi)(1/\psi)}{n/\varphi+1/\psi}(\bar{x}-\theta_0)^2$$

$$=\frac{1}{1+n\psi/\varphi}\;\frac{n(\bar{x}-\theta_0)^2}{\varphi}.$$

It follows that

$$p_1(\boldsymbol{x}\,|\,\varphi)\propto\varphi^{-n/2}\left(1+\frac{n\psi}{\varphi}\right)^{-\frac{1}{2}}\exp\left[-\tfrac{1}{2}\left(S+\frac{n(\bar{x}-\theta_0)^2}{1+n\psi/\varphi}\right)\Big/\varphi\right].$$

To go any further it is necessary to make some assumption about the relationship between φ and ψ. If it is assumed that

$$\psi=k\varphi$$

and a reference prior $p(\varphi)\propto 1/\varphi$ is used, then the predictive distribution under the alternative hypothesis becomes

$$p_1(\boldsymbol{x})=\int p(\boldsymbol{x},\varphi)\,\mathrm{d}\varphi=\int p(\varphi)p(\boldsymbol{x}\,|\,\varphi)\,\mathrm{d}\varphi$$

$$\propto(1+nk)^{-\frac{1}{2}}\int\varphi^{-(\nu+1)/2-1}\exp\left[-\tfrac{1}{2}\left(S+\frac{n(\bar{x}-\theta_0)^2}{1+nk}\right)\Big/\varphi\right]\mathrm{d}\varphi$$

$$\propto(1+nk)^{-\frac{1}{2}}\left\{1+\frac{1}{\nu}\;\frac{t^2}{1+nk}\right\}^{-(\nu+1)/2}$$

where t is the same statistic encountered in the case of the null hypothesis. It follows that the Bayes factor is

$$\frac{\{1+t^2/\nu\}^{-(\nu+1)/2}}{(1+nk)^{-\frac{1}{2}}\{1+t^2(1+nk)^{-1}/\nu\}^{-(\nu+1)/2}}$$

and hence it is possible to find p_0 and p_1.

It should be noted that as $n\to\infty$ the exponential limit shows that the Bayes factor is asymptotically

$$\frac{\exp[-\frac{1}{2}t^2]}{(1+nk)^{-\frac{1}{2}}\exp[-\frac{1}{2}t^2(1+nk)^{-1}]}=(1+nk)^{-\frac{1}{2}}\exp[-\tfrac{1}{2}t^2nk/(1+nk)]$$

which as $t\cong z$ is the same as in the known variance case.

4.6 The Doogian philosophy

Description of the method

Good (1983, Chapter 4 and elsewhere) has argued in favour of a compromise between Bayesian and non-Bayesian approaches to hypothesis testing. His technique can be summarized as follows (in his own words):

"The Bayes/non-Bayes synthesis is the following technique for synthesizing subjective and objective techniques in statistics. (i) We use the neo-Bayes/Laplace philosophy [i.e. the techniques described in Section 4.4 on point null hypotheses with prior information] in order to arrive at a factor F [which is $1/B$ in the notation used here] in favour of the non-null hypothesis. For the particular case of discrimination between two simple statistical hypotheses, the factor in favour is equal to the likelihood ratio [as was shown in Section 4.1 when hypothesis testing was first considered], but not in general. The neo-Bayes/Laplace philosophy usually works with inequalities between probabilities, but for definiteness we here assume that the initial distributions are taken as precise, though not necessarily uniform. (ii) We then use F as a statistic and try to obtain its distribution on the null hypothesis, and work out its tail area, P. (iii) Finally, we look to see if F lies in the range

$$(1/30P,\ 3/10P).$$

If it does not lie in this range we think again. (Note that F is here the factor *against* H)."

This is certainly not unworkable. For example, in the previous section we found that

$$1/F = B = \{1 + n\psi/\varphi\}^{\frac{1}{2}} \exp\left[-\tfrac{1}{2}z^2(1 + \varphi/n\psi)^{-1}\right]$$

so that B is a monotonic function of z^2, and hence the probability $B \geqslant b$ equals the (two-tailed) P-value corresponding to the value of z, which is easily found as z has a standard normal distribution.

Numerical example

Thus if, as in an example discussed in the previous section, $\pi_0 = \frac{1}{2}$ and the density under the alternative hypothesis has $\psi = \varphi$ (and so is $N(\theta_0, \varphi)$), then for

$$z = 1.96 \qquad \text{and} \qquad n = 15$$

the P-value is $P = 0.05$ and the Bayes factor is $B = 0.66 = 1/1.5$. Good's method then asks us to check whether $F = 1.5$ lies in the range

$(1/30P, 3/10P)$, that is, $(0.67, 6.0)$. Consequently, we do not in this case need to "think again".

Good attributes this approach to "the Tibetan lama K. Caj Doog", but it does not appear that thc lama has made many converts apart from Good himself.

Exercises on Chapter 4

1. Watkins (1986, Section 1.3) reports that theory predicted the existence of a Z particle of mass 93.3 ± 0.9 GeV, while first experimental results showed its mass to be 93.0 ± 1.8 GeV. Find the prior and posterior odds and the Bayes ratio for the hypothesis that its mass is less than 93.0 GeV.

2. An experimental station wishes to test whether a growth hormone will increase the yield of wheat above the average value of 100 units per plot produced under currently standard conditions. Twelve plots treated with the hormone give the yields:

$$140, 103, 73, 171, 137, 91, 81, 157, 146, 69, 121, 134.$$

Find the P-value for the hypothesis under consideration.

3. In a genetic experiment, theory predicts that if two genes are on different chromosomes, then the probability of a certain event will be 3/16. In an actual trial, the event occurs 56 times in 300. Use Lindley's method to decide whether there is enough evidence to reject the hypothesis that the genes are on the same chromosome.

4. Suppose that the standard test statistic $z = (\bar{x} - \theta_0)/\sqrt{(\varphi/n)}$ takes the value $z = 2.5$ and that the sample size is $n = 100$. How close to θ_0 does a value of θ have to be for the value of the normal likelihood function at $\bar{x}$ to be within 10% of its value at $\theta = \theta_0$?

5. Show that the Bayes factor for a test of a point null hypothesis for the normal distribution (where the prior under the alternative hypothesis is also normal) can be expanded in a power series in $\lambda = \varphi/n\psi$ as

$$B = \lambda^{-\frac{1}{2}} \exp\left(-\tfrac{1}{2}z^2\right)\{1 + \tfrac{1}{2}\lambda(z^2 + 1) + \ldots\}.$$

6. At the beginning of Section 4.5, we saw that under the alternative

hypothesis that $\theta \sim N(\theta_0, \psi)$ the predictive density for $\bar{x}$ was $N(\theta_0, \psi + \varphi/n)$, so that

$$p_1(\bar{x}) = \{2\pi(\psi + \varphi/n)\}^{-\frac{1}{2}} \exp\left[-\tfrac{1}{2}(\bar{x} - \theta_0)^2/(\psi + \varphi/n)\right].$$

Show that a maximum of this density considered as a function of ψ occurs when $\psi = (z^2 - 1)\varphi/n$, which gives a possible value for ψ if $z \geqslant 1$. Hence show that if $z \geqslant 1$ then for any such alternative hypothesis the Bayes factor satisfies

$$B \geqslant \sqrt{e}\, z \exp(-\tfrac{1}{2}z^2)$$

and deduce a bound for p_0 (depending on the value of π_0).

7. A window is broken in forcing entry to a house. The refractive index of a piece of glass found at the scene of the crime is x, which is supposed $N(\theta_1, \varphi)$. The refractive index of a piece of glass found on a suspect is y, which is supposed $N(\theta_2, \varphi)$. In the process of establishing the guilt or innocence of the suspect, we are interested in investigating whether H_0: $\theta_1 = \theta_2$ is true or not. The prior distributions of θ_1 and θ_2 are both $N(\mu, \psi)$ where $\psi \gg \varphi$. Write

$$u = x - y, \qquad z = \tfrac{1}{2}(x + y).$$

Show that if H_0 is true and $\theta_1 = \theta_2 = \theta$ then θ, $x - \theta$ and $y - \theta$ are independent and

$$\theta \sim N(\mu, \psi), \quad x - \theta \sim N(0, \varphi), \quad y - \theta \sim N(0, \varphi).$$

By writing $u = (x - \theta) - (y - \theta)$ and $z = \theta + \frac{1}{2}(x - \theta) + \frac{1}{2}(y - \theta)$, go on to show that u has an $N(0, 2\varphi)$ distribution and that z has an $N(\mu, \frac{1}{2}\varphi + \psi)$, so approximately an $N(\mu, \psi)$, distribution. Conversely, show that if H_0 is false and θ_1 and θ_2 are assumed independent, then θ_1, θ_2, $x - \theta_1$ and $y - \theta_2$ are all independent and

$$\theta_1 \sim N(\mu, \psi), \theta_2 \sim N(\mu, \psi), x - \theta_1 \sim N(0, \varphi), y - \theta_2 \sim N(0, \varphi).$$

By writing

$$u = \theta_1 - \theta_2 + (x - \theta_1) - (y - \theta_2),$$
$$z = \tfrac{1}{2}\{\theta_1 + \theta_2 + (x - \theta_1) + (y - \theta_2)\}$$

show that in this case u has an $N(0, 2(\varphi + \psi))$, so approximately an $N(0, 2\psi)$, distribution, while z has an $N(\mu, \frac{1}{2}(\varphi + \psi))$, so approximately an $N(\mu, \frac{1}{2}\psi)$, distribution. Conclude that the Bayes factor is approximately

$$B = \sqrt{(\psi/2\varphi)} \exp\left[-\tfrac{1}{2}u^2/2\varphi - \tfrac{1}{2}(z - \mu)^2/\Psi\right].$$

Suppose that the ratio $\surd(\psi/\varphi)$ of the standard deviations is 100 and that $u = 2 \times \surd(2\varphi)$, so that the difference between x and y represents two standard deviations, and that $z = \mu$, so that both specimens are of commonly occurring glass. Show that a classical test would reject H_0 at the 5% level, but that $B = 9.57$, so that the odds in *favour* of H_0 are multiplied by a factor just below 10.

[This problem is due to Lindley (1977); see also Shafer (1982). Lindley comments that, "What the test fails to take into account is the extraordinary coincidence of x and y being so close together were the two pieces of glass truly different".]

8. Lindley (1957) originally discussed his paradox under slightly different assumptions from those made in this book. Follow through the reasoning used in Section 4.5 with $\rho_1(\theta)$ representing a uniform distribution on the interval $(\theta_0 - \frac{1}{2}\tau, \theta_0 + \frac{1}{2}\tau)$ to find the corresponding Bayes factor assuming that $\tau^2 \gg \varphi/n$, so that an $N(\mu, \varphi/n)$ variable lies in this interval with very high probability. Check that your answers are unlikely to disagree with those found in Section 4.5 under the assumption that $\rho_1(\theta)$ represents a normal density.

9. Express in your own words the arguments given by Jeffreys (1939; 1948; 1961, Section 5.2) in favour of a Cauchy distribution

$$\rho_1(\theta) = \frac{1}{\pi} \frac{\surd\psi}{\psi + (\theta - \theta_0)^2}$$

in the problem discussed in the previous question.

10. Suppose that x has a binomial distribution $B(n, \theta)$ of index n and parameter θ, and that it is desired to test H_0: $\theta = \theta_0$ against the alternative hypothesis H_1: $\theta \neq \theta_0$.

(a) Find lower bounds on the posterior probability of H_0 and on the Bayes factor for H_0 versus H_1, bounds which are valid for *any* $\rho_1(\theta)$.

(b) If $n = 20$, $\theta_0 = \frac{1}{3}$ and $x = 16$ is observed, calculate the (two-tailed) *P*-value and the lower bound on the posterior probability when the prior probability π_0 of the null hypothesis is $\frac{1}{2}$.

11. Twelve observations from a normal distribution of mean θ and variance φ are available, of which the sample mean is 1.2 and the sample variance is 1.1. Compare the Bayes factors of the null

hypothesis that $\theta = \theta_0$ assuming (a) that φ is unknown and (b) that it is known that $\varphi = 1$.

12. Suppose that in testing a point null hypothesis you find a value of the usual Student's t statistic of 2.4 on 8 degrees of freedom. Would the methodology of Section 4.6 require you to "think again"?

5

Two-sample Problems

5.1 Two-sample problem—both variances known

The problem of two normal samples

We now want to consider the situation in which we have independent samples from two normal distributions, namely,

$$x_1, x_2, \ldots, x_m \sim \mathrm{N}(\lambda, \varphi)$$
$$y_1, y_2, \ldots, y_n \sim \mathrm{N}(\mu, \psi)$$

which are independent of each other, and the quantity really of interest is the posterior distribution of

$$\delta = \lambda - \mu.$$

This problem arises in comparative situations, for example, in comparing the achievement in geometry tests of boy and girl pupils.

Paired comparisons

Before proceeding further, we should beware of a possible misapplication of the model. If $m = n$ and each of the xs is in some sense paired with one of the ys, say x_i with y_i, we should define

$$w_i = x_i - y_i$$

and then investigate the ws as a sample

$$w_1, w_2, \ldots, w_n \sim \mathrm{N}(\delta, \omega)$$

for some ω. This is known as the method of *paired comparisons*. It might arise if, for example, the comparison of performance of boys and girls were restricted to pairs of twins of opposite sexes. The reason that such a situation is not to be treated as a two-sample problem in the sense described at the start is that there will be an effect common to any pair of twins, so that the observations on the boys and on the girls will not be fully independent. It is a very valuable technique which can often give a *more precise* measurement of an effect, but it is important to distinguish it from a case where the two samples are independent. There is no particular difficulty in analysing the results of a paired comparison experiment by the methods described in Chapter 1 for samples from a single normal distribution.

Example of a paired comparison problem

"Student" (1908) quotes data due to A. R. Cushny and A. R. Peebles on the extra hours of sleep gained by 10 patients using laevo (L) and dextro (D) hyoscamine hydrobromide, as follows:

Patient i	1	2	3	4	5	6	7	8	9	10
Gain x_i with L	+1.9	+0.8	+1.1	+0.1	−0.1	+4.4	+5.5	+1.6	+4.6	+3.4
Gain y_i with D	+0.7	−1.6	−0.2	−1.2	−0.1	+3.4	+3.7	+0.8	0	+2.0
Difference w_i	+1.2	+2.4	+1.3	+1.3	0	+1.0	+1.8	+0.8	+4.6	+1.4

If we are interested in the difference between the effects of the two forms of the drug, we should find the mean $\bar{w} = 1.58$ and the sample sum of squares $S = 13.616$ and hence the sample standard deviation $s = 1.23$. Assuming a standard reference prior for δ and a variance known to equal 1.23^2, the posterior distribution of the effect δ of using the L rather than the D form is $N(1.58, 1.23^2)$. We can then use this distribution, for example, to give an HDR for δ or to test a hypothesis about δ (such as H_0: $\delta = 0$ versus H_1: $\delta \neq 0$) in the ways discussed in previous sections. On the other hand, if we are interested simply in the effect of the L form, then the data about the D form are irrelevant and we can use the same methods on the x_i. It is straightforward to extend the analysis to allow for a non-trivial prior for δ or an unknown variance or both.

The case where both variances are known

In the case of the two-sample problem proper, there are three cases that can arise:

(i) φ and ψ are known;
(ii) it is known that $\varphi = \psi$ but their common value is unknown;
(iii) φ and ψ are unknown.

For the rest of this section, we shall restrict ourselves to case (i). It should, however, be noted that it is not really likely that we would know the variances exactly (although we might have some idea from past experience). The main reason for discussing this case, as in the problem of a single sample from a normal distribution, is that it involves fewer complexities than the case where the variances are known.

If λ and μ have *independent* reference priors $p(\lambda) = p(\mu) \propto 1$ then it follows from Section 2.3 on "Several normal observations with a normal prior" that the posterior for λ is $N(\bar{x}, \varphi/m)$, and similarly the posterior for μ is $N(\bar{y}, \psi/n)$ independently of λ. It follows that

$$\delta = \lambda - \mu \sim N(\bar{x} - \bar{y}, \varphi/m + \psi/n).$$

Example

The weight gains (in grammes) between the 28th and 84th days of age of $m = 12$ rats receiving high-protein diets were as follows:

Rat i	1	2	3	4	5	6	7	8	9	10	11	12
Weight gain x_i	134	146	104	119	124	161	107	83	113	129	97	123

while the weight gains for $n = 7$ rats on a low-protein diet were:

Rat i	1	2	3	4	5	6	7
Weight gain y_i	70	118	101	85	107	132	94

(cf. Armitage and Berry, 1987, Section 4.4). The sample mean and sum of squares for the high-protein group are $\bar{x} = 120$ and 5032, implying a sample variance of $5032/11 = 457$. For the low-protein group the mean and sum of squares are $\bar{y} = 101$ and 2552, implying a sample variance of $2552/6 = 425$. Although the values for the variances are derived from the samples, the method will be illustrated by proceeding *as if* they were known (perhaps from past experience). Then

$$m = 12, n = 7, \bar{x} = 120, \bar{y} = 101, \varphi = 457, \psi = 425$$

from which it follows that the posterior distribution of the parameter δ that measures the effect of using a high-protein rather than a low-protein diet is $N(120 - 101, 457/12 + 425/7)$, that is, $N(19, 99)$.

It is now possible to conclude, for example, that a 90% HDR for δ is $19 \pm 1.6449\sqrt{99}$, that is, $(3, 35)$. Also, the posterior probability that $\delta > 0$ is $\Phi(19/\sqrt{99}) = \Phi(1.91) = 0.9719$ or about 97%. Furthermore, it is possible to conduct a test of the point null hypothesis that $\delta = 0$. If the variance of δ under the alternative hypothesis is denoted ω (rather than ψ as in Section 4.5 on "Point null hypotheses for the normal distribution" since ψ now has another meaning), then the Bayes factor is

$$B = \{1 + \omega/(\varphi/m + \psi/n)\}^{\frac{1}{2}} \exp\left[-\tfrac{1}{2}z^2\{1 + (\varphi/m + \psi/n)/\omega\}^{-1}\right]$$

where z is the standardized normal variable (under the null hypothesis), namely, $z = (19 - 0)/\sqrt{99} = 1.91$. It is not wholly clear what value should be used for ω. One possibility might be to take

$\omega=\varphi+\psi=457+425=882$, and if this is done then

$$B=\{1+882/99\}^{\frac{1}{2}}\exp\left[-\tfrac{1}{2}(1.91)^2\{7+99/882\}^{-1}\right]$$
$$=9.91^{\frac{1}{2}}\exp(-1.64)=0.61.$$

If the prior probability of the null hypothesis is taken as $\pi_0=\frac{1}{2}$, then this gives a posterior probability of $p_0=(1+0.61^{-1})^{-1}=0.38$, so that it has dropped, but not dropped very much.

Non-trivial prior information

The method is easily generalized to the case where substantial prior information is available. If the prior for λ is $N(\lambda_0, \varphi_0)$ then the posterior is $\lambda\sim N(\lambda_1, \varphi_1)$ where (as was shown in Section 2.3 on "Several normal observations with a normal prior")

$$\varphi_1=\{\varphi_0{}^{-1}+(\varphi/m)^{-1}\}^{-1}$$
$$\lambda_1=\varphi_1\{\lambda_0/\varphi_0+\bar{x}/(\varphi/m)\}.$$

Similarly, if the prior for μ is $N(\mu_0, \psi_0)$ then the posterior for μ is $N(\mu_1, \psi_1)$, where ψ_1 and μ_1 are similarly defined. It follows that

$$\delta=\lambda-\mu\sim N(\lambda_1-\mu_1, \varphi_1+\psi_1)$$

and inferences can proceed much as before.

5.2 Variances unknown but equal

Solution using reference priors

We shall now consider the case where we are interested in $\delta=\lambda-\mu$ and we have independent vectors $\boldsymbol{x}=(x_1, x_2, \ldots, x_m)$ and $\boldsymbol{y}=(y_1, y_2, \ldots, y_n)$ such that

$$x_i\sim N(\lambda, \varphi) \qquad \text{and} \qquad y_i\sim N(\mu, \varphi)$$

so that the two samples have a common variance φ.

We can proceed much as we did in Section 2.12 on "Normal mean and variance both unknown". We begin by defining

$$S_x=\sum(x_i-\bar{x})^2, \qquad S_y=\sum(y_i-\bar{y})^2, \qquad S=S_x+S_y,$$
$$\nu_x=m-1, \qquad \nu_y=n-1, \qquad \nu=\nu_x+\nu_y.$$

For the moment, we take independent priors uniform in λ, μ and $\log\varphi$, that is,

$$p(\lambda, \mu, \varphi)\propto 1/\varphi.$$

With this prior, the posterior is

$$\begin{aligned} p(\lambda, \mu, \varphi | \boldsymbol{x}, \boldsymbol{y}) &\propto p(\lambda, \mu, \varphi) p(\boldsymbol{x} | \lambda, \varphi) p(\boldsymbol{y} | \mu, \varphi) \\ &\propto (1/\varphi)(2\pi\varphi)^{-(m+n)/2} \\ &\quad \times \exp\left[-\tfrac{1}{2}\{\sum(x_i - \lambda)^2 + \sum(y_i - \mu)^2\}/\varphi\right] \\ &\propto \varphi^{-(m+n)/2-1} \exp\left[-\tfrac{1}{2}\{S_x + m(\bar{x} - \lambda)^2 + S_y \right. \\ &\quad \left. + n(\bar{y} - \mu)^2\}/\varphi\right] \\ &\propto \varphi^{-\nu/2-1} \exp\left[-\tfrac{1}{2}S/\varphi\right] (2\pi\varphi/m)^{-\frac{1}{2}} \exp\left[-\tfrac{1}{2}m(\lambda - \bar{x})^2/\varphi\right] \\ &\quad \times (2\pi\varphi/n)^{-\frac{1}{2}} \exp\left[-\tfrac{1}{2}n(\mu - \bar{y})^2/\varphi\right] \\ &\propto p(\varphi | S) p(\lambda | \varphi, \bar{x}) p(\mu | \varphi, \bar{y}) \end{aligned}$$

where

$p(\varphi | S)$ is an $S\chi^{\nu - 2}$ density
$p(\lambda | \varphi, \bar{x})$ is an $\mathrm{N}(\bar{x}, \varphi/m)$ density,
$p(\mu | \varphi, \bar{y})$ is an $\mathrm{N}(\bar{y}, \varphi/n)$ density.

It follows that, for given φ, the parameters λ and μ have independent normal distributions, and hence that the joint density of $\delta = \lambda - \mu$ and φ is

$$p(\delta, \varphi | \boldsymbol{x}, \boldsymbol{y}) = p(\varphi | S) p(\delta | \bar{x} - \bar{y}, \varphi)$$

where $p(\delta | \bar{x} - \bar{y}, \varphi)$ is an $\mathrm{N}(\bar{x} - \bar{y}, \varphi(m^{-1} + n^{-1}))$ density. The variance can now be integrated out just as in Section 2.12 when we considered a single sample from a normal distribution of unknown variance, giving a very similar conclusion, that is, that if

$$t = \frac{\delta - (\bar{x} - \bar{y})}{s(m^{-1} + n^{-1})^{\frac{1}{2}}}$$

where $s^2 = S/\nu$, then $t \sim \mathrm{t}_\nu$. Note that the variance estimator s^2 is found by adding the sums of squares S_x and S_y about the observed means and dividing by the sum of the corresponding numbers of degrees of freedom, ν_x and ν_y, and that this latter sum gives the number of degrees of freedom of the resulting Student's t variable. Another way of looking at it is that s^2 is a weighted mean of the variance estimators s_x^2 and s_y^2 given by the two samples with weights proportional to the corresponding degrees of freedom.

Example

This section can be illustrated by using the data considered in the last section on the weight growth of rats, this time supposing (more realistically) that the variances are equal but unknown. We found that $S_x = 5032$, $S_y = 2552$, $\nu_x = 11$ and $\nu_y = 6$ so that $S = 7584$, $\nu = 17$, $s^2 = 7584/17 = 446$ and

$$s(m^{-1}+n^{-1})^{\frac{1}{2}} = \{446(12^{-1}+7^{-1})\}^{\frac{1}{2}} = 10.$$

Since $\bar{x} = 120$ and $\bar{y} = 101$, the posterior distribution of δ is given by

$$(\delta - 19)/10 \sim t_{17}.$$

From tables of the t distribution it follows, for example, that a 90% HDR for δ is $19 \pm 1.740 \times 10$, that is, (2, 36). This is not very different from the result in the previous section, and indeed it will not usually make a great deal of difference to assume that variances are known unless the samples are very small.

It would also be possible to do other things with this posterior distribution, for example, to find the probability that $\delta > 0$ or to test the point null hypothesis that $\delta = 0$, but this should be enough to give the idea.

Non-trivial prior information

A simple analysis is possible if we have prior information which, at least approximately, is such that the prior for φ is $S_0\chi_{\nu_0}^{-2}$ and, conditional on φ, the priors for λ and μ are such that

$$\lambda \sim N(\lambda_0, \varphi/m_0)$$
$$\mu \sim N(\mu_0, \varphi/n_0)$$

independently of one another. This means that

$$p(\lambda, \mu, \varphi) \propto \varphi^{-(\nu_0+2)/2-1} \exp[-\tfrac{1}{2}\{S_0 + m_0(\lambda-\lambda_0)^2 + n_0(\mu-\mu_0)^2\}/\varphi].$$

Of course, as in any case where conjugate priors provide a handy mathematical theory, it is a question that has to be faced up to in any particular case whether or not a prior of this form *is* a reasonable approximation to your prior beliefs, and if it is not then a more untidy analysis involving numerical integration will be necessary. The reference prior used earlier is of this form, though it results from the slightly strange choice of values $\nu_0 = -2$, $S_0 = m_0 = n_0 = 0$. With such a prior, the posterior is

$$p(\lambda, \mu, \varphi \,|\, \boldsymbol{x}, \boldsymbol{y}) \propto \varphi^{-\nu_1/2-1} \exp[-\tfrac{1}{2}S/\varphi](2\pi\varphi/m_1)^{-\frac{1}{2}} \exp[-\tfrac{1}{2}m_1(\lambda-\lambda_1)^2/\varphi]$$
$$\times (2\pi\varphi/n_1)^{-\frac{1}{2}} \exp[-\tfrac{1}{2}n_1(\mu-\mu_1)^2/\varphi]$$

where

$$\nu_1 = \nu_0 + m + n, \qquad m_1 = m_0 + m, \qquad n_1 = n_0 + n,$$

$$\lambda_1 = (m_0\lambda_0 + m\bar{x})/m_1, \qquad \mu_1 = (n_0\mu_0 + n\bar{y})/n_1,$$

$$S_1 = S_0 + S_x + S_y + (m_0^{-1} + m^{-1})^{-1}(\bar{x} - \lambda_0)^2 + (n_0^{-1} + n^{-1})^{-1}(\bar{y} - \mu_0)^2.$$

(The formula for S_1 takes a little manipulation). It is now possible to proceed as in the reference prior case, and so, for given φ, the parameters λ and μ have independent normal distributions, so that the joint density of $\delta = \lambda - \mu$ and φ is

$$p(\delta, \varphi \mid \boldsymbol{x}, \boldsymbol{y}) = p(\varphi \mid S_1)p(\delta \mid \bar{x} - \bar{y}, \varphi)$$

where $p(\delta \mid \bar{x} - \bar{y}, \varphi)$ is an $\mathrm{N}(\bar{x} - \bar{y}, \varphi(m^{-1} + n^{-1}))$ density. The variance can now be integrated out as before, giving a very similar result, namely, that if

$$t = \frac{\delta - (\lambda_1 - \mu_1)}{s_1(m_1^{-1} + n_1^{-1})^{\frac{1}{2}}}$$

where $s_1^2 = S_1/\nu_1$, then $t \sim \mathrm{t}_{\nu_1}$.

The methodology is sufficiently similar to the case where a reference prior is used that it does not seem necessary to give a numerical example. Of course, the difficulty in using it in practice lies in finding appropriate values of the parameters of the prior distribution $p(\lambda, \mu, \varphi)$.

5.3 Variances unknown and unequal (Behrens–Fisher problem)

Formulation of the problem

In this section we are concerned with the most general case of the problem of two normal samples, where neither the means nor the variances are assumed equal. Consequently, we have independent vectors $\boldsymbol{x} = (x_1, x_2, \ldots, x_m)$ and $\boldsymbol{y} = (y_1, y_2, \ldots, y_n)$ such that

$$x_i \sim \mathrm{N}(\lambda, \varphi) \qquad \text{and} \qquad y_i \sim \mathrm{N}(\mu, \psi)$$

and $\delta = \lambda - \mu$. This is known as the Behrens–Fisher problem (or sometimes as the Behrens problem or the Fisher–Behrens problem).

It is convenient to use the notation of the previous section, except that sometimes we write $\nu(x)$ and $\nu(y)$ instead of ν_x and ν_y to avoid using sub-subscripts. In addition it is useful to define

$$s_x^2 = S_x/\nu_x, \qquad s_y^2 = S_y/\nu_y.$$

For the moment we shall assume independent reference priors uniform in λ, μ, φ and ψ. Then, just as in Section 2.12 on "Normal mean and variance both unknown", it follows that the posterior distributions of λ and μ are independent and are such that

$$T_x = \frac{\lambda - \bar{x}}{s_x/\sqrt{m}} \sim t_{\nu(x)} \qquad \text{and} \qquad T_y = \frac{\mu - \bar{y}}{s_y/\sqrt{n}} \sim t_{\nu(y)}.$$

It is now useful to define T and θ by

$$T = \frac{\delta - (\bar{x} - \bar{y})}{\sqrt{(s_x^2/m + s_y^2/n)}} \qquad \text{and} \qquad \tan\theta = \frac{s_x/\sqrt{n}}{s_y/\sqrt{m}}$$

(θ can be taken in the first quadrant). It is then easy to check that

$$T = T_x \sin\theta - T_y \cos\theta.$$

Since θ is known (from the data) and the distributions of T_x and T_y are known, it follows that the distribution of T can be evaluated. This distribution is tabulated and is called Behrens' (or the Behrens–Fisher or Fisher–Behrens) distribution, and it will be denoted

$$T \sim \text{BF}(\nu_x, \nu_y, \theta).$$

It was first referred to in Behrens (1929).

Patil's approximation

Behrens' distribution turns out to have a rather difficult form, so that the density at any one point can only be found by a complicated integral, although a reasonable approximation was given by Patil (1965). To use this approximation, we need to find

$$f_1 = \left(\frac{m-1}{m-3}\right)\sin^2\theta + \left(\frac{n-1}{n-3}\right)\cos^2\theta,$$

$$f_2 = \frac{(m-1)^2}{(m-3)^2(m-5)}\sin^4\theta + \frac{(n-1)^2}{(n-3)^2(n-5)}\cos^4\theta,$$

$$b = 4 + (f_1^2/f_2) \qquad \text{and} \qquad a = \sqrt{\{f_1(b-2)/b\}}.$$

Then approximately

$$T/a \sim t_b.$$

Because b is not necessarily an integer, interpolation in tables of the t distribution may be needed to use this approximation.

A rather limited table of percentage points of the Behrens distribution based on this approximation is to be found in the tables at the end of the book; this will often be enough to give some idea as to what is going on. If more percentage points are required or the tables are not available, Patil's approximation has to be used.

Example

Yet again we shall consider the data on the weight growth of rats as in the two previous sections. Recall that $m = 12$, $n = 7$ (so $\nu_x = 11$, $\nu_y = 6$), $\bar{x} = 120$, $\bar{y} = 101$, $S_x = 5032$, $S_y = 2552$, and hence $s_x^2 = 457$, $s_y^2 = 425$. Therefore

$$\tan\theta = \frac{s_x/\sqrt{m}}{s_y/\sqrt{n}} = \left(\frac{457/12}{425/7}\right)^{\frac{1}{2}} = 0.8; \qquad \sqrt{(s_x^2/m + s_y^2/n)} = 9.9$$

so that $\theta = 39° = 0.67$ radians, and thus $T \sim \text{BF}(11, 6, 39°)$. From the tables in the Appendix the 95% point of BF(12, 6, 30°) is 1.91 and that of BF(12, 6, 45°) is 1.88, so the 95% point of BF(11, 6, 39°) must be about 1.89. Consequently, a 90% HDR for δ is given by $|T| \leqslant 1.89$ and so is $(120 - 101) \pm 1.89 \times 9.9$, that is, (0, 38). This is slightly wider than was obtained in the previous section, as is reasonable, because we have made fewer assumptions and can only expect to reach less precise conclusions.

The same result can be obtained directly from Patil's approximation. The required numbers turn out to be $f_1 = 1.39$, $f_2 = 0.44$, $b = 8.39$, $a = 1.03$, so that $T/1.03 \sim t_{8.39}$. Interpolating between the 95% percentage points for t_8 and t_9 (which are 1.860 and 1.833 respectively), the required percentage point for $t_{8.39}$ is 1.849, and hence a 90% HDR for δ is $(120 - 101) \pm 1.03 \times 1.849 \times 9.9$, giving a very similar answer to that obtained from the tables.

Of course it would need more extensive tables to find, for example, the posterior probability that $\delta > 0$, but there is no difficulty in principle in doing so. On the other hand, it would be quite complicated to find the Bayes factor for a test of a point null hypothesis such as $\delta = 0$, and since such tests are only to be used with caution in special cases, it would not be likely to be worthwhile.

Substantial prior information

If we do happen to have substantial prior information about the parameters which can be approximated reasonably well by independent normal/chi-squared distributions for (λ, φ) and (μ, ψ), then the method of this section can usually be extended to include it. All that will happen is that T_x and T_y will be replaced by slightly different quantities with independent t distributions, derived as in Section 2.12 on "Normal mean and variance both unknown". It should be fairly clear how to carry out the details, so no more will be said about this case.

5.4 The Behrens–Fisher controversy

The Behrens–Fisher problem from a classical standpoint

As pointed out in Section 2.6 on "Highest density regions", in the case of a single normal observation of known variance there is a close relationship between classical results and Bayesian results using a reference prior, which can be summarized in terms of the "tilde" notation by saying that, in classical statistics, results depend on saying that

$$(\theta - \tilde{x})/\sqrt{\varphi} \sim \mathrm{N}(0, 1)$$

while Bayesian results depend on saying that

$$(\tilde{\theta} - x)/\sqrt{\varphi} \sim \mathrm{N}(0, 1).$$

As a result of this, if $\varphi = 1$ then the observation $x = 5$, say, leads to the same interval, 5 ± 1.96, which is regarded as a 95% confidence interval for θ by classical statisticians and as a 95% HDR for θ by Bayesians (at least if they are using a reference prior). It is not hard to see that very similar relationships exist if we have a sample of size n and replace x by $\bar{x}$, and also when the variance is unknown (provided that the normal distribution is replaced by the t distribution).

There is also no great difficulty in dealing with the case of a two-sample problem in which the variances are known. If they are unknown but equal (that is, $\varphi = \psi$), it was shown that if

$$t = \frac{\delta - (\bar{x} - \bar{y})}{s(m^{-1} + n^{-1})^{\frac{1}{2}}}$$

then the posterior distribution of t is Student's t on $\nu = \nu_x + \nu_y$ degrees of freedom. A classical statistician would say that this "pivotal quantity" has the same distribution *whatever* (λ, μ, φ) are, and so would be able to give confidence intervals for δ which were exactly the same as HDRs derived by a Bayesian statistician (always assuming that the latter used a reference prior).

This seems to suggest that there is always likely to be a way of interpreting classical results in Bayesian terms and vice versa, provided that a suitable prior distribution is used. One of the interesting aspects of the Behrens–Fisher problem is that no such correspondence exists in this case. To see why, recall that the Bayesian analysis led us to conclude that

$$T \sim \mathrm{BF}(\nu_x, \nu_y, \theta)$$

where

$$T = \frac{\delta - (\bar{x} - \bar{y})}{\sqrt{(s_x^2/m + s_y^2/n)}} \qquad \text{and} \qquad \tan\theta = \frac{s_x/\sqrt{m}}{s_y/\sqrt{n}}$$

Moreover, changing the prior inside the conjugate family would only alter the parameters slightly, but would still give results of the same general character. So if there is to be a classical analogue to the Bayesian result, then if T is regarded as a function of the data $\boldsymbol{x} = (x_1, x_2, \ldots, x_m)$ and $\boldsymbol{y} = (y_1, y_2, \ldots, y_n)$ for fixed values of the parameters $(\lambda, \mu, \varphi, \psi)$, it must have Behrens' distribution over repeated samples $\boldsymbol{x}$ and $\boldsymbol{y}$. There is an obvious difficulty in this, in that the parameter θ depends on the samples, whereas there is no such parameter in the normal or t distributions. However, it is still possible to investigate whether the sampling distribution of T depends on the parameters $(\lambda, \mu, \varphi, \psi)$.

It turns out that its distribution over repeated sampling does not just depend on the sample sizes m and n—it also depends on the ratio φ/ψ (which is not in general known to the statistician). It is easiest to see this when $m = n$ and so $\nu_x = \nu_y = \nu/2$ (say). We first suppose that (unknown to the statistician) it is in fact the case that $\varphi/\psi = 1$. Then the sampling distribution found in the previous section for the case where the statistician did happen to know that $\varphi = \psi$ must still hold (his or her ignorance can scarcely affect what happens in repeated sampling). Because if $m = n$ then

$$\sqrt{(s_x^2/m + s_y^2/n)} = \sqrt{\{2(S_x + S_y)/\nu m\}} = s\sqrt{(2/m)} = s(m^{-1} + n^{-1})^{\frac{1}{2}}$$

where $s_x^2 = S_x/\nu_x$, $s_y^2 = S_y/\nu_y$ and $s^2 = S/\nu$ as in the previous section, it follows that

$$T = \frac{\delta - (\bar{x} - \bar{y})}{s(m^{-1} + n^{-1})^{\frac{1}{2}}} \sim \mathrm{t}_\nu .$$

On the other hand if $\psi = 0$, then necessarily $S_y = 0$ and so $s_y^2 = 0$, and hence $\sqrt{(s_x^2/m + s_y^2/n)} = \sqrt{(s_x^2/m)}$. Since it must also be the case that $\bar{y} = \mu$ and so $\delta - (\bar{x} - \bar{y}) = \lambda - \bar{x}$, the distribution of T is given by

$$T = \frac{\lambda - \bar{x}}{s_x/\sqrt{m}} \sim \mathrm{t}_{\nu(x)}$$

that is, T has a t distribution on $\nu_x = \nu/2$ degrees of freedom. For intermediate values of φ/ψ the distribution over repeated samples is intermediate between these forms (but is not, in general, a t distribution).

Example

Bartlett (1936) quotes an experiment in which the yields x_i (in pounds per acre) on m plots for early hay were compared with the yields y_i for ordinary hay on another n plots. It turned out that $m = n = 7$ (so $\nu_x = \nu_y = 6$), $\bar{x} = 448.6$, $\bar{y} = 408.8$, $s_x{}^2 = 308.6$, $s_y{}^2 = 1251.3$. It follows that $\bar{x} - \bar{y} = 39.8$, $\sqrt{((s_x{}^2/m) + (s_y{}^2/n))} = 14.9$ and $\tan\theta = \sqrt{(308.6/1251.3)}$ so that $\theta = 26° = 0.46$ radians. The Bayesian analysis now proceeds by saying that $(\delta - 39.8)/14.9 \sim \mathrm{BF}(6, 6, 26°)$. By interpolation in tables of Behrens' distribution a 50% HDR for δ is $39.8 \pm 0.75 \times 14.9$, that is, (28.6, 51.0).

A classical statistician who was willing to assume that $\theta = \psi$ would use tables of t_{12} to conclude that a 50% confidence interval was $39.8 \pm 0.695 \times 14.9$, that is, (29.4, 50.2). This interval is different, although not very much so, from the Bayesian's HDR. Without some assumption such as $\varphi = \psi$ he or she would not be able to give any exact answer.

The controversy

The Bayesian solution was championed by Fisher (1935, 1937, 1939). Fisher had his own theory of *fiducial inference* which does not have many adherents nowadays, and did not in fact support the Bayesian arguments put forward here. In an introduction to Fisher (1939) reprinted in his *Collected Papers* (1971–74), Fisher said that

"Pearson and Neyman have laid it down axiomatically that the level of significance of a test must be equal to the frequency of a wrong decision 'in repeated samples from the same population'. The idea was foreign to the development of tests of significance given by the author in *Statistical Methods for Research Workers*, for the experimenter's experience does not consist in repeated samples from the same population, although in simple cases the numerical values are often the same; and it was, I believe, this coincidence which misled Pearson and Neyman, who were not very familiar with the ideas of 'Student' and the author."

Although Fisher was not a Bayesian, the above quotation does put one of the objections which any Bayesian must have to classical tests of significance.

In practice, classical statisticians can at least give intervals which, while they may not have an exact significance level, have a significance level between two reasonably close bounds. A recent review of the problem is given by Robinson (1976, 1982).

5.5 Inferences concerning a variance ratio

Statement of the problem

In this section we are concerned with the data of the same form as we met in the Behrens–Fisher problem. That is to say, we have independent vectors $\boldsymbol{x} = (x_1, x_2, \ldots, x_m)$ and $\boldsymbol{y} = (y_1, y_2, \ldots, y_n)$ such that

$$x_i \sim \mathrm{N}(\lambda, \varphi) \qquad \text{and} \qquad y_i \sim \mathrm{N}(\mu, \psi)$$

where all of $(\lambda, \mu, \varphi, \psi)$ are unknown. The difference is that in this case the quantity of interest is the ratio

$$\kappa = \varphi/\psi$$

of the two unknown variances, so that the intention is to discover how much more (or less) variable the one population is than the other. We shall use the same notation as before, and in addition we will find it useful to define

$$k = \frac{s_x^2}{s_y^2} = \frac{S_x/\nu_x}{S_y/\nu_y} = \frac{\nu_y}{\nu_x}\frac{S_x}{S_y},$$

$$\eta = \kappa/k.$$

Again, we shall begin by assuming a reference prior

$$p(\lambda, \mu, \varphi, \psi) \propto 1/\varphi\psi.$$

As was shown in Section 2.12 on "Normal mean and variance both unknown", the posterior distributions of φ and ψ are independent and such that $\varphi \sim S_x \chi_{\nu(x)}^{-2}$ and $\psi \sim S_y \chi_{\nu(y)}^{-2}$ so that

$$p(\varphi, \psi \mid \boldsymbol{x}, \boldsymbol{y}) \propto \varphi^{-\nu(x)/2-1} \psi^{-\nu(y)/2-1} \exp\left(-\tfrac{1}{2}S_x/\varphi - \tfrac{1}{2}S_y/\psi\right).$$

It turns out that η has (Snedecor's) F distribution on ν_y and ν_x degrees of freedom (or equivalently that its reciprocal has an F distribution on ν_x and ν_y degrees of freedom). The proof of this fact, which is not of great importance and can be omitted if you are prepared to take it for granted, is in the next subsection.

The result is of the same type, although naturally the parameters are slightly different, if the priors for φ and ψ are from the conjugate family. Even if, by a fluke, we happened to know the means but not the variances, the only change would be an increase of 1 in each of the degrees of freedom.

Derivation of the F distribution

In order to find the distribution of κ, we need first to change variables to (κ, φ), noting that

$$\frac{\partial(\kappa, \varphi)}{\partial(\varphi, \psi)} = \begin{vmatrix} 1/\psi & -\varphi/\psi^2 \\ 1 & 0 \end{vmatrix} = \varphi/\psi^2 = \kappa^2/\varphi.$$

It follows that

$$p(\kappa, \varphi \mid \boldsymbol{x}, \boldsymbol{y}) = p(\varphi, \psi \mid \boldsymbol{x}, \boldsymbol{y}) \left| \frac{\partial(\kappa, \varphi)}{\partial(\varphi, \psi)} \right|^{-1} = p(\varphi, \psi \mid \boldsymbol{x}, \boldsymbol{y}) \kappa^{-2} \varphi$$
$$= \kappa^{\nu(y)/2-1} \varphi^{-(\nu(x)+\nu(y))/2-1} \exp\{-\tfrac{1}{2}(S_x + \kappa S_y)/\varphi\}.$$

It is now easy enough to integrate φ out by substituting $x = \frac{1}{2}A/\varphi$ where $A = S_x + \kappa S_y$ and thus reducing the integral to a standard gamma function integral (cf. Section 2.12 on "Normal mean and variance both unknown"). Hence

$$p(\kappa \mid \boldsymbol{x}, \boldsymbol{y}) \propto \kappa^{\nu(y)/2-1} \{S_x + \kappa S_y\}^{-\{\nu(x)+\nu(y)\}/2}$$
$$\propto \kappa^{\nu(y)/2-1} \{(S_x/S_y) + \kappa\}^{-\{\nu(x)+\nu(y)\}/2}.$$

Defining k and η as above, and noting that $d\eta/d\kappa$ is constant, this density can be transformed to give

$$p(\eta \mid \boldsymbol{x}, \boldsymbol{y}) \propto \eta^{\nu(y)/2-1} \{\nu_x + \nu_y \eta\}^{-\{\nu(x)+\nu(y)\}/2}.$$

From the Appendix, it can be seen that this is an F distribution on ν_y and ν_x degrees of freedom, so that

$$\eta \sim F_{\nu(y), \nu(x)}.$$

We note that by symmetry

$$\eta^{-1} \sim F_{\nu(x), \nu(y)}.$$

For most purposes, it suffices to think of an F distribution as being, by definition, the distribution of the ratio of two chi-squared (or inverse chi-squared) variables divided by their respective degrees of freedom.

Example

Jeffreys (1939, 1948, 1961, Section 5.4) quotes the following data (due to Lord Rayleigh) on the masses x_i (in grammes) of $m = 12$ samples of nitrogen obtained from air (A) and the masses y_i of $n = 8$ samples obtained by chemical method (C) within a given container at standard temperature and pressure.

A 2.31035 2.31026 2.31024 2.31012 2.31027 2.31017 2.30986 2.31010
2.31001 2.31024 2.31010 2.31028

C 2.30143 2.29890 2.29816 2.30182 2.29869 2.29940 2.29849 2.29889

It turns out that $\bar{x} = 2.31017$, $\bar{y} = 2.29947$, $s_x^2 = 19.02 \times 10^{-9}$, $s_y^2 = 1902 \times 10^{-9}$, so that $k = 19/1902 = 0.010$. Hence the posterior of κ is such that

$$\eta = \kappa/k = 100\,\kappa \sim \mathrm{F}_{7,11}$$

or equivalently

$$\eta^{-1} = k/\kappa = 0.01/\kappa \sim \mathrm{F}_{11,7}.$$

This makes it possible to give an interval in which we can be reasonably sure that the ratio κ of the variances lies. For reasons similar to those for which we chose to use intervals corresponding to HDRs for $\log \chi^2$ in Section 2.8 on "HDRs for the normal variance", it seems sensible to use intervals corresponding to HDRs for log F. From the tables in the Appendix, such an interval of probability 90% for $\mathrm{F}_{11,7}$ is (0.32, 3.46), so that κ lies in the interval from 0.01/0.32 to 0.01/3.46, i.e. (0.003, 0.031), with a posterior probability of 90%. Because the distribution is markedly asymmetrical, it may also be worth finding the mode of κ, which (from the mode of $\mathrm{F}_{\nu(y),\,\nu(x)}$ as given in the Appendix) is

$$k \frac{\nu_x}{\nu_x + 1} \frac{\nu_y - 2}{\nu_y} = 0.010 \frac{11}{12} \frac{5}{7} = 0.0065.$$

5.6 Comparison of two proportions; the 2 × 2 table

Methods based on the log odds-ratio

In this section we are concerned with another two-sample problem, but this time one arising from the binomial rather than the normal distribution. Suppose

$$x \sim \mathrm{B}(m, \pi) \qquad \text{and} \qquad y \sim \mathrm{B}(n, \rho)$$

and that we are interested in the relationship between π and ρ. Another way of describing this situation is in terms of a 2 × 2 table

	Population I	Population II
Success	$a = x$	$c = y$
Failures	$b = m - x$	$d = n - y$
Total	m	n

We shall suppose that the priors for π and ρ are such that $\pi \sim \text{Be}(\alpha_0, \beta_0)$ and $\rho \sim \text{Be}(\gamma_0, \delta_0)$, independently of one another. It follows that the posteriors are also beta distributions, and, more precisely, if

$$\alpha = \alpha_0 + x, \qquad \beta = \beta_0 + m - x, \qquad \gamma = \gamma_0 + y, \qquad \delta = \delta_0 + n - y$$

then

$$\pi \sim \text{Be}(\alpha, \beta) \qquad \text{and} \qquad \rho \sim \text{Be}(\gamma, \delta).$$

We recall from Section 3.1 on the binomial distribution that if

$$\Lambda = \log \lambda = \log \{\pi/(1-\pi)\}, \qquad \Lambda' = \log \lambda' = \log \{\rho/(1-\rho)\}$$

then $\frac{1}{2}\Lambda + \frac{1}{2}\log(\beta/a) \sim z_{2\alpha, 2\beta}$, so that

$$\begin{aligned} \mathsf{E}\Lambda &\cong \log \{(\alpha - \tfrac{1}{2})/(\beta - \tfrac{1}{2})\}, \\ \mathscr{V}\Lambda &\cong \alpha^{-1} + \beta^{-1} \end{aligned}$$

and similarly for Λ'. Now the z distribution is approximately normal (this is the reason why Fisher preferred to use the z distribution rather than the F distribution, which is not so near to normality), and so Λ and Λ' are approximately normal with these means and variances. Hence the *log odds-ratio*

$$\Lambda - \Lambda' = \log (\lambda/\lambda')$$

is also approximately normal, that is,

$$\Lambda - \Lambda' \sim \text{N}(\log\{(\alpha - \tfrac{1}{2})(\delta - \tfrac{1}{2})/ (\beta - \tfrac{1}{2})(\gamma - \tfrac{1}{2})\}, \alpha^{-1} + \beta^{-1} + \gamma^{-1} + \delta^{-1}),$$

or more approximately

$$\Lambda - \Lambda' \sim \text{N}(\log\{\alpha\delta/\beta\gamma\}, \alpha^{-1} + \beta^{-1} + \gamma^{-1} + \delta^{-1}).$$

If the Haldane reference priors are used, so that $\alpha_0 = \beta_0 = \gamma_0 = \delta_0 = 0$, then $\alpha = a$, $\beta = b$, $\gamma = c$ and $\delta = d$, and so

$$\Lambda - \Lambda' \sim \text{N}(\log\{ad/bc\}, a^{-1} + b^{-1} + c^{-1} + d^{-1}).$$

The quantity ab/cd is sometimes called the *cross-ratio*, and there are good grounds for saying that any measure of association in the 2×2 table should be a function of the cross-ratio (cf. Edwards, 1963).

The log odds-ratio is a sensible measure of the degree to which the two populations are identical, and in particular $\pi > \rho$ if and only if $\Lambda - \Lambda' > 0$. On the other hand, knowledge of the posterior distribution of the log odds-ratio does not in itself imply knowledge of the posterior distribution of the difference $\pi - \rho$ or the ratio π/ρ. The approximation is likely to be reasonable provided that all of the entries in the 2×2 table are at least 5.

Example

The table below [quoted from Di Raimondo (1951)] relates to the effect on mice of bacterial inoculum (*Staphyloccus aureus*). Two different types of injection were tried, a standard one and one with 0.15 U of penicillin per ml.

	Standard	Penicillin
Alive	8	48
Dead	12	62
Total	20	110

The cross-ratio is $ad/bc = (8\times 62)/(12\times 48) = 0.861$ so its logarithm is -0.150 and $a^{-1}+b^{-1}+c^{-1}+d^{-1} = 0.245$, and so the posterior distribution of the log odds-ratio is $\Lambda - \Lambda' \sim \mathrm{N}(-0.150, 0.245)$. Allowing for the $\frac{1}{2}$s in the more exact form for the mean does not make much difference; in fact -0.150 becomes -0.169. The posterior probability that $\pi > \rho$, that is, that the log odds ratio is positive, is

$$\Phi(-0.169/\sqrt{0.245}) = \Phi(-0.341) = 0.3665.$$

The data thus show no great difference between the injections with and without the penicillin.

The inverse root-sine transformation

In Section 1.5 on "Means and variances" we saw that if $x \sim \mathrm{B}(m, \pi)$, then the transformation $z = \sin^{-1}\sqrt{(x/m)}$ resulted in $\mathsf{E}z = \sin^{-1}\sqrt{\pi} = \psi$, say, and $\mathcal{V}z = 1/4m$, and in fact it is approximately true that $z \sim \mathrm{N}(\psi, 1/4m)$. This transformation was also mentioned in Section 3.2 on "Reference prior for the binomial likelihood", and it was pointed out there that one of the possible reference priors for π was $\mathrm{Be}(\frac{1}{2}, \frac{1}{2})$, and that this prior was equivalent to a uniform prior in $\sin^{-1}\sqrt{\pi} = \psi$. Now if we use such a prior, then clearly the posterior for ψ is approximately $\mathrm{N}(z, 1/4m)$, that is,

$$\psi = \sin^{-1}\sqrt{\pi} \sim \mathrm{N}(\sin^{-1}\sqrt{(x/m)}, 1/4m).$$

This is of no great use if there is only a single binomial variable, but when there are two it can be used to conclude that approximately

$$\sin^{-1}\sqrt{\pi} - \sin^{-1}\sqrt{\rho} \sim \mathrm{N}(\sin^{-1}\sqrt{(x/m)} - \sin^{-1}\sqrt{(y/n)}, 1/4m + 1/4n)$$

and so to give another approximation to the probability that $\pi > \rho$.

Thus with the same data as above, $\sin^{-1}\surd(x/m) = \sin^{-1}\surd(8/20) = 0.685$ radians, $\sin^{-1}\surd(48/110) = 0.721$ radians, and $1/4m + 1/4n = 0.0148$, so that the posterior probability that $\pi > \rho$ is about $\Phi(-0.036/\surd 0.0148) = \Phi(-0.296) = 0.3936$. The two methods do not give precisely the same answer, but it should be borne in mind that the numbers are not very large, so the approximations involved are not very good, and also that we have assumed slightly different reference priors in deriving the two answers.

If there is non-trivial prior information, it can be incorporated in this method as well as in the previous method. The approximations involved are reasonably accurate provided that $x(1 - x/n)$ and $y(1 - y/n)$ are both at least 5.

Other methods

If all the entries in the 2×2 table are at least 10, then the posterior beta distributions are reasonably well approximated by normal distributions of the same means and variances. This is quite useful in that it gives rise to an approximation to the distribution of $\pi - \rho$ which is much more likely to be of interest than some function of π minus the same function of ρ. It will therefore allow us to give an approximate HDR for $\pi - \rho$ or to approximate the probability that $\pi - \rho$ lies in a particular interval.

In quite a different case, where the values of π and ρ are small, which will be reflected in small values of x/m and y/n, then the binomial distributions can be reasonably well approximated by Poisson distributions, which means that the posteriors of π and ρ are multiples of chi-squared distributions (cf. Section 3.4 on "The Poisson distribution"). It follows from this that the posterior of π/ρ is a multiple of an F distribution (cf. the previous section). Again, this is quite useful because π/ρ is a quantity of interest in itself. The Poisson approximation to the binomial is likely to be reasonable if $n > 10$ and either $x/n < 0.05$ or $x/n > 0.95$ (in the latter case, π has to be replaced by $1 - \pi$).

The exact probability that $\pi < \rho$ can be worked out in terms of hypergeometric probabilities (cf. Altham, 1969), although the resulting expression is not usually useful for hand computation. It is even possible to give an expression for the posterior probability that $\pi/\rho \leqslant c$ (cf. Weisberg, 1972), but this is even more unwieldy.

Exercises on Chapter 5

1. Two analysts measure the percentage of ammonia in a chemical process over 9 days and find the following discrepancies between their results:

Day	1	2	3	4	5	6	7	8	9
Analyst A	12.04	12.37	12.35	12.43	12.34	12.36	12.48	12.33	12.33
Analyst B	12.18	12.37	12.38	12.36	12.47	12.48	12.57	12.28	12.42

Investigate the mean discrepancy θ between their results and in particular give an interval in which you are 90% sure that it lies.

2. With the same data as in the previous question, test the hypothesis that there is no discrepancy between the two analysts.

3. Two analysts in the same laboratory made repeated determinations of the percentage of fibre in soya cotton cake, the results being as shown below.

Analyst A	12.38	12.53	12.25	12.37	12.48	12.58	12.43	12.43	12.30
Analyst B	12.25	12.45	12.31	12.31	12.30	12.20	12.25	12.25	12.26
	12.42	12.17	12.09						

Investigate the mean discrepancy θ between their mean determinations and in particular give an interval in which you are 90% sure that it lies

(a) assuming that it is known from past experience that the standard deviation of both sets of observations is 0.1, and

(b) assuming simply that it is known that the standard deviations of the two sets of observations are equal.

4. A random sample $\boldsymbol{x} = (x_1, x_2, \ldots, x_m)$ is available from an $N(\lambda, \varphi)$ distribution and a second independent random sample $\boldsymbol{y} = (y_1, y_2, \ldots, y_n)$ is available from an $N(\mu, 2\varphi)$ distribution. Obtain, under the usual assumptions, the posterior distributions of $\lambda - \mu$ and of φ.

5. Verify the formula for S_1 given towards the end of Section 5.2.

6. The following data consist of the lengths of mm of cuckoo's eggs found in nests belonging to the dunnock and to the reed warbler:

Dunnock	22.0	23.9	20.9	23.8	25.0	24.0	21.7	23.8	22.8	23.1
Reed warbler	23.2	22.0	22.2	21.2	21.6	21.9	22.0	22.9	22.8	

Investigate the difference θ between these lengths without making any particular assumptions about the variances of the two populations, and in particular give an interval in which you are 90% sure that it lies.

7. Suppose that T_x, T_y and θ are defined as in Section 5.3 and that

$$T = T_x \sin\theta - T_y \cos\theta, \qquad U = T_x \cos\theta + T_y \sin\theta$$

Show that the transformation from (T_x, T_y) to (T, U) has unit Jacobian and hence show that the density of T satisfies

$$p(T|\boldsymbol{x}, \boldsymbol{y}) \propto \int_0^\infty [1 + (T\sin\theta + U\cos\theta)^2/\nu_x]^{-(\nu(x)+1)/2}$$
$$\times [1 + (-T\cos\theta + U\sin\theta)^2/\nu_y]^{-(\nu(y)+1)/2}\,\mathrm{d}U.$$

8. Two different microscopic methods, A and B, are available for the measurement of very small dimensions in microns. As a result of several such measurements on the same object, estimates of variance are available as follows:

Method	A	B
No. of observations	$m = 15$	$n = 25$
Estimated variance	$s_1^2 = 7.533$	$s_2^2 = 1.112$

Give an interval in which you are 95% sure that the ratio of the variances lies.

9. The table below [quoted from Jeffreys (1939; 1948; 1961, Section 5.1)] gives the relationship between grammatical gender in Welsh and psychoanalytical symbolism according to Freud:

Psycho. \ Gram.	M	F
M	45	30
F	28	29
Total	73	59

Find the posterior probability that the log odds-ratio is positive and compare it with the comparable probability found by using the inverse root-sine transformation.

10. Show that if $\pi \cong \rho$ then the log odds-ratio is such that

$$\Lambda - \Lambda' \cong (\pi - \rho)/\{\pi(1-\pi)\}.$$

11. A report issued in 1966 about the effect of radiation on patients with inoperable lung cancer compared the effect of radiation treatment

with placebos. The numbers surviving after a year were:

	Radiation	Placebos
No. of cases	308	246
No. surviving	56	34

What arc the approximate posterior odds that the one-year survival rate of irradiated patients is at least 0.01 greater than that of those who were not irradiated?

12. Suppose that $x \sim \mathrm{P}(8.5)$, that is, x is Poisson of mean 8.5, and $y \sim \mathrm{P}(11.0)$. What is the approximate distribution of $x - y$?

6

Correlation, Regression and the Analysis of Variance

6.1 Theory of the correlation coefficient

Definitions

The standard measure of association between two random variables, which was first mentioned in Section 1.5 on "Means and variances", is the *correlation coefficient*

$$\rho(x, y) = \frac{\mathscr{C}(x, y)}{\sqrt{(\mathscr{V} x)(\mathscr{V} y)}}.$$

It is used to measure the strength of linear association between two variables, most commonly in the case where it might be expected that both have, at least approximately, a normal distribution. It is most important in cases where it is not thought that either variable is dependent on the other. One example of its use would be an investigation of the relationship between the height and the weight of individuals in a population, and another would be in finding how closely related barometric gradients and wind velocities were. You should, however, be warned that it is very easy to conclude that measurements are closely related because they have a high correlation, when in fact the relationship is due to their having a common time trend or a common cause and there is no close relationship between the two (see, for example, the relationship between the growth of money supply and Scottish dysentery as pointed out in a letter to *The Times* dated 6 April 1977). We should also be aware that two closely related variables can have a low correlation if the relationship between them is highly non-linear.

We suppose, then, that we have a set of n ordered pairs of observations, the pairs being independent of one another but members of the same pair being, in general, not independent. We shall denote these observations (x_i, y_i) and, as usual, we shall write $\boldsymbol{x} = (x_1, x_2, \ldots, x_n)$ and $\boldsymbol{y} = (y_1, y_2, \ldots, y_n)$. Further, we suppose that these pairs have a *bivariate normal distribution* with

$$\mathsf{E}x_i = \lambda, \quad \mathsf{E}y_i = \mu, \quad \mathscr{V}x_i = \varphi, \quad \mathscr{V}y_i = \psi, \quad \rho(x_i, y_i) = \rho$$

and we shall use the notation

$$\bar{x} = \sum x_i/n, \quad \bar{y} = \sum y_i/n, \quad S_{xx} = \sum (x_i - \bar{x})^2, \quad S_{yy} = \sum (y_i - \bar{y})^2$$

(S_{xx} and S_{yy} have previously been denoted S_x and S_y), and

$$S_{xy} = \sum (x_i - \bar{x})(y_i - \bar{y}).$$

It is also useful to define the *sample correlation coefficient* r by

$$r = \frac{\sum (x_i - \bar{x})(y_i - \bar{y})}{\sqrt{[\{\sum (x_i - \bar{x})^2\}\{\sum (y_i - \bar{y})^2\}]}} = \frac{S_{xy}}{\sqrt{(S_{xx}S_{yy})}}$$

so that $S_{xy} = r\sqrt{(S_{xx}S_{yy})}$.

We shall show that, with standard reference priors for λ, μ, φ and ψ, a reasonable approximation to the posterior density of ρ is given by

$$p(\rho \mid \boldsymbol{x}, \boldsymbol{y}) \propto p(\rho) \frac{(1-\rho^2)^{(n-1)/2}}{(1-\rho r)^{n-(3/2)}}$$

where $p(\rho)$ is its prior density. Making the substitution

$$\rho = \tanh \zeta, \qquad r = \tanh z$$

we will go on to show that after another approximation

$$\zeta \sim \mathrm{N}(z, 1/n).$$

These results will be derived after quite a complicated series of substitutions [due to Fisher (1915, 1921)]. Readers who are prepared to take these results for granted can omit the rest of this section.

Approximate posterior distribution of the correlation coefficient

As before, we shall have use for the formulae

$$\sum (x_i - \lambda)^2 = S_{xx} + n(\bar{x} - \lambda)^2, \qquad \sum (y_i - \mu)^2 = S_{yy} + n(\bar{y} - \mu)^2$$

and also for a similar one not used before:

$$\sum (x_i - \lambda)(y_i - \mu) = S_{xy} + n(\bar{x} - \lambda)(\bar{y} - \mu).$$

Now the (joint) density function of a single pair (x, y) of observations from a bivariate normal distribution is

$$p(x, y \mid \lambda, \mu, \varphi, \psi, \rho) = \frac{1}{2\pi\sqrt{\{\varphi\psi(1-\rho^2)\}}} \exp\left(-\frac{1}{2(1-\rho)} Q_0\right)$$

where

$$Q_0 = \varphi^{-1}(x-\lambda)^2 - 2\rho(\varphi\psi)^{-\frac{1}{2}}(x-\lambda)(y-\mu) + \psi^{-1}(y-\mu)^2$$

and hence the joint density of the vector $(\boldsymbol{x}, \boldsymbol{y})$ is

$$\begin{aligned} p(\boldsymbol{x}, \boldsymbol{y} \mid \lambda, \mu, \varphi, \psi, \rho) \propto & \{\varphi\psi(1-\rho^2)\}^{-(n-1)/2} \exp\{-\tfrac{1}{2}Q/(1-\rho^2)\} \\ & \times [2\pi n^{-1}\sqrt{\{\varphi\psi(1-\rho^2)]^{-1}} \exp\{-\tfrac{1}{2}nQ'/(1-\rho^2)\} \end{aligned}$$

where

$$\begin{aligned} Q &= \varphi^{-1}S_{xx} - 2\rho(\varphi\psi)^{-\frac{1}{2}}S_{xy} + \psi^{-1}S_{yy}, \\ Q' &= \varphi^{-1}(\bar{x}-\lambda)^2 - 2\rho(\varphi\psi)^{-\frac{1}{2}}(\bar{x}-\lambda)(\bar{y}-\mu) + \psi^{-1}(\bar{y}-\mu)^2. \end{aligned}$$

It follows that the vector $(\bar{x}, \bar{y}, S_{xx}, S_{yy}, r)$ is sufficient for $(\lambda, \mu, \varphi, \psi, \rho)$. For the moment, we shall use independent priors of a simple form. For λ, μ, φ and ψ, we shall take the standard reference priors, and for the moment we shall use a perfectly general prior for ρ, so that

$$p(\lambda, \mu, \varphi, \psi, \rho) \propto p(\rho)/\varphi\psi$$

and hence

$$\begin{aligned} p(\lambda, \mu, \varphi, \psi, \rho \mid \boldsymbol{x}, \boldsymbol{y}) \propto & \, p(\rho)(1-\rho^2)^{-(n-1)/2}\{\varphi\psi\}^{-(n+1)/2} \\ & \times \exp\{-\tfrac{1}{2}Q/(1-\rho^2)\} \\ & \times [2\pi n^{-1}\sqrt{\{\varphi\psi(1-\rho^2)]^{-1}} \exp\{-\tfrac{1}{2}nQ'/(1-\rho^2)\}. \end{aligned}$$

The last factor is evidently the (joint) density of λ and μ considered as bivariate normal with means $\bar{x}$ and $\bar{y}$, variances φ/n and ψ/n and correlation ρ. Consequently, it integrates to unity, and so as the remaining factors do not depend on λ or μ

$$p(\varphi, \psi, \rho \mid \boldsymbol{x}, \boldsymbol{y}) \propto p(\rho)(1-\rho^2)^{-(n-1)/2}\{\varphi\psi\}^{-(n+1)/2} \exp\{-\tfrac{1}{2}Q/(1-\rho^2)\}.$$

To integrate φ and ψ out, it is convenient to define

$$\xi = \sqrt{\{(\varphi\psi)/(S_{xx}S_{yy})\}}, \quad \omega = \sqrt{\{(\varphi S_{yy})/(\psi S_{xx})\}}$$

so that $\varphi = \omega\xi S_{xx}$ and $\psi = \omega^{-1}\xi S_{yy}$. The Jacobian is

$$\frac{\partial(\varphi, \psi)}{\partial(\xi, \omega)} = \begin{vmatrix} \omega S_{xx} & \xi S_{xx} \\ \omega^{-1}S_{yy} & -\omega^{-2}\xi S_{yy} \end{vmatrix} = 2\omega^{-1}\xi S_{xx}S_{yy}$$

and hence

$$p(\xi, \omega, \rho \mid \boldsymbol{x}, \boldsymbol{y}) \propto p(\varphi, \psi, \rho \mid \boldsymbol{x}, \boldsymbol{y}) \frac{\partial(\varphi, \psi)}{\partial(\xi, \omega)}$$

$$\propto p(\rho)(1-\rho^2)^{-(n-1)/2}\omega^{-1}\xi^{-n}\exp\left(-\tfrac{1}{2}R/\xi\right)$$

where

$$R = (\omega + \omega^{-1} - 2\rho r)/(1-\rho^2).$$

The substitution $x = \frac{1}{2}R/\xi$ (so that $\xi = \frac{1}{2}R/x$) reduces the integral over ξ to a standard gamma function integral, and hence we can deduce that

$$p(\omega, \rho \mid \boldsymbol{x}, \boldsymbol{y}) \propto p(\rho)(1-\rho^2)^{-(n-1)/2}\omega^{-1}R^{-(n-1)}$$

$$\propto p(\rho)(1-\rho^2)^{+(n-1)/2}\omega^{-1}(\omega+\omega^{-1}-2\rho r)^{-(n-1)}.$$

Finally, integrating over ω

$$p(\rho \mid \boldsymbol{x}, \boldsymbol{y}) \propto p(\rho)(1-\rho^2)^{(n-1)/2}\int_0^\infty \omega^{-1}(\omega+\omega^{-1}-2\rho r)^{-(n-1)}\,\mathrm{d}\omega.$$

By substituting $1/\omega$ for ω it is easily checked that the integral from 0 to 1 is equal to that from 1 to ∞, so that as constant multiples are irrelevant, the lower limit of the integral can be taken to be 1 rather than 0.

By substituting $\omega = \exp(t)$, the integral can be put in the alternative form

$$p(\rho \mid \boldsymbol{x}, \boldsymbol{y}) \propto p(\rho)(1-\rho^2)^{(n-1)/2}\int_0^\infty (\cosh t - \rho r)^{-(n-1)}\,\mathrm{d}t.$$

The exact distribution corresponding to $p(\rho) \propto 1$ has been tabulated in David (1954), but for most purposes it suffices to use an approximation. The usual way to proceed is by yet a further substitution, in terms of u where $\cosh t - \rho r = (1-\rho r)/(1-u)$, but this is rather messy and gives more than is necessary for a first-order approximation. Instead, note that for small t

$$\cosh t \cong 1 + \tfrac{1}{2}t^2$$

while the contribution to the integral from values where t is large will, at least for large n, be negligible. Using this approximation

$$\int_0^\infty (\cosh t - \rho r)^{-(n-1)}\,\mathrm{d}t = \int_0^\infty \left(1+\tfrac{1}{2}t^2-\rho r\right)^{-(n-1)}\mathrm{d}t$$

$$= (1-\rho r)^{-(n-1)}\int_0^\infty \left\{1+\tfrac{1}{2}t^2/(1-\rho r)\right\}^{-(n-1)}\mathrm{d}t.$$

On substituting

$$t = \sqrt{\{2(1-\rho r)\}} \tan \theta$$

the integral is seen to be proportional to

$$(1-\rho r)^{-n+(3/2)} \int_0^{\frac{1}{2}\pi} \cos^{n-3} \theta \, \mathrm{d}\theta.$$

Since the integral in this last expression does not depend on ρ, we can conclude that

$$p(\rho \mid \boldsymbol{x}, \boldsymbol{y}) \propto p(\rho) \frac{(1-\rho^2)^{(n-1)/2}}{(1-\rho r)^{n-(3/2)}}.$$

Although evaluation of the constant of proportionality would still require the use of numerical methods, it is much simpler to calculate the distribution of ρ using this expression than to have to evaluate an integral for every value of ρ. In fact the approximation is quite good [some numerical comparisons can be found in Box and Tiao (1973, Section 8.4.8)].

The hyperbolic tangent substitution

Although the exact mode does not usually occur at $\rho = r$, it is easily seen that for plausible choices of the prior $p(\rho)$, the approximate density derived above is greatest when ρ is near r. However, except when $r = 0$, this distribution is asymmetrical. Its asymmetry can be reduced by writing

$$\rho = \tanh \zeta, \qquad r = \tanh z$$

so that $\mathrm{d}\rho/\mathrm{d}\zeta = \operatorname{sech}^2 \zeta$ and

$$1-\rho^2 = \operatorname{sech}^2 \zeta, \qquad 1-\rho r = \frac{\cosh(\zeta - z)}{\cosh \zeta \cosh z}.$$

It follows that

$$p(\zeta \mid \boldsymbol{x}, \boldsymbol{y}) \propto p(\zeta) \cosh^{-5/2} \zeta \cosh^{(3/2)-n}(\zeta - z).$$

If n is large, since the factor $p(\zeta) \cosh^{-5/2} \zeta \cosh^{(3/2)} (\zeta - z)$ does not depend on n, it may be regarded as approximately constant over the range over which $\cosh^{-n} (\zeta - z)$ is appreciably different from zero, so that

$$p(\zeta \mid \boldsymbol{x}, \boldsymbol{y}) \propto \cosh^{-n} (\zeta - z).$$

Finally, we put

$$\xi = (\zeta - z)\sqrt{n}$$

and note that if ζ is close to z then $\cosh(\zeta - z) \cong 1 + \frac{1}{2}\xi^2/n + \ldots$ Putting this into the expression for $p(\zeta \mid \boldsymbol{x}, \boldsymbol{y})$ and using the exponential limit

$$p(\xi \mid \boldsymbol{x}, \boldsymbol{y}) \propto \exp(-\tfrac{1}{2}\xi^2)$$

so that approximately $\xi \sim \mathrm{N}(0, 1)$, or equivalently

$$\zeta \sim \mathrm{N}(z, 1/n).$$

A slightly better approximation to the mean and variance can be found by using approximations based on the likelihood as in Section 3.10. If we take a uniform prior for ζ or at least assume that the prior does not vary appreciably over the range of values of interest, we obtain

$$\begin{aligned} L(\zeta) &= -(5/2)\log\cosh\zeta + \{(3/2) - n\}\log\cosh(\zeta - z) \\ L'(\zeta) &= -(5/2)\tanh\zeta + \{(3/2) - n\}\tanh(\zeta - z) \\ &= -(5/2)\rho + \{(3/2) - n\}\{(\zeta - z) + \ldots\}. \end{aligned}$$

We can now approximate ρ by r (we could write $\tanh\zeta = \tanh\{z + (\zeta - z)\}$ and so get a better approximation, but this is not worthwhile). We can also approximate $n - (3/2)$ by n, so obtaining the root of the likelihood equation as

$$\zeta = z - 5r/2n.$$

Further

$$\begin{aligned} L''(\zeta) &= -(5/2)\,\mathrm{sech}^2\,\zeta + \{(3/2) - n\}\,\mathrm{sech}^2(\zeta - z) \\ &= -(5/2)(1 - \tanh^2\zeta) + \{(3/2) - n\}\{1 + \ldots\} \end{aligned}$$

so that again approximating ρ by r, we have at $\zeta = \hat{\zeta}$

$$L''(\hat{\zeta}) = -(5/2)(1 - r^2) + (3/2) - n.$$

It follows that the distribution of ζ is given slightly more accurately by

$$\zeta \sim \mathrm{N}(z - 5r/2n, \quad \{n - (3/2) + (5/2)(1 - r^2)\}^{-1}).$$

This approximation differs a little from that usually given by classical statisticians, who usually quote the variance as $(n-3)^{-1}$, but the difference is not of great importance.

Reference prior

Clearly the results will be simplest if the prior used has the form

$$p(\rho) \propto (1 - \rho^2)^c$$

for some c. The simplest choice is to take $c = 0$, that is, $p(\rho) \propto 1$, and it seems quite a reasonable choice. It is possible to use the multi-parameter version of Jeffreys' rule to find a prior for (φ, ψ, ρ), although this is not wholly simple. The easiest way is to write $\kappa = \rho\varphi\psi$ for the covariance and to work in terms of the *inverse* of the variance/covariance matrix, that is, in terms of (α, β, γ), where

$$\begin{pmatrix} \alpha & \gamma \\ \gamma & \beta \end{pmatrix} = \begin{pmatrix} \varphi & \kappa \\ \kappa & \psi \end{pmatrix}^{-1}.$$

It turns out that $p(\alpha, \beta, \gamma) \propto \Delta^{3/2}$ where Δ is the determinant $\varphi\psi - \kappa^2$, and that the Jacobian determinant

$$\frac{\partial(\alpha, \beta, \gamma)}{\partial(\varphi, \psi, \kappa)} = -\Delta^{-3}$$

so that $p(\varphi, \psi, \kappa) \propto \Delta^{-3/2}$. Finally, transforming to the parameters (φ, ψ, ρ) that are really of interest, it transpires that

$$p(\varphi, \psi, \rho) \propto (\varphi\psi)^{-1}(1-\rho^2)^{-3/2}$$

which corresponds to the choice $c = -3/2$ and the usual reference priors for φ and ψ.

Incorporation of prior information

It is not difficult to adapt the above analysis to the case where prior information from the conjugate family [that is, inverse chi-squared for φ and ψ and of the form $(1-\rho^2)^c$ for ρ] is available. In practice, this information will usually be available in the form of previous measurements of a similar type and in this case it is best dealt with by transforming all the information about ρ into statements about $\zeta = \tanh^{-1}\rho$ so that the theory we have built up for the normal distribution can be used.

6.2 Examples on the use of the correlation coefficient

Use of the hyperbolic tangent transformation

The following data are a small subset of a much larger quantity of data on the length and breadth (in mm) of the eggs of cuckoos (*C. canorus*).

Egg no. (i):	1	2	3	4	5	6	7	8	9
Length (x_i):	22.5	20.1	23.3	22.9	23.1	22.0	22.3	23.6	24.7
Breadth (y_i):	17.0	14.9	16.0	17.4	17.4	16.5	17.2	17.2	18.0

Here $n = 9$, $\bar{x} = 22.7\dot{2}$, $\bar{y} = 16.8\dot{4}$, $S_{xx} = 12.816$, $S_{yy} = 6.842$, $S_{xy} = 7.581$, $r = 0.810$ and so $z = \tanh^{-1} 0.810 = 1.127$ and $1/n = 1/9$.

We can conclude that with 95% posterior probability ζ is in the interval $1.127 \pm 1.960 \times \surd(1/9)$, that is, (0.474, 1.780), giving rise to (0.441, 0.945) as a corresponding interval for ρ, using Neave (1978, Table 6.3).

Combination of several correlation coefficients

One of the important uses of the hyperbolic tangent transformation lies in the way in which it makes it possible to combine different observations of the correlation coefficient. Suppose, for example, that on one occasion we observe that $r = 0.7$ on the basis of 19 observations and on another we observe that $r = 0.9$ on the basis of 25 observations. Then after the first set of observations, our posterior for ζ is $\mathrm{N}(\tanh^{-1} 0.7, 1/19)$. The second set of observations now puts us into the situation of a normal prior and likelihood, so the posterior after all the observations is still normal, with variance

$$(19 + 25)^{-1} = 0.0227$$

and mean

$$0.0227(19 \times \tanh^{-1} 0.7 + 25 \times \tanh^{-1} 0.9) = 1.210$$

that is, is $\mathrm{N}(1.210, 0.0227)$, suggesting a point estimate of $\tanh 1.210 = 0.8367$ for ρ.

The transformation also allows one to investigate whether or not the correlations on the two occasions really were from the same population or at least from reasonably similar populations.

The squared correlation coefficient

There is a temptation to take r as such too seriously and to think that if it is very close to 1 then the two variables are closely related, but we will see shortly when we come to consider regression that r^2, which measures the proportion of the variance of one variable that can be accounted for by the other variable, is in many ways at least as useful a quantity to consider.

6.3 Regression and the bivariate normal model

The model

The problem we will consider in this section is that of using the values of one variable to explain or predict values of another. We shall refer to an *explanatory* and a *dependent* variable, although it is conventional to refer to an independent and a dependent variable. An important reason for preferring the phrase explanatory variable is that the word "independent" if used in this context has nothing to do with the use of the word in the phrase "independent random variables". Some authors for example, Novick and Jackson (1974, Section 9.11), refer to the dependent variable as the criterion variable. The theory can be applied, for example, to finding a way of predicting the weight (the dependent variable) of typical individuals in terms of their height (the explanatory variable). It should be noted that the relationship which best predicts weight in terms of height will not necessarily be the best relationship for predicting height in terms of weight.

The basic situation and notation are the same as in the last two sections, although in this case there is not the symmetry between the two variables that existed there. We shall suppose that the xs represent the explanatory variable and ys the dependent variables.

There are two slightly different situations. In the first the experimenters are free to set the values of x_i, whereas in the second both values are random, although one is thought of as having a causal or explanatory relationship with the other. The analysis, however, turns out to be the same in both cases.

The most general model is

$$p(\boldsymbol{x}, \boldsymbol{y} \mid \boldsymbol{\theta}_1, \boldsymbol{\theta}_2) \propto p(\boldsymbol{x} \mid \boldsymbol{\theta}_1) p(\boldsymbol{y} \mid \boldsymbol{x}, \boldsymbol{\theta}_2)$$

where in the first situation described above $\boldsymbol{\theta}_1$ is a null vector and the distribution of $\boldsymbol{x}$ is degenerate. If it is assumed that $\boldsymbol{\theta}_1$ and $\boldsymbol{\theta}_2$ have independent priors, so that $p(\boldsymbol{\theta}_1 \, \boldsymbol{\theta}_2) = p(\boldsymbol{\theta}_1) p(\boldsymbol{\theta}_2)$, then

$$p(\boldsymbol{\theta}_1, \boldsymbol{\theta}_2 \mid \boldsymbol{x}, \boldsymbol{y}) \propto p(\boldsymbol{\theta}_1) p(\boldsymbol{\theta}_2) p(\boldsymbol{x} \mid \boldsymbol{\theta}_1) p(\boldsymbol{y} \mid \boldsymbol{x}, \boldsymbol{\theta}_2).$$

It is now obvious that we can integrate over $\boldsymbol{\theta}_1$ to obtain

$$p(\boldsymbol{\theta}_2 \mid \boldsymbol{x}, \boldsymbol{y}) \propto p(\boldsymbol{\theta}_2) p(\boldsymbol{y} \mid \boldsymbol{x}, \boldsymbol{\theta}_2).$$

Technically, given $\boldsymbol{\theta}_2$, the vector $\boldsymbol{x}$ is *sufficient* for $\boldsymbol{\theta}_1$ and, given $\boldsymbol{\theta}_1$, the vector $\boldsymbol{x}$ is *ancillary* for $\boldsymbol{\theta}_2$. It follows that in so far as we wish to make inferences about $\boldsymbol{\theta}_2$, we may *act as if* $\boldsymbol{x}$ were constant.

Bivariate linear regression

We now move on to a very important particular case. Suppose that *conditional* on $\boldsymbol{x}$ we have

$$y_i \sim \mathrm{N}(\eta_0 + \eta_1 x_i, \phi).$$

Thus

$$\boldsymbol{\theta}_2 = (\eta_0, \eta_1, \phi)$$

unless one or more of η_0, η_1 and φ are known, in which case the ones that are known can be dropped from $\boldsymbol{\theta}_2$. Thus we are supposing that, on average, the dependence of the *y*s on the *x*s is *linear*. It would be necessary to use rather different methods if there were grounds for thinking, for example, that $\mathsf{E}(y_i|x_i) = \eta_0 + \eta_1 x_i + \eta_2 x_i^2$ or that $\mathsf{E}(y_i|x_i) = \gamma_0 \cos(x_i + \gamma_1)$. It is also important to suppose that the *y*s are *homoscedastic*, that is, that the variance $\mathscr{V}(y_i|x_i)$ has the same constant value φ whatever the value of x_i; modifications to the analysis would be necessary if it were thought that, for example, $y_i \sim \mathrm{N}(\eta_0 + \eta_1 x_i, x_i)$ so that the variance increased with x_i.

It simplifies some expressions to write $\eta_0 + \eta_1 x_i$ as $\alpha + \beta(x_i - \bar{x})$ where, of course, $\bar{x} = \sum x_i/n$, so that $\eta_0 = \alpha - \beta\bar{x}$ and $\eta_1 = \beta$, hence $\alpha = \eta_0 + \eta_1 \bar{x}$. The model can now be written as

$$y_i \sim \mathrm{N}(\alpha + \beta(x_i - \bar{x}), \varphi).$$

Because a key feature of the model is the *regression line* $y = \alpha + \beta(x - \bar{x})$ on which the expected values lie, the parameter β is usually referred to as the slope and α is sometimes called the intercept, although this term is also sometimes applied to η_0. For the rest of this section we shall take a reference prior that is independently uniform in α, β and $\log\varphi$, so that

$$p(\alpha, \beta, \varphi) \propto 1/\varphi.$$

In addition to the notation of the last two sections, it is helpful to define

$$S_{ee} = S_{yy} - S_{xy}^2/S_{xx} = S_{yy}(1 - r^2),$$
$$a = \bar{y}, \quad b = S_{xy}/S_{xx}, \quad e_0 = \bar{y} - b\bar{x}.$$

It then turns out that

$$\begin{aligned} p(\alpha, \beta, \varphi | \boldsymbol{x}, \boldsymbol{y}) &\propto p(\alpha, \beta, \gamma) p(\boldsymbol{y} | \boldsymbol{x}, \alpha, \beta, \varphi) \\ &\propto \varphi^{-1}(2\pi\varphi)^{-n/2} \exp[-\tfrac{1}{2}\textstyle\sum\{y_i - \alpha - \beta(x_i - \bar{x})\}^2/\varphi]. \end{aligned}$$

Now since $\bar{y} - \alpha$ is a constant and $\sum(x_i - \bar{x}) = \sum(y_i - \bar{y}) = 0$ the sum of squares can be written as

$$\begin{aligned} \textstyle\sum\{y_i - \alpha - \beta(x_i - \bar{x})\}^2 &= \textstyle\sum\{(y_i - \bar{y}) + (\bar{y} - \alpha) - \beta(x_i - \bar{x})\}^2 \\ &= S_{yy} + n(\bar{y} - \alpha)^2 + \beta^2 S_{xx} - 2\beta S_{xy} \\ &= S_{yy} - S_{xy}^2/S_{xx} + n(\alpha - \bar{y})^2 + S_{xx}(\beta - S_{xy}/S_{xx})^2 \\ &= S_{ee} + n(\alpha - a)^2 + S_{xx}(\beta - b)^2. \end{aligned}$$

Thus the joint posterior is

$$p(\alpha, \beta, \varphi | \boldsymbol{x}, \boldsymbol{y}) \propto \varphi^{-(n+2)/2} \exp[-\tfrac{1}{2}\{S_{ee} + n(\alpha - a)^2 + S_{xx}(\beta - b)^2\}/\varphi].$$

It is now clear that *for given b and* φ the posterior for β is $N(b, \varphi/S_{xx})$, and so we can integrate β out to obtain)

$$p(\alpha, \varphi | \boldsymbol{x}, \boldsymbol{y}) \propto \varphi^{-(n+1)/2} \exp[-\tfrac{1}{2}\{S_{ee} + n(\alpha - a)^2\}/\varphi]$$

(note the change in the exponent of φ).

In Section 2.12 on "Normal mean and variance both unknown", we saw that if

$$p(\theta, \varphi) \propto \varphi^{-(\nu+1)/2-1} \exp[-\tfrac{1}{2}\{S + n(\theta - \bar{x})^2\}/\varphi]$$

and $s^2 = S/\nu$ then

$$t = \frac{(\theta - \bar{x})}{s/\sqrt{n}} \sim t_\nu \quad \text{and} \quad \varphi \sim S\chi_\nu^{-2}.$$

It follows from just the same argument that in this case the posterior for α given $\boldsymbol{x}$ and $\boldsymbol{y}$ is such that if $s^2 = S_{ee}/(n-2)$ then

$$\frac{(\alpha - a)}{s/\sqrt{n}} \sim t_{n-2}.$$

Similarly, the posterior of β can be found to be integrating α out to show that

$$\frac{(\beta - b)}{s/\sqrt{S_{xx}}} \sim t_{n-2}.$$

Finally, note that

$$\varphi \sim S_{ee} \chi_{n-2}^{-2}.$$

It should, however, be noted that the posteriors for α and β are *not* independent, although they are independent for given φ.

It may be noted that the posterior means of α and β are a and b and that these are the values that minimize the sum of squares

$$\sum\{y_i - \alpha - \beta(x_i - \bar{x})\}^2$$

and that S_{ee} is the minimum sum of squares. This fact is clear because the sum is

$$S_{ee} + n(\alpha - a)^2 + S_{xx}(\beta - b)^2$$

and it constitutes the *principle of least squares*, for which reason a and b are referred to as the *least squares estimates* of α and β. The regression line

$$y = a + b(x - \bar{x}),$$

which can be plotted for all x as opposed to just those x_i observed, is called the *line of best fit* for y on x. The principle is very old; it was probably first published by Legendre but first discovered by Gauss; for its history see Harter (1974, 1975, 1976). It should be noted that the line of best fit for y on x is not, in general, the same as the line of best fit for x on y.

Example

This example goes to show that what I naïvely thought to be true of York's weather is in fact false. I guessed that if November was wet, the same thing would be true in December, and so I thought I would try to see how far this December's rainfall could be predicted in terms of November's. It turns out that the two are in fact *negatively* correlated, so that if November is very wet there is a slight indication that December will be on the dry side. However, the data (given in mm) serves quite as well to indicate the method.

Year (i)	1971	1972	1973	1974	1975	1976	1977	1978	1979	1980
Nov. (x_i)	23.9	43.3	36.3	40.6	57.0	52.5	46.1	142.0	112.6	23.7
Dec. (y_i)	41.0	52.0	18.7	55.0	40.0	29.2	51.0	17.6	46.6	57.0

It turns out that $\bar{x} = 57.8$, $\bar{y} = 40.8$, $S_{xx} = 13{,}539$, $S_{yy} = 1889$ and $S_{xy} = -2178$ so that

$$a = \bar{y} = 40.8, \quad b = S_{xy}/S_{xx} = -0.161, \quad r = S_{xy}/\sqrt{(S_{xx}S_{yy})} = -0.431$$
$$S_{ee} = S_{yy} - S_{xy}{}^2/S_{xx} = S_{yy}(1-r^2) = 1538.$$

It follows that

$$s^2 = S_{ee}/(n-2) = 192, \quad s/\sqrt{n} = 4.38, \quad s/\sqrt{S_{xx}} = 0.119.$$

Since the 75th percentile of t_8 is 0.706, it follows that a 50% HDR for the intercept α is $40.8 \pm 0.706 \times 4.38$, that is, (37.7, 43.9). Similarly, a 50% HDR for the slope β is $-0.161 \pm 0.706 \times 0.119$, that is, $(-0.245, -0.077)$. Further, from tables of values of chi-squared corresponding to HDRs for $\log \chi_8{}^2$, an interval of posterior probability 50% for the variance φ is from 1538/11.079 to 1538/5.552, that is, (139, 277).

Very often the slope β is of more importance than the intercept α. Thus in the above example, the fact that the slope is negative with high probability corresponds to the conclusion that high rainfall in November indicates that there is less likely to be high rainfall in December, as was mentioned earlier.

Case of known variance

If, which is not very likely, we should happen to *know* the variance φ, the problem is even simpler. In this case, it is easy to deduce that (with the same priors for α and β)

$$\begin{aligned} p(\alpha, \beta \mid \boldsymbol{x}, \boldsymbol{y}) &\propto \exp[-\tfrac{1}{2}\sum\{y_i - \alpha - \beta(x_i - \bar{x})\}^2/\varphi] \\ &\propto \exp[-\tfrac{1}{2}\{n(\alpha - a)^2 + S_{xx}(\beta - b)^2\}/\varphi]. \end{aligned}$$

It is clear that in this case the posteriors for α and β *are* independent and such that $\alpha \sim \mathrm{N}(a, \varphi/n)$ and $\beta \sim \mathrm{N}(b, \varphi/S_{xx})$.

The mean value at a given value of the explanatory variable

Sometimes there are other quantities of interest than α, β and φ. For example, you might want to know what the expected value of y is at a given value of x. A particular case would arise if you wanted to estimate the average weight of women of a certain height on the basis of data on the heights and weights of n individuals. Similarly, you might want to know about the value of the parameter η_0 in the original formulation (which corresponds to the particular value $x = 0$). Suppose that the parameter of interest is

$$\gamma = \alpha + \beta(x_0 - \bar{x}).$$

Now we know that for given $\boldsymbol{x}$, $\boldsymbol{y}$, x_0 and φ

$$\alpha \sim \mathrm{N}(a, \varphi/n) \qquad \text{and} \qquad \beta \sim \mathrm{N}(b, \varphi/S_{xx})$$

independently of one another. It follows that, given the same values,

$$\gamma \sim \mathrm{N}(a + b(x_0 - \bar{x}),\ \varphi\{n^{-1} + (x_0 - \bar{x})^2/S_{xx}\}).$$

It is now easy to deduce $p(\gamma, \varphi \mid \boldsymbol{x}, \boldsymbol{y}, x_0)$ from the fact that φ has a (multiple of an) inverse chi-squared distribution. The same arguments as used in Section 2.12 on "Normal mean and variance both unknown" can be used to deduce that

$$\frac{\gamma - a - b(x_0 - \bar{x})}{s\sqrt{\{n^{-1} + (x_0 - \bar{x})^2/S_{xx}\}}} \sim \mathrm{t}_{n-2}.$$

In particular, setting $x_0 = 0$ and writing $e_0 = a - b\bar{x}$ we obtain

$$\frac{\eta_0 - e_0}{s\sqrt{\{n^{-1} + \bar{x}^2/S_{xx}\}}} \sim \mathrm{t}_{n-2}.$$

Prediction of observations at a given value of the explanatory variable

It should be noted that if you are interested in the distribution of a potential observation at a value $x = x_0$, that is, the *predictive distribution*, then the result is slightly different. The mean of such observations conditional on $\boldsymbol{x}$, $\boldsymbol{y}$ and x_0 is still $a+b(x_0-\bar{x})$, but since

$$y_0-\alpha-\beta(x_0-\bar{x})\sim \mathrm{N}(0,\ \varphi)$$

in addition to the distribution found above for $\gamma=\alpha+\beta(x_0-\bar{x})$, it follows that

$$y_0\sim \mathrm{N}(a+b(x_0-\bar{x}),\ \varphi\{1+n^{-1}+(x_0-x)^2/S_{xx}\})$$

and so on integrating φ out

$$\frac{y_0-a-b(x_0-\bar{x})}{s\surd\{1+n^{-1}+(x_0-\bar{x})^2/S_{xx}\}}\sim \mathrm{t}_{n-2}.$$

Continuation of the example

To find the mean rainfall to be expected in December in a year when the rainfall $x_0=46.1$ mm in November, we first find $a+b(x_0-\bar{x})$ = 42.7 and $n^{-1}+(x_0-\bar{x})^2/S_{xx}=0.110$, and hence $s\surd\{n^{-1}+(x_0-\bar{x})^2/S_{xx}\}$ = 4.60. Then the expected value γ at $x=x_0$ has a t distribution which is approximately $\mathrm{N}(42.7,\ 4.60^2)$. On the other hand, in single years in which the rainfall in November is 46.1 there is a greater variation in the December rainfall than the variance for the mean of $4.60^2=21.2$ implies—in fact $s\surd\{1+n^{-1}+(x_0-\bar{x})^2/S_{xx}\}=14.6$ and the corresponding variance is $14.6^2=213.2$.

Multiple regression

Very often there is more than one independent variable, and we want to predict the value of y using the values of two or more variables $x^{(1)}$, $x^{(2)}$, etc. It is not difficult to adapt the method described above to estimate the parameters in a model such as

$$y_i\sim \mathrm{N}(\alpha+\beta^{(1)}(x_i^{(1)}-\bar{x}^{(1)})+\beta^{(2)}(x_i^{(2)}-\bar{x}^{(2)}),\ \varphi)$$

although we will find some complications unless it happens that

$$\sum(x_i^{(1)}-\bar{x}^{(1)})(x_i^{(2)}-\bar{x}^{(2)})=0.$$

For this reason, it is best to deal with such *multiple regression* problems by using matrix analysis. Readers who are interested will find a brief introduction to this topic in the context of the general linear model in Section 6.7, while a full account in Box and Tiao (1973).

Polynomial regression

A difficult problem which will not be discussed in any detail is that of *polynomial regression*, that is, of fitting a model

$$y \sim \mathrm{N}(\eta_0 + \eta_1 x + \eta_2 x^2 + \ldots + \eta_k x^k, \varphi)$$

where all the parameters, *including the degree k* of the polynomial are unknown *a priori*. Some relevant references are Jeffreys (1939; 1948; 1961, Sections 5.9–5.92) and Sprent (1969, Sections 5.4, 5.5). There is also an interesting discussion in Meyer and Collier (1970, p. 114 *et seq.*) in which Lindley starts by remarking:

"I agree the problem of fitting a polynomial to the data is one that at the moment I can't fit very conveniently to the Bayesian analysis. I have prior beliefs in the smoothness of the polynomial. We need to express this idea quantitatively, but I don't know how to do it. We could bring in our prior opinion that some of the regression coefficients are very small."

Subsequently, a Bayesian approach to this problem has been developed by A. S. Young (1977).

6.4 Conjugate prior for the bivariate regression model

The problem of updating a regression line

In the last section we saw that with the regression line in the standard form $y_i = \alpha + \beta(x_i - \bar{x})$ the joint posterior is

$$p(\alpha, \beta, \varphi \mid \boldsymbol{x}, \boldsymbol{y}) \propto \varphi^{-(n+2)/2} \exp\left[-\tfrac{1}{2}\{S_{ee} + n(\alpha - a)^2 + S_{xx}(\beta - b)^2\}/\varphi\right].$$

For reasons that will soon emerge, we denote the quantities derived from the data with a prime as n', $\bar{x}'$, $\bar{y}'$, a', b', S_{xx}', S_{ee}', etc. In the example on rainfall, we found that $n' = 10$ and

$$\bar{x}' = 57.8, \quad a' = \bar{y}' = 40.8, \quad b' = -0.161, \quad S_{xx}' = 13{,}539, \quad S_{ee}' = 1538.$$

Now suppose that we collect further data, thus:

Year (i)	1981	1982	1983	1984	1985	1986
Nov. (x_i)	34.1	62.0	106.9	34.1	68.3	81.0
Dec. (y_i)	12.3	90.4	28.8	106.2	62.3	50.5

If this had been all the data available, we would have constructed a regression line based on data for $n'' = 6$ years, with

$$\bar{x}'' = 64.4, \quad a'' = \bar{y}'' = 58.4, \quad b'' = -0.381, \quad S_{xx}'' = 3939, \quad S_{ee}'' = 5815.$$

If, however, we had information about all 16 years, then the regression line would have been based on data for $n = 16$ years resulting in

$$\bar{x} = 60.3, \quad a = \bar{y} = 47.4, \quad b = -0.184, \quad S_{xx} = 17{,}641, \quad S_{ee} = 8841.$$

Formulae for recursive construction of a regression line

By the sufficiency principle it must be possible to find $\bar{x}$, $\bar{y}$, b, etc., from $\bar{x}'$, $\bar{y}'$, b', etc., and $\bar{x}''$, $\bar{y}''$, b'', etc. It is in fact not too difficult to show that n, $\bar{x}$ and $a = \bar{y}$ are given by

$$
\begin{aligned}
n &= n' + n'' \\
\bar{x} &= (n'\bar{x}' + n''\bar{x}'')/n \\
\bar{y} &= (n'\bar{y}' + n''\bar{y}'')/n
\end{aligned}
$$

and that if we define

$$
\begin{aligned}
n^h &= (n'^{-1} + n''^{-1})^{-1} \\
S_{xx}{}^c &= n^h(\bar{x}' - \bar{x}'')^2 \\
S_{xy}{}^c &= n^h(\bar{x}' - \bar{x}'')(\bar{y}' - \bar{y}'') \\
b^c &= S_{xy}{}^c / S_{xx}{}^c
\end{aligned}
$$

then S_{xx}, b and S_{ee} are given by

$$
\begin{aligned}
S_{xx} &= S_{xx}' + S_{xx}'' + S_{xx}{}^c \\
b &= (b'S_{xx}' + b''S_{xx}'' + b^cS_{xx}{}^c)/S_{xx} \\
S_{ee} &= S_{ee}' + S_{ee}'' + [(b' - b'')^2 S_{xx}'S_{xx}'' \\
&\quad + (b'' - b^c)^2 S_{xx}''S_{xx}{}^c + (b^c - b')^2 S_{xx}{}^cS_{xx}']/S_{xx}.
\end{aligned}
$$

Of these formulae, the only one that is at all difficult to deduce is the last, which is established thus

$$
\begin{aligned}
S_{xx} &= S_{yy} - b^2S_{xx} \\
&= S_{yy}' + S_{yy}'' + S_{yy}{}^c - b^2S_{xx} \\
&= S_{ee}' + S_{ee}'' + b'^2S_{xx}' + b''S_{xx}'' + b^{c2}S_{xx}{}^c - b^2S_{xx}
\end{aligned}
$$

(it is easily checked that there is no term $S_{ee}{}^c$). However,

$$
\begin{aligned}
&S_{xx}(b'^2S_{xx}' + b''S_{xx}'' + b^{c2}S_{xx}{}^c - b^2S_{xx}) \\
&\quad = (S_{xx}' + S_{xx}'' + S_{xx}{}^c)(b'^2S_{xx}' + b''^2S_{xx}'' + b^{c2}S_{xx}{}^c) \\
&\qquad - (b'S_{xx}' + b''S_{xx}'' + b^cS_{xx}{}^c)^2 \\
&\quad = b'^2S_{xx}'S_{xx}'' + b''^2S_{xx}'' - 2b'b''S_{xx}'S_{xx}'' \\
&\qquad + b''^2S_{xx}''S_{xx}{}^c + b^{c2}S_{xx}''S_{xx}{}^c - 2b''b^cS_{xx}''S_{xx}{}^c \\
&\qquad + b^{c2}S_{xx}{}^cS_{xx}' + b'^2S_{xx}{}^cS_{xx}' - 2b^cb'S_{xx}{}^cS_{xx}' \\
&\quad = (b' - b'')^2S_{xx}'S_{xx}'' + (b'' - b^c)^2S_{xx}''S_{xx}{}^c + (b^c - b')^2S_{xx}{}^cS_{xx}'
\end{aligned}
$$

giving the result.

With the data in the example, it turns out that $n = n' + n'' = 16$ and $\bar{x}$ (a weighted mean of $\bar{x}'$ and $\bar{x}''$) is 60.3, and similarly $\bar{y} = 47.4$.

Moreover,

$$n^h = 3.75, \quad S_{xx}{}^c = 163.35, \quad S_{xy}{}^c = 435.6, \quad b^c = 2.667$$

so that

$$S_{xx} = 17{,}641, \quad b = -0.184, \quad S_{ee} = 8841,$$

in accordance with the results quoted earlier obtained by considering all 16 years together.

Finding an appropriate prior

In the analysis above, our prior knowledge could be summarized by saying that if the regression line is put in the form

$$y_i = \alpha + \beta(x_i - \bar{x}')$$

then

$$p(\alpha, \beta, \varphi) \propto \varphi^{-(n'+2)/2} \exp\left[-\tfrac{1}{2}\{S_{ee}' + n'(\alpha - a')^2 + S_{xx}'(\beta - b')^2\}/\varphi\right].$$

We then had observations (denoted $\boldsymbol{x}''$ and $\boldsymbol{y}''$) that resulted in a posterior which is such that if the regression line is put in the form

$$y_i = \alpha + \beta(x_i - \bar{x})$$

then

$$p(\alpha, \beta, \varphi \mid \boldsymbol{x}'', \boldsymbol{y}'') \propto \varphi^{-(n+2)/2} \exp\left[-\tfrac{1}{2}\{S_{ee}' + n(\alpha - a)^2 + S_{xx}(\beta - b)^2\}/\varphi\right].$$

This, of course, gives a way of incorporating prior information into a regression model provided that it can be put into the form which occurs above. It is, however, often difficult to specify prior knowledge about a regression line unless, as in the case above, it is explicitly the result of previous data. Appropriate questions to ask to fix which prior of this class to use are as follows:

(1) What number of observations is your present knowledge worth? Write the answer as n'.

(2) What single point is the regression line most likely to go through? Write the answer as $(\bar{x}', \bar{y}')$.

(3) What is the best guess as to the slope of the regression line? Write the answer as b'?

(4) What is the best guess as to the variance of the observation y_i about the regression line? Write the answer as s'^2 and find S_{ee}' as $(n' - 2)s'^2$.

(5) Finally make S_{xx}' such that the estimated variances for the slope β and the intercept α are in the ratio n' to S_{xx}'.

As noted above, it is difficult to believe that the process can be carried out in a very convincing manner, although the first three steps do not present as much difficulty as the last two. However, the case where information is received and then more information of the same type is used to update the regression line can be useful.

It is of course possible (and indeed simpler) to do similar things with the correlation coefficient.

6.5 Comparison of several means—the one way model

Description of the one way layout

Sometimes we want to compare more than two samples. We might, for example, wish to compare the performance of children from a number of schools at a standard test. The usual model for such a situation is as follows. We suppose that $\boldsymbol{\theta} = (\theta_1, \theta_2, \ldots, \theta_I)$ is a vector of unknown parameters and that there are $N = \sum K_i$ independent observations

$$x_{ik} \sim \mathrm{N}(\theta_i, \varphi) \qquad (i = 1, 2, \ldots, I;\ k = 1, 2, \ldots, K_i)$$

from I independent populations with, however, a common variance φ. For simplicity we shall assume independent reference priors uniform in $\theta_1, \theta_2, \ldots, \theta_I$ and $\log \varphi$, that is,

$$p(\boldsymbol{\theta}, \varphi) \propto 1/\varphi.$$

The likelihood is

$$(2\pi\varphi)^{-N/2} \exp(-\tfrac{1}{2}S/\varphi)$$

where

$$S = \sum_i \sum_k (x_{ik} - \theta_i)^2$$

and so the posterior is

$$p(\boldsymbol{\theta}, \varphi \mid \boldsymbol{x}) \propto \varphi^{-N/2-1} \exp(-\tfrac{1}{2}S/\varphi).$$

It is useful to define the following notation:

$$x_{i.} = \sum_k x_{ik}/K_i$$

$$x_{..} = \sum_i \sum_k x_{ik}/N$$

$$\lambda = \sum_i \theta_i/I$$

$$\mu_i = \theta_i - \lambda$$

$$\hat{\lambda} = x_{..}$$

$$\hat{\mu} = x_{i.} - x_{..}.$$

The reason for thinking of the μ_i is that we are often concerned as to whether all the θ_i are equal. If, for example, the x_{ik} represent yields of wheat on fields on which I different fertilizers have been used, then we are likely to be interested in whether the yields are on average all equal (or nearly so), that is, $\theta_1 = \theta_2 = \ldots = \theta_I$ or equivalently whether or not

$$\mu_1 = \mu_2 = \ldots = \mu_I = 0.$$

The μ_i satisfy the condition

$$\sum_i K_i \mu_i = 0$$

so that if we know the values of $\mu_1, \ldots, \mu_{I-1}$ we automatically know

$$\boldsymbol{\mu} = (\mu_1, \ldots, \mu_I).$$

Similarly, the $\hat{\mu}_i$ satisfy $\sum K_i \hat{\mu}_i = 0$.

Integration over the nuisance parameters

Since the Jacobian determinant of the transformation from $\boldsymbol{\theta}$, φ to λ, $\mu_1, \mu_2, \ldots, \mu_{I-1}, \varphi$ consists of entries all of which are $1/n$, 1 or 0, its value is a constant, and so

$$\begin{aligned} p(\lambda, \boldsymbol{\mu}, \varphi \mid \boldsymbol{x}) &\propto p(\lambda, \mu_1, \ldots, \mu_{I-1}, \varphi \mid \boldsymbol{x}) \\ &= p(\boldsymbol{\theta}, \varphi \mid \boldsymbol{x}). \end{aligned}$$

Now we must re-express S in terms of λ, $\boldsymbol{\mu}$, φ. Since $x_{i.} = \hat{\lambda} + \hat{\mu}_i$ and $\theta_i = \lambda + \mu_i$ it follows that

$$-(x_{ik} - \theta_i) = (\lambda - \hat{\lambda}) + (\mu_i - \hat{\mu}) + (x_{i.} - x_{ik}).$$

It is easily checked that sums of products of terms on the right vanish, and so it follows that

$$\begin{aligned} S &= \sum\sum (x_{ik} - \theta_i)^2 \\ &= N(\lambda - \hat{\lambda})^2 + S_t(\boldsymbol{\mu}) + S_e \end{aligned}$$

where

$$S_t(\boldsymbol{\mu}) = \sum K_i(\mu_i - \hat{\mu})^2, \qquad S_e = \sum\sum (x_{i.} - x_{ik})^2.$$

It is also useful to define

$$\nu = N - I, \qquad s^2 = S_e/\nu.$$

It follows that the posterior may be written in the form

$$p(\lambda, \boldsymbol{\mu}, \varphi \mid \boldsymbol{x}) \propto \varphi^{-N/2-1} \exp\left[-\tfrac{1}{2}\{N(\lambda - \hat{\lambda})^2 + S_t(\boldsymbol{\mu}) + S_e\}/\varphi\right].$$

As explained above, the value of λ is not usually of any great interest, and it is easily integrated out to give

$$p(\boldsymbol{\mu}, \varphi \mid \boldsymbol{x}) \propto \varphi^{-(N+1)/2} \exp\left[-\tfrac{1}{2}\{S_t(\boldsymbol{\mu}) + S_e\}/\varphi\right].$$

The variance φ can now be integrated out in just the same way as it was in Section 2.12 on "Normal mean and variance both unknown" by reducing to a standard gamma function integral. The result is that

$$\begin{aligned} p(\boldsymbol{\mu} \mid \boldsymbol{x}) &\propto \{S_t(\boldsymbol{\mu}) + S_e\}^{-(N-1)/2} \\ &\propto \{1 + (I-1)F(\boldsymbol{\mu})/\nu\}\varphi^{-(N-1)/2} \end{aligned}$$

where

$$F(\boldsymbol{\mu}) = \frac{S_t(\boldsymbol{\mu})/(I-1)}{S_e/\nu} = \frac{\sum K_i(\mu_i - \hat{\mu}_i)^2/(I-1)}{s^2}.$$

This is similar to a result obtained in one dimension (see Section 2.12 again; the situation there is not quite that obtained by setting $I = 1$ here because here λ has been integrated out). In that case we deduced that

$$p(\mu \mid \boldsymbol{x}) \propto \{1 + t^2/\nu\}^{-(\nu+1)/2}$$

where

$$t^2 = \frac{K(\mu - \bar{x})^2}{s^2}.$$

By analogy with that situation, the posterior distribution for $\boldsymbol{\mu}$ is called the multivariate t distribution. It was discovered independently by Cornish (1954, 1955) and by Dunnett and Sobel (1954). The constant of proportionality can be evaluated, but we will not need to use it.

It should be clear that the density is a maximum when $\boldsymbol{\mu} = \hat{\boldsymbol{\mu}}$ and decreases as the distance from $\boldsymbol{\mu}$ to $\hat{\boldsymbol{\mu}}$, and indeed an HDR for $\boldsymbol{\mu}$ is clearly a hyperellipsoid centred on $\hat{\boldsymbol{\mu}}$, that is, it is of the form

$$\mathrm{E}(F) = \{\boldsymbol{\mu}; F(\boldsymbol{\mu}) \leqslant F\}$$

in which the length of each of the axes is in a constant ratio to $\sqrt{F}$.

To find an HDR of any particular probability it therefore suffices to find the distribution of $F(\boldsymbol{\mu})$, and since $F(\boldsymbol{\mu})$ is a ratio of sums of squares divided by appropriate numbers of degrees of freedom it seems reasonable to conjecture that

$$F(\boldsymbol{\mu}) \sim \mathrm{F}_{I-1,\nu}$$

which is indeed so.

Derivation of the F distribution

It is not really necessary to follow this proof that $F(\boldsymbol{\mu})$ really does have an F distribution, but it is included for completeness.

$$\begin{aligned}
\mathrm{P}\{F \leqslant F(\boldsymbol{\mu}) \leqslant F+\mathrm{d}F\} &= \int_{E(F+\mathrm{d}F)\setminus E(F)} p(\boldsymbol{\mu}\,|\,\boldsymbol{x})\,\mathrm{d}\boldsymbol{\mu} \\
&= \int_{E(F+\mathrm{d}F)\setminus E(F)} \{1+(I-1)F(\boldsymbol{\mu})/\nu\}^{-(N-1)/2}\,\mathrm{d}\boldsymbol{\mu} \\
&= \{1+(I-1)F(\boldsymbol{\mu})/\nu\}^{-(N-1)/2} \int_{E(F+\mathrm{d}F)\setminus E(F)} \mathrm{d}\boldsymbol{\mu} \\
&= \{1+(I-1)F(\boldsymbol{\mu})/\nu\}^{-(N-1)/2}[\mathrm{V}(F+\mathrm{d}F)-\mathrm{V}(F)]
\end{aligned}$$

where $\mathrm{V}(F)$ is the volume of the hyperellipsoid $\mathrm{E}(F)$. At first sight it appears that this is I-dimensional, but because $\sum K_i(\mu_i-\hat{\mu}_i)=0$ it represents the intersection of a hyperellipsoid in I dimensions with a hyperplane through its centre, which is a hyperellipsoid in $(I-1)$ dimensions. If this is not clear, it may help to note that an ordinary (3-dimensional) sphere cuts a plane in a circle, that is, a sphere in $3-1=2$ dimensions. It follows that

$$\mathrm{V}(F) \propto (\sqrt{F})^{I-1} = F^{(I-1)/2}$$

and hence

$$\mathrm{V}(F+\mathrm{d}F)-\mathrm{V}(F) \propto (F+\mathrm{d}F)^{(I-1)/2} - F^{(I-1)/2} \propto F^{(I-1)/2-1}\,\mathrm{d}F.$$

It follows that the density of $F(\boldsymbol{\mu})$ is proportional to

$$F^{(I-1)/2-1}\{1+(I-1)F/\nu\}^{-(N-1)/2}.$$

Comparing this with the standard form in the Appendix, and noting that $I-1+\nu=N-1$, it can be seen that indeed $F(\boldsymbol{\mu}) \sim \mathrm{F}_{I-1,\nu}$ as asserted.

Relationship to the analysis of variance

This relates to the classical approach to the one way layout. Note that if

$$F(\mathbf{0}) = \frac{\sum K_i \hat{\mu}_i^2/(I-1)}{s^2} = \frac{\sum K_i(x_{i.}-x_{..})^2/(I-1)}{\sum\sum (x_{ik}-x_{i.})^2/\nu}$$

then $F(\boldsymbol{\mu}) = F(\mathbf{0})$ at the point $\boldsymbol{\mu}=\mathbf{0}$ which represents no treatment effect. Consequently, if

$$\pi(\boldsymbol{\mu}_0) = \mathrm{P}\{F(\boldsymbol{\mu}) \leqslant F(\boldsymbol{\mu}_0)\}$$

then $\pi(\mathbf{0})$ is the probability of an HDR which just includes $\boldsymbol{\mu}=\mathbf{0}$. It is thus possible to carry out a significance test at level α of the hypothesis that $\boldsymbol{\mu} = \mathbf{0}$ in the sense of Section 4.3 on "Lindley's method" by rejecting if and only if $\pi(\mathbf{0}) > 1 - \alpha$.

This procedure corresponds exactly to the classical analysis of variance (ANOVA) procedure in which a table is consructed as follows. First we find

$$S_T = \sum\sum (x_{ik} - x_{..})^2,$$
$$S_t(\mathbf{0}) = \sum K_i(x_{i.} - x_{..})^2.$$

It is convenient to write S_t for $S_t(\mathbf{0})$. Then we find S_e by subtraction, as it is easily shown that

$$S_T = S_t + S_e.$$

In computing, it should be noted that it makes no difference if a constant is subtracted from each of the x_{ik} and that S_T and S_t can be found by

$$S_T = \sum\sum x_{ik}{}^2 - Nx_{..}{}^2 = \sum x_{ik}{}^2 - C,$$
$$S_t = \sum K_i x_{i.}{}^2 - Nx_{i..}{}^2 = \sum T_i^2/K_i - C$$

where $T_i = \sum x_{ik} = K_i x_{i.}$ is the total for treatment i, $G = \sum\sum x_{ik} = Nx_{..}$ is the grand total, and $C = G^2/N$ is the "correction for error". (Note that these formulae are subject to rounding error if used incautiously).

The value of $F(\mathbf{0})$ is then found easily by setting out a table as below:

ANOVA Table

Source	Sum of squares	Degrees of freedom	Mean square	Ratio
Treatments	S_t	$I-1$	$S_t/(I-1)$	$F(\mathbf{0})$
Error	S_e	$\nu = N - I$	$s^2 = S_e/\nu$	
Total	S_T	$N-1$		

We will now consider an example.

Example

Cochran and Cox (1957, Section 4.13) quote the following data from an experiment on the effect of sulphur in reducing scab disease in potatoes. In addition to untreated plots which serve as a control, three amounts of dressing were compared—300, 600 and 1200 pounds per acre. Both an autumn and a spring application of each treatment were tried, so that in all there were seven distinct treatments. The effectiveness of the treatments was measured by the "scab index", which is (roughly speaking) the average percentage of the area of 100 potatoes taken at random from each plot that is affected with scab. The data are as follows.

i	Treatment	K_i	Scab indices x_{ik}								T_i	$x_{i.}$	$\hat{\mu}_i$
1	0	8	12	30	10	18	24	32	29	26	181	22.6	7.0
2	A3	4	9	9	16	4					38	9.5	−6.2
3	S3	4	30	7	21	9					67	16.8	1.1
4	A6	4	16	10	18	18					62	15.5	−0.2
5	S6	4	18	24	12	19					73	18.2	2.6
6	A12	4	10	4	4	5					23	5.8	−9.9
7	S12	4	17	7	16	17					57	14.2	−1.2

There are $I = 7$ treatments and $N = \sum K_i = 32$ observations, the grand total being $G = 501$ (and the grand average $x_{..}$ being 15.66), the crude sum of squares being $\sum\sum x_{ik}{}^2 = 9939$ and the correction for error $C = G^2/N = 7844$. Further

$$S_T = 9939 - 7844 = 2095$$
$$S_t = 181^2/8 + (38^2 + 67^2 + 62^2 + 73^2 + 23^2 + 57^2)/4 - 7844 = 972$$

and hence the analysis of variance table is as follows:

ANOVA Table

Source	Sum of squares	Degrees of freedom	Mean square	Ratio
Treatments	972	6	162	3.60
Error	1123	25	45	
Total	2095	31		

From tables of the F distribution, an $F_{6,25}$ variable exceeds 3.63 with probability 0.01. Consequently, a 99% HDR is

$$E(3.63) = \{\boldsymbol{\mu};\, F(\boldsymbol{\mu}) \leqslant 3.63\}$$

so that $\pi(\mathbf{0}) \cong 0.99$ and, according to Lindley's method, as described in Section 4.3, the data are very nearly enough to cause the null hypothesis of no treatment to be rejected at the 1% level.

The 99% HDR can be re-expressed by noting that $\boldsymbol{\mu}$ is in it if and only if $F(\boldsymbol{\mu}) \leqslant 3.63$ or

$$\sum K_i(\mu_i - \hat{\mu}_i)^2 \leqslant 3.63 s^2 = 3.63 \times 45 = 163$$

that is, if and only if

$$2(\mu_1 - \hat{\mu}_1)^2 + \sum_{i=2}^{7} (\mu_i - \hat{\mu}_i)^2 \leqslant 41.$$

It is of course difficult to visualize such sets, which is one reason why the significance test mentioned earlier is helpful in giving some ideas as to what is going on. However, as was explained when significance tests were first introduced, they should not be taken too seriously—in most cases we would expect to see a treatment effect, even if only a small one. One point to note is that we can obtain some idea of the size of the treatment effect from the significance level.

Relationship to a simple linear regression model

A way of visualizing the analysis of variance in terms of the simple linear regression model was pointed out by Kelley (1927, p. 178); see also Novick and Jackson (1974, Sections 4–7).

Kelley's work is relevant to a *random-effects model* (sometimes known as a components of variance model or Model II for the analysis of variance). An idea of what this is can be gained by considering an example quoted by Scheffé (1959, Section 7.2). We suppose a machine is used by different workers on different days, being used by worker i on K_i days for $i = 1, 2, \ldots, I$, and that the output when worker i uses it on day k is x_{ik}. Then it might be reasonable to suppose that

$$x_{ik} = m_i + e_{ik}$$

where m_i is the "true" mean for the ith worker and e_{ik} is his "error" on the kth day. We could then assume that the I workers are a random sample from a large labour pool, rather than being fixed if unknown effects. In such a case all of our knowledge of the x_{ik} contributes to knowledge of the distribution of the m_i, and so if we want to estimate a particular m_i we should take into account the observations $x_{i'k}$ for $i' \neq i$ as well as the observations x_{ik}. Kelley's suggestion is that we treat the individual measurements x_{ik} as the explanatory variable and the treatment means $x_{i.}$ as the dependent variable, so that the model to be fitted is

$$x_{i.} = \eta_0 + \eta_1 x_{ik} + \varepsilon_{ik}$$

where the ε_{ik} are error terms of mean zero, or equivalently

$$x_{i.} = \alpha + \beta(x_{ik} - x_{..}).$$

In terms of the notation we used in connection with simple linear regression,

$$\begin{aligned}
S_{xx} &= \sum\sum(x_{ik} - x_{..})^2 = S_T \\
S_{yy} &= \sum\sum(x_{i.} - x_{..})^2 = \sum K_i(x_{i.} - x_{..})^2 = S_t \\
S_{xy} &= \sum\sum(x_{ik} - x_{..})(x_{i.} - x_{..}) = \sum K_i(x_{i.} - x_{..})^2 = S_t
\end{aligned}$$

$$r = \frac{S_{xy}}{\surd(S_{xx}S_{yy})} = \surd(S_t/S_T)$$

$$1 - r^2 = (S_T - S_t)/S_T = S_e/S_T$$

$$S_{ee} = S_{yy}(1 - r^2) = S_t S_e / S_T.$$

In accordance with the theory of simple linear regression we estimate α and β by, respectively,

$$a = x_{..} \qquad \text{and} \qquad b = S_{xy}/S_{xx} = S_t/S_T$$

so that the regression line takes the form

$$x_{i.} = (S_t x_{ik} + S_e x_{..})/S_T.$$

The point of this formula is that if we were to try one single replicate with another broadly similar treatment to those already tried, we could estimate the overall mean for that treatment not simply by the one observation we have for that treatment, but by a weighted mean of that observation and the overall mean of all observations available to date.

Investigation of contrasts

Often in circumstances where the treatment effect does not appear substantial we may want to make further investigations. Thus in the example above about sulphur treatment for potatoes, we might want to see how the effect of *any* sulphur treatment compares with that of none, that is, we might like an idea of the size of

$$d = \sum_{i=2}^{7} \theta_i/6 - \theta_1 = \sum_{i=2}^{7} \mu_i/6 - \mu_1.$$

More generally, it may be of interest to investigate any *contrast*, that is, any linear combination

$$d = \sum \delta_i \mu_i \qquad \text{where} \qquad \sum \delta_i = 0.$$

If we then write $\hat{d} = \sum \delta_i \hat{\mu}_i$ and

$$K_d = (\sum \delta_i^2/K_i)^{-1}$$

then it is not difficult to show that we can write

$$S_t(\boldsymbol{\mu}) = \sum K_i(\mu_i - \hat{\mu}_i)^2$$

$$= K_d(d - \hat{d})^2 + S_t'(\boldsymbol{\mu}')$$

where S_t' is a quadratic much like S_t except that has one less dimension and $\boldsymbol{\mu}'$ consists of linear combinations of $\boldsymbol{\mu} - \hat{\boldsymbol{\mu}}$. It follows that

$$p(\boldsymbol{\mu}, \phi | \boldsymbol{x}) \propto \phi^{-(N+1)/2} \exp [-\tfrac{1}{2}\{K_d(d - \hat{d})^2 + S_t'(\boldsymbol{\mu}') + S_e\}/\phi].$$

It is then possible to integrate over the $I-2$ linearly independent components of $\boldsymbol{\mu}'$ to obtain

$$p(d, \varphi\,|\,\boldsymbol{x}) \propto \varphi^{-(N-I+3)/2} \exp\left[-\tfrac{1}{2}\{K_d(d-\hat{d})^2+S_e\}/\varphi\right]$$

and then to integrate φ out to give

$$\begin{aligned} p(d, \varphi\,|\,\boldsymbol{x}) &\propto \{K_d(d-\hat{d})^2+S_e\}^{-(N-I+1)/2} \\ &\propto \{1+t^2/\nu\}^{-(\nu+1)/2} \end{aligned}$$

(remembering that $\nu = N-I$), where

$$t = \sqrt{K_d}(d-\hat{d})/s.$$

It follows that $t \sim \mathrm{t}_\nu$.

For example, in the case of the contrast concerned with the main effect of sulphur, $\hat{d} = -14/6-7 = -9.3$ and $K_d = \{6(1/6)^2/4 + 1^2/8\}^{-1} = 6$, so that

$$\frac{\sqrt{K_d}(d-\hat{d})}{s} = \frac{\sqrt{6}(d+9.33)}{\sqrt{45}} = \frac{d+9.33}{2.74} \sim \mathrm{t}_\nu$$

so that, for example, as a t_{25} random variable is less than 2.060 in modulus with probability 0.95, a 95% HDR for d is between $-9.33 \pm 2.74 \times 2.060$, that is, $(-15.0, -3.7)$.

6.6 The two way layout

Notation

Sometimes we come across observations which can be classified in two ways. We shall consider a situation in which each of a number of treatments is applied to a number of plots in each of a number of blocks. However, the terminology need not be taken literally; the terms "treatments" and "blocks" are purely conventional and could be interchanged in what follows. A possible application is to the yields on plots on which different varieties of wheat have been sown and to which different fertilizers are applied, and in this case either the varieties or the fertilizers could be termed treatments or blocks. Another application is to an analysis of rainfall per hour, in which case the months of the year might be treated as the blocks and the hours of the day as the treatments (or vice versa).

We consider the simplest situation in which we have $N = IJK$ observations

$$x_{ijk} \sim \mathrm{N}(\theta_{ij}, \varphi) \quad (i = 1, 2, \ldots, I;\, j = 1, 2, \ldots, J;\, k = 1, 2, \ldots, K)$$

to which I treatments have been applied in J blocks, the observations having a common variance φ. For simplicity we assume independent reference priors uniform in the θ_{ij} and $\log\varphi$, that is,

$$p(\boldsymbol{\theta}, \varphi) \propto 1/\varphi.$$

The likelihood is

$$(2\pi\varphi)^{-N/2} \exp\left(-\tfrac{1}{2}S/\varphi\right)$$

where

$$S = \sum_i \sum_j \sum_k (x_{ijk} - \theta_{ij})^2$$

and so the posterior is

$$p(\boldsymbol{\theta}, \varphi \mid \boldsymbol{x}) \propto \varphi^{-N/2-1} \exp\left(-\tfrac{1}{2}S/\varphi\right).$$

As in the previous section, we use dots to indicate averaging over suffices, so, for example, $x_{.j.}$ is the average of x_{ijk} over i and k for fixed j. We write

$$\begin{aligned}
\lambda &= \theta_{..} & \hat{\lambda} &= x_{...} \\
\tau_i &= \theta_{i.} - \theta_{..} & \hat{\tau}_i &= x_{i..} - x_{...} \\
\beta_j &= \theta_{.j} - \theta_{..} & \hat{\beta}_j &= x_{.j.} - x_{...} \\
\kappa_{ij} &= \theta_{ij} - \theta_{i.} - \theta_{.j} + \theta_{..} & \hat{\kappa}_{ij} &= x_{ij.} - x_{i..} - x_{.j.} + x_{...}
\end{aligned}$$

It is conventional to refer to τ_i as the *main effect* of treatment i and to β_j as the main effect of block j. If $\tau_i = 0$ for all i there is said to be no main effect due to treatments; similarly if $\beta_j = 0$ for all j there is said to be no main effect due to blocks. Further, κ_{ij} is referred to as the *interaction* of the ith treatment with the jth block, and if $\kappa_{ij} = 0$ for all i and j there is said to be no interaction between treatments and blocks. We note that the parameters satisfy the conditions

$$\sum \tau_i = \sum \beta_j = 0$$

so that $\boldsymbol{\tau}$ is $(I-1)$-dimensional and β is $(J-1)$-dimensional. Similarly, for all i

$$\sum_i \kappa_{ij} = 0$$

and for all j

$$\sum_i \kappa_{ij} = 0.$$

Because both of these imply $\sum\sum \kappa_{ij} = 0$, there are only $I+J-1$ linearly independent constraints, so that $\boldsymbol{\kappa}$ *is* $IJ-I-J+1=(I-1)(J-1)$-dimensional.

Marginal posterior distributions

Because

$$-(x_{ijk}-\theta_{ij})=(\lambda-\hat{\lambda})+(\tau_i-\hat{\tau}_i)+(\beta_j-\hat{\beta}_j)+(\kappa_{ij}-\hat{\kappa}_{ij})+(x_{ij.}-x_{...})$$

the sum of squares S can be split up in a slightly more complicated way than in the previous section as

$$S=N(\lambda-\hat{\lambda})^2+S_t(\boldsymbol{\tau})+S_b(\boldsymbol{\beta})+S_{tb}(\boldsymbol{\kappa})+S_e$$

where

$$\begin{aligned}S_t(\boldsymbol{\tau})&=JK\sum(\tau_i-\hat{\tau}_i)^2\\ S_b(\boldsymbol{\beta})&=IK\sum(\beta_j-\hat{\beta}_j)^2\\ S_{tb}(\boldsymbol{\kappa})&=K\sum\sum(\kappa_{ij}-\kappa_{ij})^2\\ S_e&=\sum\sum\sum(x_{ij.}-x_{...})^2.\end{aligned}$$

It is also useful to define

$$\nu=IJ(K-1),\qquad s^2=S_e/\nu.$$

After a change of variable as in the last section, the posterior can be written as

$$p(\lambda,\boldsymbol{\tau},\boldsymbol{\beta},\boldsymbol{\kappa},\varphi\,|\,\boldsymbol{x})\propto\varphi^{-N/2-1}$$
$$\exp[-\tfrac{1}{2}\{N(\lambda-\hat{\lambda})^2+S_t(\boldsymbol{\tau})+S_b(\boldsymbol{\beta})+S_{tb}(\boldsymbol{\kappa})+S_e\}/\varphi]$$

so that on integrating λ out

$$p(\boldsymbol{\tau},\boldsymbol{\beta},\boldsymbol{\kappa},\varphi)\propto\varphi^{-N/2-1}\exp[-\tfrac{1}{2}\{S_t(\boldsymbol{\tau})+S_b(\boldsymbol{\beta})+S_{tb}(\boldsymbol{\kappa})+S_e\}/\varphi].$$

To investigate the effects of the treatments, we can now integrate over the $\{(J-1)+(I-1)(J-1)\}$-dimensional space of values of $(\boldsymbol{\beta},\boldsymbol{\kappa})$ to obtain

$$p(\boldsymbol{\tau})\propto\varphi^{-(\nu+I+1)/2-1}\exp[-\tfrac{1}{2}\{S_t(\boldsymbol{\tau})+S_e\}/\varphi].$$

We can now integrate φ out just as in the last section to obtain

$$\begin{aligned}p(\boldsymbol{\tau})&\propto\{S_t(\boldsymbol{\tau})+S_e\}^{-(\nu+I-1)/2}\\ &\propto\{1+(I-1)F_t(\boldsymbol{\tau})/\nu\}^{-(\nu+I-1)/2}\end{aligned}$$

where

$$F_t(\boldsymbol{\tau})=\frac{S_t(\boldsymbol{\tau})/(I-1)}{S_e/\nu}.$$

Again, as in the last section, it can be shown that

$$F_t(\boldsymbol{\tau}) \sim \mathrm{F}_{I-1,\nu}$$

and so it is possible to conduct significance tests by Lindley's method and to find HDRs for $\boldsymbol{\tau}$.

Similarly, it can be shown that $F_b(\boldsymbol{\beta})$ (defined in the obvious way) is distributed as $F_b(\boldsymbol{\beta}) \sim \mathrm{F}_{J-1,\nu}$. Moreover, if

$$F_{tb}(\boldsymbol{\kappa}) = \frac{S_{tb}(\boldsymbol{\kappa})/(I-1)(J-1)}{S_e/\nu}$$

then $F_{tb}(\boldsymbol{\kappa}) \sim \mathrm{F}_{(I-1)(J-1),\nu}$. Thus it is also possible to investigate the blocks effect and the interaction. It should be noted that it would rarely make sense to believe in the existence of an interaction unless both main effects were there.

Analysis of variance

The analysis is helped by defining

$$\begin{aligned}
S_T &= \sum\sum(x_{ijk} - x_{...})^2 = \sum\sum\sum x_{ijk}{}^2 - C, \\
S_t &= S_t(\mathbf{0}) = JK\sum T_i^2 - C, \\
S_b &= S_b(\mathbf{0}) = IK\sum B_j^2 - C, \\
S_{tb} &= S_{tb}(\mathbf{0}) = K\sum C_{ij}{}^2 - C - S_t - S_b
\end{aligned}$$

in which T_i is the treatment total $JKx_{i..}$, while B_j is the block total $IKx_{.j.}$, and C_{ij} is the cell total $Kx_{ij.}$. Finally, G is the grand total $IJKx_{...}$, and C is the correction factor G^2/N.

ANOVA Table

Source	Sum of squares	Degrees of freedom	Mean square	Ratio
Treatments	S_t	$I-1$	$S_t/(I-1)$	$F_t(\mathbf{0})$
Blocks	S_b	$J-1$	$S_b/(J-1)$	$F_b(\mathbf{0})$
Interaction	S_{tb}	$(I-1)(J-1)$	$S_{tb}/(I-1)(J-1)$	$F_{tb}(\mathbf{0})$
Error	S_e	$\nu = IJ(K-1)$	$s^2 = S_e/\nu$	
Total	S_T	$N-1$		

Other matters, such as the exploration of treatments or blocks contrasts, can be pursued as in the case of the one way model.

6.7 The general linear model

Formulation of the general linear model

All of the last few sections have been concerned with particular cases of the so-called general linear model. It is possible to treat them all at once in an approach using matrix theory. In most of this book substantial use of matrix theory has been avoided, but if the reader has same knowledge of matrices this section may be helpful, in that the intention here is to put some of the models already considered into the form of the general linear model. An understanding of how these models can be put into such a framework will put the reader in a good position to approach the theory in its full generality, as it is dealt with in such works as Box and Tiao (1973).

It is important to distinguish row vectors from column vectors. We write $\boldsymbol{x}$ for a column vector and $\boldsymbol{x}^{\mathrm{T}}$ for its transpose; similarly, if A is an $n \times k$ matrix then $\boldsymbol{A}^{\mathrm{T}}$ is its $k \times n$ transpose. Consider a situation in which we have a column vector $\boldsymbol{x}$ of observations, so that $\boldsymbol{x} = (x_1, x_2, \ldots, x_n)^{\mathrm{T}}$ (the equation is written thus to save the excessive space taken up by column vectors). We suppose that the x_i are independently normally distributed with common variance φ and a vector $\mathsf{E}\boldsymbol{x}$ of means satisfying

$$\mathsf{E}\boldsymbol{x} = \boldsymbol{A\theta}$$

where $\boldsymbol{\theta} = (\theta_1, \theta_2, \ldots, \theta_k)^{\mathrm{T}}$ is a vector of unknown parameters and A is a known $n \times k$ matrix.

In the case of the original formulation of the bivariate linear regression model in which, conditional on x_i, we have $y_i \sim \mathrm{N}(\eta_0 + \eta_1 x_i, \varphi)$ than $\boldsymbol{y}$ takes the part of $\boldsymbol{x}$, $k = 2$, $\eta = (\eta_0, \eta_1)^{\mathrm{T}}$ takes the part of $\boldsymbol{\theta}$ and $\boldsymbol{A}_0$ takes the part of $\boldsymbol{A}$ where

$$\boldsymbol{A}_0 = \begin{pmatrix} 1 & x_1 \\ 1 & x_2 \\ \cdots & \\ 1 & x_n \end{pmatrix} \qquad \boldsymbol{A}_0\eta = \begin{pmatrix} \eta_0 + \eta_1 x_1 \\ \eta_0 + \eta_1 x_2 \\ \cdots\cdots \\ \eta_0 + \eta_1 x_n \end{pmatrix}$$

This model is reformulated in terms of $\boldsymbol{\beta} = (\alpha, \beta)^{\mathrm{T}}$ and

$$\boldsymbol{A} = \begin{pmatrix} 1 & x_1 - \bar{x} \\ 1 & x_2 - \bar{x} \\ \cdots & \\ 1 & x_n - \bar{x} \end{pmatrix} \qquad \boldsymbol{A\beta} = \begin{pmatrix} \alpha + \beta(x_1 - \bar{x}) \\ \alpha + \beta(x_2 - \bar{x}) \\ \cdots\cdots \\ \alpha + \beta(x_n - \bar{x}) \end{pmatrix}.$$

In the case of the one way model (where, for simplicity, we restrict ourselves to the case where $K_i = K$ for all i) $n = N$, $k = I$,

$\boldsymbol{\theta} = (\theta_1, \theta_2, \ldots, \theta_I)^{\mathrm{T}}$ and

$$\boldsymbol{x} = \begin{pmatrix} x_{11} \\ x_{12} \\ \cdots \\ x_{1K} \\ x_{21} \\ x_{22} \\ \cdots \\ x_{IK} \end{pmatrix} \quad \boldsymbol{A} = \begin{pmatrix} 1 & 0 & \ldots & 0 \\ 1 & 0 & \ldots & 0 \\ . & . & . & . \\ 1 & 0 & \ldots & 0 \\ 0 & 1 & \ldots & 0 \\ 0 & 1 & \ldots & 0 \\ . & . & . & . \\ 0 & 0 & \ldots & 1 \end{pmatrix} \quad \boldsymbol{A\theta} = \begin{pmatrix} \theta_1 \\ \theta_1 \\ \cdots \\ \theta_1 \\ \theta_2 \\ \theta_2 \\ \cdots \\ \theta_I \end{pmatrix}$$

The two way layout can be expressed similarly using a matrix of 0s and 1s. It is also possible to write the multiple regression model

$$y_i \sim \mathrm{N}(\eta_0 + x_{i1}\eta_1 + \ldots + x_{ik}\eta_k, \varphi)$$

(the x_{ij} being treated as known) as a case of the general linear model.

Derivation of the posterior

Noting that for any vector $\boldsymbol{u}$ we have $\sum u_i^2 = \boldsymbol{u}^{\mathrm{T}}\boldsymbol{u}$, we can write the likelihood function for the general linear model in the form

$$(2\pi\varphi)^{-n/2} \exp\{-\tfrac{1}{2}(\boldsymbol{x} - \boldsymbol{A\theta})^{\mathrm{T}}(\boldsymbol{x} - \boldsymbol{A\theta})\}.$$

Taking standard reference priors, that is, $p(\theta, \varphi) \propto 1/\varphi$, the posterior is

$$p(\boldsymbol{\theta}, \varphi \,|\, \boldsymbol{x}) \propto \varphi^{-n/2-1} \exp\{-\tfrac{1}{2}(\boldsymbol{x} - \boldsymbol{A\theta})^{\mathrm{T}}(\boldsymbol{x} - \boldsymbol{A\theta})\}.$$

Now as $(\boldsymbol{AB})^{\mathrm{T}} = \boldsymbol{B}^{\mathrm{T}}\boldsymbol{A}^{\mathrm{T}}$ and scalars equal their own transposes

$$(\boldsymbol{x} - \boldsymbol{A\theta})^{\mathrm{T}}(\boldsymbol{x} - \boldsymbol{A\theta}) = \boldsymbol{\theta}^{\mathrm{T}}\boldsymbol{A}^{\mathrm{T}}\boldsymbol{A\theta} - 2\boldsymbol{\theta}^{\mathrm{T}}\boldsymbol{A}^{\mathrm{T}}\boldsymbol{x} + \boldsymbol{x}^{\mathrm{T}}\boldsymbol{x}$$

so that if $\hat{\boldsymbol{\theta}}$ is such that

$$\boldsymbol{A}^{\mathrm{T}}\boldsymbol{A}\hat{\boldsymbol{\theta}} = \boldsymbol{A}^{\mathrm{T}}\boldsymbol{x}$$

(so that $\boldsymbol{\theta}^{\mathrm{T}}\boldsymbol{A}^{\mathrm{T}}\boldsymbol{A}\hat{\boldsymbol{\theta}} = \boldsymbol{\theta}^{\mathrm{T}}\boldsymbol{A}^{\mathrm{T}}\boldsymbol{x}$), that is, assuming $\boldsymbol{A}^{\mathrm{T}}\boldsymbol{A}$ is non-singular,

$$\hat{\boldsymbol{\theta}} = (\boldsymbol{A}^{\mathrm{T}}\boldsymbol{A})^{-1}\boldsymbol{A}^{\mathrm{T}}\boldsymbol{x}$$

we have

$$\begin{aligned} S = (\boldsymbol{x} - \boldsymbol{A\theta})^{\mathrm{T}}(\boldsymbol{x} - \boldsymbol{A\theta}) &= \boldsymbol{\theta}^{\mathrm{T}}\boldsymbol{A}^{\mathrm{T}}\boldsymbol{A\theta} - 2\boldsymbol{\theta}^{\mathrm{T}}\boldsymbol{A}^{\mathrm{T}}\boldsymbol{x} + \boldsymbol{x}^{\mathrm{T}}\boldsymbol{x} \\ &= (\boldsymbol{\theta} - \hat{\boldsymbol{\theta}})^{\mathrm{T}}\boldsymbol{A}^{\mathrm{T}}\boldsymbol{A}(\boldsymbol{\theta} - \hat{\boldsymbol{\theta}}) + \boldsymbol{x}^{\mathrm{T}}\boldsymbol{x} - \hat{\boldsymbol{\theta}}^{\mathrm{T}}\boldsymbol{A}^{\mathrm{T}}\boldsymbol{A}\hat{\boldsymbol{\theta}} \\ &= S_t(\boldsymbol{\theta}) + S_e \end{aligned}$$

where

$$S_t(\boldsymbol{\theta}) = (\boldsymbol{\theta} - \hat{\boldsymbol{\theta}})^{\mathrm{T}}\boldsymbol{A}^{\mathrm{T}}\boldsymbol{A}(\boldsymbol{\theta} - \hat{\boldsymbol{\theta}}), \qquad S_e = \boldsymbol{x}^{\mathrm{T}}\boldsymbol{x} - \hat{\boldsymbol{\theta}}^{\mathrm{T}}\boldsymbol{A}^{\mathrm{T}}\boldsymbol{A}\hat{\boldsymbol{\theta}}.$$

It is also useful to define

$$\nu = n - k, \qquad s^2 = S_e/\nu.$$

Because $S_t(\boldsymbol{\theta})$ is of the form $\boldsymbol{u}^{\mathrm{T}}\boldsymbol{u}$, it is always non-negative, and it clearly vanishes if $\boldsymbol{\theta} = \hat{\boldsymbol{\theta}}$. Further, S_e is the minimum value of the sum of squares S and so is positive. It is sometimes worth noting that

$$S_e = (\boldsymbol{x} - \boldsymbol{A}\hat{\boldsymbol{\theta}})^{\mathrm{T}}(\boldsymbol{x} - \boldsymbol{A}\hat{\boldsymbol{\theta}})$$

as is easily shown.

It follows that the posterior can be written as

$$p(\boldsymbol{\theta}, \varphi \,|\, \boldsymbol{x}) \propto \varphi^{-n/2-1} \exp\left[-\tfrac{1}{2}\{S_t(\boldsymbol{\theta}) + S_e\}/\varphi\right].$$

In fact this means that for given φ the vector $\boldsymbol{\theta}$ has a multivariate normal distribution of mean $\boldsymbol{\theta}$ and variance–covariance matrix $\boldsymbol{A}^{\mathrm{T}}\boldsymbol{A}$.

If we are now interested in $\boldsymbol{\theta}$ as a whole we can integrate with respect to φ to obtain

$$\begin{aligned} p(\boldsymbol{\theta} \,|\, \boldsymbol{x}) &\propto \{S_t(\boldsymbol{\theta}) + S_e\}^{-n/2} \\ &\propto \{1 + kF(\boldsymbol{\theta})/\boldsymbol{\nu}\}^{-n/2} \end{aligned}$$

where

$$F(\boldsymbol{\theta}) = \frac{S_t(\boldsymbol{\theta})/k}{S_e/\nu} = \frac{S_t(\boldsymbol{\theta})/k}{s^2}.$$

It may also be noted that the set

$$\mathrm{E}(F) = \{\boldsymbol{\theta};\, F(\boldsymbol{\theta}) \leqslant F\}$$

is a hyperellipsoid in k-dimensional space in which the length of each of the axes is in a constant ratio to $\sqrt{F}$. The argument of Section 6.5 on the one way layout can now be adapted to show that $F(\boldsymbol{\theta}) \sim \mathrm{F}_{k,\nu}$, so that $\mathrm{E}(F)$ is an HDR for $\boldsymbol{\theta}$ of probability p if F is the appropriate percentage point of $\mathrm{F}_{k,\nu}$.

Inference for a subset of the parameters

However, it is often the case that most of the interest centres on a subset of the parameters, say on $\theta_1, \theta_2, \ldots, \theta_j$. If so, then it is convenient to write $\boldsymbol{\theta}' = (\theta_1, \theta_2, \ldots, \theta_j)$ and $\boldsymbol{\theta}'' = (\theta_{j+1}, \ldots, \theta_k)$. If it happens that $S_t(\boldsymbol{\theta})$ splits into a sum

$$S_t(\boldsymbol{\theta}) = S_t'(\boldsymbol{\theta}') + S_t''(\boldsymbol{\theta}'')$$

then it is easy to integrate

$$p(\boldsymbol{\theta}', \boldsymbol{\theta}'', \varphi \,|\, \boldsymbol{x}) \propto \varphi^{-n/2-1} \exp\left[-\tfrac{1}{2}\{S_t'(\boldsymbol{\theta}') + S_t''(\boldsymbol{\theta}'') + S_e\}/\varphi\right]$$

to obtain

$$p(\boldsymbol{\theta}', \varphi \mid \boldsymbol{x}) \propto \varphi^{-(n-k+j)/2-1} \exp\left[-\tfrac{1}{2}\{S_t'(\boldsymbol{\theta}') + S_e\}/\varphi\right]$$

and thus as $n-k=\nu$

$$\begin{aligned} p(\boldsymbol{\theta}' \mid \boldsymbol{x}) &\propto \{S_t'(\boldsymbol{\theta}') + S_e\}^{-(j+\nu)/2} \\ &\propto \{1 + jF_t'(\boldsymbol{\theta}')/\nu\}^{-(j+\nu)/2} \end{aligned}$$

where

$$F'(\boldsymbol{\theta}') = \frac{S_t'(\boldsymbol{\theta})/j}{S_e/\nu}.$$

It is now easy to show that $F'(\boldsymbol{\theta}') \sim \mathrm{F}_{j,\nu}$ and hence to make inferences for $\boldsymbol{\theta}'$.

Unfortunately, the quadratic form $S_t(\boldsymbol{\theta})$ does in general contain terms $\theta_r\theta_s$ where $r \leqslant j$ but $s > j$ and hence it does not in general split into $S_t'(\boldsymbol{\theta}) + S_t''(\boldsymbol{\theta}'')$. Without going into details, we can say that by completing squares it will always be possible to express $S_t(\boldsymbol{\theta})$ in the form

$$S_t(\boldsymbol{\theta}) = S_t'(\boldsymbol{\theta}) + S_t''(\boldsymbol{\theta}'' - \boldsymbol{B}\boldsymbol{\theta}')$$

for a suitable matrix $\boldsymbol{B}$ and then proceed much as above.

Application to bivariate linear regression

The theory can be illustrated by considering the simple linear regression model. Consider first the reformulated version in terms of $\boldsymbol{A}$ and $\boldsymbol{\beta} = (\alpha, \beta)^{\mathrm{T}}$. In this case

$$\boldsymbol{A}^{\mathrm{T}}\boldsymbol{A} = \begin{pmatrix} n & 0 \\ 0 & S_{xx} \end{pmatrix}$$

and the fact that this matrix is easy to invert is one of the underlying reasons why this reformulation was sensible. Also

$$\hat{\boldsymbol{\theta}} = (\boldsymbol{A}^{\mathrm{T}}\boldsymbol{A})^{-1}\boldsymbol{A}^{\mathrm{T}}\boldsymbol{y} = \begin{pmatrix} \bar{y} \\ S_{xy}/S_{xx} \end{pmatrix},$$

$$S_t(\boldsymbol{\theta}) = (\boldsymbol{\theta} - \hat{\boldsymbol{\theta}})^{\mathrm{T}}\boldsymbol{A}^{\mathrm{T}}\boldsymbol{A}(\boldsymbol{\theta} - \hat{\boldsymbol{\theta}}) = n(\alpha - \bar{y})^2 + S_{xx}(\beta - S_{xy}/S_{xx})^2,$$

$$S_e = \mathbf{y}^{\mathrm{T}}\mathbf{y} - \hat{\boldsymbol{\theta}}^{\mathrm{T}}\boldsymbol{A}^{\mathrm{T}}\boldsymbol{A}\hat{\boldsymbol{\theta}} = \sum y_i^2 - n\bar{y}^2 - S_{xx}(S_{xy}/S_{xx})^2 = S_{yy} - S_{xy}{}^2/S_{xx}$$

so that S_e is what was denoted S_{ee} in Section 6.4 on bivariate linear regression.

If we are particularly interested in α, then in this case we should note that the quadratic form splits with $S_t'(\boldsymbol{\theta}) = n(\alpha - \bar{y})^2$ and $S_t''(\boldsymbol{\theta}'') =$

$S_{xx}(\beta - S_{xy}/S_{xx})^2$. Consequently, the posterior distribution of α is given by

$$\frac{\nu(\alpha - \bar{y})^2}{s^2} = \frac{S_t{}'(\boldsymbol{\theta}')/1}{S_e/(n-2)} \sim F_{1,n-2}.$$

Since the square of a t_{n-2} variable can easily be shown to have an $\mathrm{F}_{1,n-2}$ distribution, this conclusion is equivalent to that of Section 6.4.

The greater difficulties that arise when $\boldsymbol{A}^{\mathrm{T}}\boldsymbol{A}$ is non-diagonal can be seen by following the same process through for the original formulation of the bivariate linear regression model in terms of $\boldsymbol{A_0}$ and $\boldsymbol{\eta} = (\eta_0, \eta_1)^{\mathrm{T}}$. In this case it is easy to find the posterior distribution of η_1, but it involves some rearrangement to obtain that of η_0.

Exercises on Chapter 6

1. The sample correlation coefficient between length and weight of a species of frog was determined at each of a number of sites. The results were as follows:

Site	1	2	3	4	5
Number of frogs	12	45	23	19	30
Correlation	0.631	0.712	0.445	0.696	0.535

Find an interval in which you are 95% sure that the correlation coefficient lies.

2. Three groups of children were given two tests. The numbers of children and the sample correlation coefficients between the two test scores in each group were as follows:

Number of children	45	34	49
Correlation	0.489	0.545	0.601

Is there any evidence that the association between the two tests differs in the three groups?

3. From the approximation

$$p(\rho\,|\,\boldsymbol{x}, \boldsymbol{y}) \propto (1-\rho^2)^{n/2}(1-\rho r)^{-n}$$

which holds for large n, deduce an expression for the log-likelihood

$L(\rho|\boldsymbol{x}, \boldsymbol{y})$ and hence show that the maximum likelihood occurs when $\rho = r$. An approximation to the information can now be made by replacing r by ρ in the second derivative of the likelihood, since ρ is near r with high probability. Show that this approximation suggests a prior density of the form

$$p(\rho) \propto (1 - \rho^2)^{-1}.$$

4. Use the fact that

$$\int_0^\infty (\cosh t + \cos\theta)^{-1}\,\mathrm{d}t = \theta/\sin\theta$$

(cf. Edwards, 1921, article 180) to show that

$$p(\rho|\boldsymbol{x}, \boldsymbol{y}) \propto p(\rho)(1-\rho)^{(n-1)/2}\frac{\mathrm{d}^{n-2}}{\mathrm{d}(\rho r)^{n-2}}\left(\frac{\operatorname{arc}\cos(-\rho r)}{\sqrt{(1-\rho^2 r^2)}}\right).$$

5. Show that in the special case where the sample correlation coefficient $r = 0$ and the prior takes the special form $p(\rho) \propto (1-\rho^2)^k$ the variable

$$\sqrt{(k+n+1)}\rho/\sqrt{(1-\rho^2)}$$

has a Student's t distribution on $k+n+1$ degrees of freedom.

6. By writing

$$\omega^{-1}(\omega+\omega^{-1}-2\rho r)^{-(n-1)} = \omega^{n-2}(1-\rho^2)^{-(n-1)} \times [1+(\omega-\rho r)^2(1-\rho^2 r^2)^{-1}]^{-(n-1)}$$

and using repeated integration by parts, show that the posterior distribution of ρ can be expressed as a finite series involving powers of

$$\sqrt{\{(1-\rho r)/(1+\rho r)\}}$$

and Student's t integrals (cf. Box and Tiao (1973, Section 8.5)).

7. By substituting

$$\cosh t - \rho r = \frac{1-\rho r}{1-u}$$

in the form

$$p(\rho|\boldsymbol{X}, \boldsymbol{Y}) \propto p(\rho)(1-\rho^2)^{(n-1)/2}\int_0^\infty (\cosh t - \rho r)^{-(n-1)}\,\mathrm{d}t$$

for the posterior density of the correlation coefficient and then expanding

$$[1-\tfrac{1}{2}(1+\rho r)u]^{-\frac{1}{2}}$$

as a power series in u, show that the integral can be expressed as a series of beta functions. Hence deduce that

$$p(\rho\,|\,X, Y) \propto p(\rho)(1-\rho^2)^{(n-1)/2}(1-\rho r)^{-n+(3/2)}S_n(\rho r)$$

where

$$S_n(\rho r) = 1 + \sum_{l=1}^{\infty} \frac{1}{l!}\left(\frac{1+\rho r}{8}\right)^l \prod_{s=1}^{l} \frac{(2s-1)^2}{(n-(3/2)+s)}.$$

8. Fill in the details of the derivation of the prior

$$p(\varphi, \psi, \rho) \propto (\varphi\psi)^{-1}(1-\rho^2)^{-3/2}$$

from Jeffreys' rule as outlined at the end of Section 6.1.

9. The following data consist of the estimated gestational ages (in weeks) and weights (in grammes) of 12 female babies:

Age	40	36	40	38	42	39	40	37	36	38	39	40
Weight	3317	2729	2935	2754	3210	2817	3126	2539	2412	2991	2875	3231

Give an interval in which you are 90% sure that the gestational age of a particular baby will lie if its weight is 3000 grammes, and give a similar interval in which the mean weight of all such babies lies.

10. Show directly from the definitions that, in the notation of Section 6.3,

$$S_{ee} = \sum\{y_i - a - b(x_i - \bar{x})\}^2.$$

11. Observations y_i for $i = -m, -m+1, \ldots, m$ are available which satisfy the regression model

$$y_i \sim \mathrm{N}(\alpha + \beta u_i + \gamma v_i, \varphi)$$

where $u_i = i$ and $v_i = i^2 - \frac{1}{3}m(m+1)$. Adopting the standard reference prior $p(\alpha, \beta, \gamma, \varphi) \propto 1/\varphi$, show that the posterior distribution of α is such that

$$\frac{\alpha - \bar{y}}{s/\sqrt{n}} \sim \mathrm{t}_{n-3}$$

where $n = 2m + 1$, $s^2 = S_{ee}/(n-3)$ and

$$S_{ee} = S_{yy} - S_{uy}^2/S_{uu} - S_{vy}^2/S_{vv}$$

in which S_{yy}, S_{uy}, etc., are defined by

$$S_{yy} = \sum (y_i - \bar{y})^2, \quad S_{uy} = \sum (u_i - \bar{u})(y_i - \bar{y}).$$

[*Hint:* Note that $\sum u_i = \sum v_i = \sum u_i v_i = 0$, and hence $\bar{u} = \bar{v} = 0$ and $S_{uv} = 0$.]

12. Fisher (1925b, Section 41) quotes an experiment on the accuracy of counting soil bacteria. In it, a soil sample was divided into four parallel samples, and from each of these after dilution seven plates were inoculated. The number of colonies on each plate is shown below. Do the results from the four samples agree within the limits of random sampling?

Plate \ Sample	A	B	C	D
1	72	74	78	69
2	69	72	74	67
3	63	70	70	66
4	59	69	58	64
5	59	66	58	64
6	53	58	56	58
7	51	52	56	54

7

Other Topics

7.1 The likelihood principle

Introduction

This section would logically come much earlier in the book than it is placed, but it is important to have some examples of Bayesian procedures firmly in place before considering this material. The basic result is due to Birnbaum (1962), and a more detailed consideration of these issues can be found in Berger and Wolpert (1984).

The nub of the argument here is that in drawing any conclusion from an experiment only the actual observation x made (and not the other possible outcomes that might have occurred) is relevant. This is in contrast to methods by which, for example, a null hypothesis is rejected because the probability of a value as large as *or larger than* that actually observed is small, an approach which leads to Jeffreys' criticism that was mentioned in Section 4.1 when we first considered hypothesis tests, namely, that "a hypothesis that may be true may be rejected because it has not predicted observable results that have not occurred". Virtually all of the ideas discussed in this book abide by this principle, which is known as the *likelihood principle* (there are some exceptions, for example Jeffreys' rule is not in accordance with it). We shall show that it follows from two other principles, called the conditionality principle and the sufficiency principle, both of which are hard to argue against.

In this section we write x for a particular piece of data, not necessarily one-dimensional, the density $p(x \mid \theta)$ of which depends on an unknown parameter θ. For simplicity, we suppose that x and θ are discrete (although they may be more than one-dimensional). The triple $E = \{\tilde{x}, \theta, p(x \mid \theta)\}$ represents the essential features of an experiment to gain information about θ, and accordingly we refer to E as an experiment. Note that the random variable $\tilde{x}$ is a feature of the experiment, not the particular value x that may be observed when the experiment is carried out on a particular occasion. If such an experiment is carried out and the value x is observed, then we write

$\text{Ev}\{E, x, \theta\}$ for the evidence provided by carrying out experiment E and observing the value x about the value of θ. This "evidence" is not presumed to be in any particular form. To a Bayesian it would normally be the posterior distribution of θ or some feature of it, but for the moment we are not restricting ourselves to Bayesian inference, and a classical statistician might consider evidence to be made up of significance levels and confidence intervals, etc., while the notation does not rule out some form of evidence that would be new to both.

For example, we might be interested in the proportion θ of defective articles coming from a factory. A possible experiment E would consist in observing a fixed number n of articles chosen at random and observing the number x of defective articles, so that $p(x|\theta)$ is a family of binomial densities. To have a definite experiment, it is necessary to fix the value of n, so that, for example, $n = 100$; once n is known E is fully determined. If we then observe that $x = 3$, then $\text{Ev}\{E, 3, \theta\}$ denotes the conclusions we arrive at about the value of θ.

The conditionality principle

The conditionality principle can be explained as the assertion that if you have decided which of two experiments you performed by tossing a coin, then once you tell me the end-result of the experiment, it will not make any difference to any inferences I make about an unknown parameter θ whether or not I know which way the coin landed and hence which way the coin landed (assuming that the probability of the coin's landing "heads" does not in any way depend on θ). For example, if we are told that an analyst has reported on the chemical composition of a sample, then it is irrelevant whether we had always intended to ask him or her to analyse the sample or had tossed a coin to decide whether to ask that scientist or the one in the laboratory next door to analyse it. Put this way, the principle should seem plausible, and we shall now try to formalize it.

We first need to define a *mixed* experiment. Suppose that there are two experiments, $E_1 = \{\tilde{y}, \theta, p(y|\theta)\}$ and $E_2 = \{\tilde{z}, \theta, p(z|\theta)\}$ and that the random variable $\tilde{k}$ is such that $p(k = 1) = p(k = 2) = \frac{1}{2}$, whatever θ is and independently of y and z. Then the mixed experiment E^* consists of carrying out E_1 if $k = 1$ and E_2 if $k = 2$. It can also be defined as the triple $\{\tilde{x}, \theta, p(x|\theta)\}$ where

$$x = \begin{cases} (1, \ y) & \text{if } k = 1 \\ (2, \ z) & \text{if } k = 2 \end{cases}$$

and

$$p(x \mid \theta) = \begin{cases} \frac{1}{2}p(y \mid \theta) & \text{if } k = 1, \text{ so } x = (1, y) \\ \frac{1}{2}p(z \mid \theta) & \text{if } k = 2, \text{ so } x = (2, z) \end{cases}$$

We only need to assume the following rather weak form of the principle:

Weak conditionality principle. If E_1, E_2 and E^* are as defined above, then

$$\text{Ev}\{E^*, x, \theta\} = \begin{cases} \text{Ev}\{E_1, y, \theta\} & \text{if } k = 1, \text{ so } x = (1, y) \\ \text{Ev}\{E_2, z, \theta\} & \text{if } k = 2, \text{ so } x = (2, z) \end{cases}$$

that is, the evidence about θ from E^* is just the evidence from the experiment actually performed.

The sufficiency principle

The *sufficiency principle* says that if $t(x)$ is sufficient for θ given x, then any inference we may make about θ may be based on the value of t, and once we know that we have no need of the value of x. We have already seen in Section 2.9 that Bayesian inference satisfies the sufficiency principle. The form in which the sufficiency principle will be used in this section is as follows.

Weak sufficiency principle. We consider the experiment $E = \{\tilde{x}, \theta, p(x \mid \theta)\}$ and suppose that $t = t(x)$ is sufficient for θ given x. Then if $t(x_1) = t(x_2)$

$$\text{Ev}\{E, x_1, \theta\} = \text{Ev}\{E, x_2, \theta\}.$$

The likelihood principle

For the moment, we shall state what the likelihood principle is—its implications will be explored later.

Likelihood principle. Consider two experiments $E_1 = \{\tilde{y}, \theta, p(y \mid \theta)\}$ and $E_2 = \{\tilde{z}, \theta, p(z \mid \theta)\}$ where θ is the same quantity in each experiment. Suppose that there are particular possible outcomes y^* of experiment E_1 and z^* of E_2 such that

$$p(y^* \mid \theta) = cp(z^* \mid \theta)$$

for some constant c, that is, the likelihoods of θ as given by these possible outcomes of the two experiments are proportional, so that

$$l(\theta \mid y^*) \propto l(\theta \mid z^*).$$

Then

$$\mathrm{Ev}\{E_1, y^*, \theta\} = \mathrm{Ev}\{E_2, z^*, \theta\}.$$

The following theorem [due to Birnbaum (1962)] shows that the likelihood principle follows from the other two principles described above.

Theorem. The likelihood principle follows from the weak conditionality principle and the weak sufficiency principle.

Proof. If E_1 and E_2 are the two experiments about θ figuring in the statement of the likelihood principle, consider the mixed experiment E^* which arose in connection with the weak conditionality principle. We define a statistic t by

$$t = t(x) = \begin{cases} (1,\ y^*) & \text{if } k = 2 \text{ and } z = z^* \\ x & \text{otherwise.} \end{cases}$$

We now note that if $t \neq (1, y^*)$ then

$$p(x \mid t, \theta) = \begin{cases} 1 & \text{if } t = t(x) \\ 0 & \text{otherwise} \end{cases}$$

whereas if $t = x = (1, y^*)$ then

$$p(x \mid t, \theta) = \frac{\frac{1}{2}p(y^* \mid \theta)}{\frac{1}{2}p(y^* \mid \theta) + \frac{1}{2}p(z^* \mid \theta)} = \frac{c}{1+c}$$

and if $t = (1, y^*)$ but $x = (2, z^*)$ then

$$p(x \mid t, \theta) = \frac{\frac{1}{2}p(z^* \mid \theta)}{\frac{1}{2}p(y^* \mid \theta) + \frac{1}{2}p(z^* \mid \theta)} = \frac{1}{1+c}$$

while for $t = (1, y^*)$ and all other x we have $p(x \mid t, \theta) = 0$. In no case does $p(x \mid t, \theta)$ depend on θ and hence, from the definition given when sufficiency was first introduced in Section 2.9, t is sufficient for θ given x. It follows from the weak sufficiency principle that $\mathrm{Ev}\{E^*, (1, y^*), \theta\} = \mathrm{Ev}\{E^*, (2, z^*), \theta\}$. But the weak conditionality principle now ensures that

$$\begin{aligned} \mathrm{Ev}\{E_1, y^*, \theta\} &= \mathrm{Ev}\{E^*, (1, y^*), \theta\} = \mathrm{Ev}\{E^*, (2, z^*), \theta\} \\ &= \mathrm{Ev}\{E_2, z^*, \theta\} \end{aligned}$$

establishing the likelihood principle.

Corollary. If $E = \{\tilde{x}, \theta, p(x \mid \theta)\}$ is an experiment, then $\mathrm{Ev}\{E, x, \theta\}$ should depend on E and x only through the likelihood

$$l(\theta \mid x) \propto p(x \mid \theta).$$

Proof. For any one particular value x_1 of x, we define

$$y = \begin{cases} 1 & \text{if } x = x_1 \\ 0 & \text{otherwise} \end{cases}$$

so that $\mathrm{P}(y = 1 \mid \theta) = p(x_1 \mid \theta)$ (since we have assumed for simplicity that everything is discrete this will not, in general, be zero). Now we let the experiment E_1 consist simply of observing y, that is, of noting whether or not $x = x_1$. Then the likelihood principle ensures that $Ev\{E, x_1, \theta\} = \mathrm{Ev}\{E_1, 1, \theta\}$, and E_1 depends solely on $p(x, \mid \theta)$ and hence solely on the likelihood of the observation actually made.

Converse. If the likelihood principle holds, then so do the weak conditionality principle and the weak sufficiency principle.

Proof. Using the notation introduced earlier for the mixed experiment, we see that if $x = (1, y)$ then

$$p(x \mid \theta) = \tfrac{1}{2} p(y \mid \theta)$$

and so by the likelihood principle $\mathrm{Ev}\{E^*, x, \theta\} = \mathrm{Ev}\{E_1, y, \theta\}$, implying the weak conditionality principle. Moreover, if t is a sufficient statistic and $t(x_1) = t(x_2)$, then x_1 and x_2 have proportional likelihood functions, so that the likelihood principle implies the weak sufficiency principle.

Discussion

From the formulation of Bayesian inference as "posterior is proportional to prior times likelihood", it should be clear that Bayesian inference obeys the likelihood principle. It is not logically necessary that if you find the arguments for the likelihood principle convincing, you have to accept Bayesian inference, and there are some authors, for example, Edwards (1972), who have argued for a non-Bayesian form of inference based on the likelihood. Nevertheless, I think that Savage was right in saying in the discussion on Birnbaum (1962) that "... I suspect that once the likelihood principle is widely recognized, people will not long stop at that halfway house but will go forward and accept the implications of personalistic probability for statistics."

Conversely, much of classical statistics notably fails to obey the likelihood principle—any use of tail areas (for example, the probability of observing a value as large as that seen *or greater*) evidently involves matters other than the likelihood of the observations actually made. Another quotation from Savage, this time from Savage *et al.* (1962), may help to point to some of the difficulties that arise in connection with confidence intervals. "Imagine, for example, that two Meccans carefully drawn at random differ from each other in height by only 0.01 mm. Would you offer 19 to 1 odds that the standard deviation of the height of Meccans is less than 1.13 mm? That is the 95 per cent upper confidence limit computed with one degree of freedom. No, I think you would not have enough confidence in that limit to offer odds of 1 to 1".

In fact, the likelihood principle has serious consequences for both classical and Bayesian statisticians and some of these consequences will be discussed in the next three sections. For classical statisticians one of the most serious is the stopping rule principle, while for Bayesians one of the most serious is that Jeffreys' rule for finding reference priors is incompatible with the likelihood principle.

7.2 The stopping rule principle

Definitions

We shall restrict ourselves to a simple situation, but it is possible to generalize the following account considerably; see Berger and Wolpert (1984, Section 4.2). Basically, in this section we will consider a sequence of experiments which can be terminated at any stage in accordance with a rule devised by the experimenter (or forced upon him or her).

Suppose that the observations $x_1, x_2, \ldots$ are independently and identically distributed with density $p(x \mid \theta)$ and let

$$\boldsymbol{x}^{(m)} = (x_1, x_2, \ldots, x_m)$$
$$\bar{x}_m = (x_1 + x_2 + \ldots + x_m)/m.$$

We say that s is a *stopping time* if it is a random variable whose values are finite natural numbers (1, 2, 3, ...) with probability one, and is such that whether or not $s > m$ depends solely on $\boldsymbol{x}^{(m)}$. In a sequential experiment E we observe the values $x_1, x_2, \ldots, x_s$, where s is such a stopping rule, and then stop. The restriction on the distribution of s

means simply that whether or not we decide to stop cannot depend on future observations (unless we are clairvoyant), but only on the ones we have available to date.

Examples

(1) Fixed sample size from a sequence of Bernoulli trials. We suppose that the x_i are independently 1 with probability π (representing "success") or 0 with probability $1-\pi$ (representing "failure"). If $s = n$ where n is a constant, then we have the usual situation which gives rise to the binomial distribution for the total number of successes.

(2) Stopping after the first success in Bernoulli trials. With the x_i as in the previous example, we could stop after the first success, so that

$$s = \min\{m; x_m = 1\}.$$

Because the probability that $s > n$ is $(1-\pi)^n$, which tends to 0 as $n \to \infty$, this is finite with probability 1.

(3) A compromise between the first two examples. With the x_i as in the previous two examples, we could stop after the first success if that occurs at or before the nth trial, but if there has not been a success by then, stop at the nth trial, so that

$$s = \min\{n, \min\{m; x_m = 1\}\}.$$

(4) Fixed size sample from the normal distribution. If the x_i are independently $N(0, 1)$ and $s = n$ where n is a constant, then we have a case which has arisen often before of a sample of fixed size from a normal distribution.

(5) Stopping when a fixed number of standard deviations from the mean. Still taking $x_i \sim N(0, 1)$, we could have

$$s = \min\{m; |\bar{x}_m| > c/\sqrt{m}\}$$

which, as $\bar{x}_m \sim N(0, 1/m)$, means stopping as soon as we observe a value of $\bar{x}_m$, that is at least c standard deviations from the mean. It is not obvious in this case that s is finite with probability 1, but it follows from the law of the iterated logarithm, a proof of which can be found in any standard text on probability.

The stopping rule principle

The *stopping rule principle* is that in a sequential experiment, if the observed value of the stopping time is m, then the evidence $\mathrm{Ev}\{E, \boldsymbol{x}^{(m)}, \theta\}$ provided by the experiment about the value of θ should not depend on the stopping rule.

Before deciding whether it is valid, we must consider what it means. It asserts, for example, that if we observe 10 Bernoulli trials, nine of which result in failure and only the last in success, then any inference about the probability π of successes cannot depend on whether the experimenter had all along intended to carry out 10 trials and had, in fact, observed one success, or whether he or she had intended to stop the experiment immediately the first success was observed. Thus it amounts to an assertion that all that matters is what actually happened and not the intentions of the experimenter if something else had happened.

Theorem. The stopping rule principle follows from the likelihood principle, and hence is a logical consequence of the Bayesian approach.

Proof. If the x_i are discrete random variables, then it suffices to note that the likelihood.

$$l(\theta \mid \boldsymbol{x}^{(s)}) \propto p(x_1 \mid \theta) p(x_2 \mid \theta) \ldots p(x_s \mid \theta)$$

which clearly does not depend on the stopping rule. There are some slight complications in the continuous case, which are largely to do with measure theoretic complications, and in particular with events of probability zero, but a general proof from the so-called relative likelihood principle is more or less convincing; for details, see Berger and Wolpert (1984, Sections 3.4.3, 4.2.6).

Discussion

The key point about stopping rules is as follows. A classical statistician is supposed to choose the stopping rule *before* the experiment and then follow it exactly. In actual practice, the ideal is often not adhered to; an experiment can end because the data already look good enough, or because there is no more time or money, and yet the experiment is often analysed *as if* it had a fixed sample size. Although stopping for some reasons would be harmless, statisticians who stop "when the data look good", a process which is sometimes described as *optional* (or *optimal*) stopping, can produce serious errors if they use a classical analysis.

It is often argued that a single number which is a good representation of our knowledge of a parameter should be unbiased, that is, should be such that its expectation over repeated sampling should be equal to that parameter. Thus if we have a sample of fixed size from a Bernoulli distribution [example (1) above], then $\mathsf{E}\bar{x}_n = \pi$, so that $\bar{x}_n$ is in that sense a good estimator of π. However, if the stopping rule in example (2) or that in example (3), is used, then the proportion $\bar{x}_s$ will, on average, be more than π. If, for instance, we take example (3) with $n = 2$, then

$$\begin{aligned}\mathsf{E}\bar{x}_s &= \mathrm{P}(x_1 = 1)\mathsf{E}(\bar{x}_1 | x_1 = 1, \pi) + \mathrm{P}(x_1 = 0 | \pi)\mathsf{E}(\bar{x}_2 | x_1 = 0, \pi) \\ &= \pi \times 1 + (1-\pi) \times \{\tfrac{1}{2} \times \pi + 0 \times (1-\pi)\} \\ &= \pi + \tfrac{1}{2}\pi(1-\pi).\end{aligned}$$

Thus a classical statistician who used the proportion of successes actually observed as an estimator of the probability π of success, would be accused of "making the probability of success look larger than it is".

The stopping rule principle also plays havoc with classical significance tests. A particular case can be constructed from example (5) above with, for example, $c = 2$. If a classical statistician were to consider data from an $\mathrm{N}(\theta, 1)$ population in which (unknown to him or her) $\theta = 0$, then because s is so constructed that, necessarily, the value of $|\bar{x}_s|$ is at least c standard deviations from the mean, a single sample of a fixed size equal to this would necessarily lead to a rejection of the null hypothesis that $\theta = 0$ at the 5% level. By taking other values of c, it can be seen that a crafty classical statistician could arrange to reject a null hypothesis that was, in fact, true, at any desired significance level.

It can thus be seen that the stopping rule principle is very hard to accept from the point of view of classical statistics. It is for these reasons that Savage said that, "I learned the stopping rule principle from Professor Barnard, in conversation in the summer of 1952. Frankly, I then thought it a scandal that anyone in the profession could advance an idea so patently wrong, even as today I can scarcely believe some people can resist an idea so patently right" (Savage *et al.*, 1962, p. 76).

From a Bayesian viewpoint, there is nothing to be said for unbiased estimates, while a test of a sharp null hypothesis would be carried out in quite a different way, and if (as is quite likely if in fact $\theta = 0$) the sample size resulting in example (5) were very large, then the posterior probability that $\theta = 0$ would remain quite large. It can thus be seen that if the stopping rule is seen to be plausible, and it is difficult to avoid this in view of the arguments for the likelihood principle in the last section, then Bayesian statisticians are not embarrassed in the way that classical statisticians are.

7.3 Informative stopping rules

An example on capture and recapture of fish

A stopping rule s is said to be *informative* if its distribution depends on θ in such a way that it conveys information about θ in addition to that available from the values of $x_1, x_2, \ldots, x_s$. The point of this section is to give a non-trivial example of an informative stopping rule; the example is due to Roberts (1967).

Consider a capture–recapture situation for a population of fish in a lake. The total number N of fish is unknown and is the parameter of interest (that is, it is the θ of the problem). It is known that R of the fish have been captured, tagged and released, and we write S for the number of untagged fish. Because $S = N - R$ and R is known, we can treat S as the unknown parameter instead of N, and it is convenient to do so. A random sample of n fish is then drawn (without replacement) from the lake. The sample yields r tagged fish and $s = n - r$ untagged ones.

We assume that there is an unknown probability π of catching each fish independently of each other. Then the stopping rule is given by the binomial distribution as

$$p(n \mid R, S, \pi) = \binom{R+S}{n} \pi^n (1-\pi)^{R+S-n} \qquad (n = 0, 1, 2, \ldots, R+S)$$

so that π is a nuisance parameter such that $0 \leqslant \pi \leqslant 1$. We note that this stopping rule *is* informative, because it depends on $N = R + S$.

Conditional on R, N, π and n, the probability of catching r tagged fish out of $n = r + s$ is given by the hypergeometric distribution

$$p(r \mid R, S, \pi, n) = \frac{\binom{R}{r}\binom{S}{s}}{\binom{R+S}{r+s}}.$$

Because we know r and s if and only if we know r and n, it follows that

$$\begin{aligned} p(r, s \mid R, S, \pi) &= p(r \mid R, S, \pi, n) p(n \mid R, S, \pi) \\ &= \frac{\binom{R}{r}\binom{S}{s}}{\binom{R+S}{r+s}} \binom{R+S}{r+s} \pi^{r+s} (1-\pi)^{R+S-r-s} \\ &= \binom{R}{r}\binom{S}{s} \pi^{r+s} (1-\pi)^{R+S-r-s}. \end{aligned}$$

Choice of prior and derivation of posterior

We assume that not much is known about the number of the fish in the lake *a priori*, and we can represent this by an improper prior

$$p(S) \propto 1.$$

On the other hand, in the process of capturing the first sample R for tagging, some knowledge will have been gained about the probability π of catching a fish. Suppose that this knowledge can be represented by a beta prior, so that $\pi \sim \mathrm{Be}(r', R'-r')$, that is,

$$p(\pi) \propto \pi^{r'-1}(1-\pi)^{R'-r'-1}$$

independently of S. It follows that

$$\begin{aligned} p(S, \pi \mid R, r, s) &\propto p(S)p(\pi)p(r, s \mid R, S, \pi) \\ &= \pi^{r'-1}(1-\pi)^{R'-r'-1}\tbinom{R}{r}\tbinom{S}{s}\pi^{r+s}(1-\pi)^{R+S-r-s} \\ &\propto \pi^{(r''-1)-1}(1-\pi)^{(R''-r'')-1}\left[\tbinom{S}{s}\pi^{s+1}(1-\pi)^{S-s}\right] \end{aligned}$$

where

$$r'' = r + r', \qquad R'' = R + R'.$$

It follows that *for given* π the distribution of S is such that $S-s$ has a negative binomial distribution $\mathrm{NB}(s+1, \pi)$ (see the Appendix). Summing over S from s to ∞, it can also be seen that

$$p(S, \pi \mid R, r, s) \propto \pi^{(r''-1)-1}(1-\pi)^{(R''-r'')-1}$$

so that the posterior for π is $\mathrm{Be}(r''-1, R''-r'')$.

To find the unconditional distribution of S, it is necessary to integrate the joint posterior for S and π over π. It can be shown without great difficulty that the result is that

$$p(S \mid R, r, s) = \binom{S}{s}\frac{\mathrm{B}(r''+s, R''-r''+S-s)}{\mathrm{B}(r''-1, R''-r'')}$$

where $\mathrm{B}(\alpha, \beta)$ is the usual beta function. This distribution is sometimes known as the *beta–Pascal distribution*, and its properties have been investigated by Raiffa and Schlaifer (1961, Section 7.11). It follows from there that the posterior mean of S is

$$\mathsf{E}S = (s+1)\left(\frac{R''-2}{r''-2}\right) - 1$$

from which the posterior mean of N follows as $N = R + S$.

The maximum likelihood estimator

A standard classical approach would seek to estimate S or equivalently N by the maximum likelihood estimator, that is, by the value of N which maximizes

$$p(r, s \mid R, S, \pi) = \frac{\tbinom{R}{r}\tbinom{S}{s}}{\tbinom{R+S}{r+s}}$$

Now it is easily shown that

$$p(r, s|R, S, \pi)/p(r, s|R, S-1), \pi) = (R+S-r-s)S/(S-s)(R+S)$$

and this increases as a function of S until it reaches unity when $(r+s)S = (R+S)s$ and then decreases, so that the maximum likelihood estimator of S is

$$\hat{S} = Rs/r.$$

Numerical example

As a numerical example, let us suppose that the original catch was $R = 41$ fish and that the second sample results in $r = 8$ tagged and $s = 24$ untagged fish. We suppose further that the prior for the probability π of catching a fish is Be(2, 23), so that

$$R = 41, \quad r = 8, \quad s = 24, \quad R' = 25, \quad r' = 2$$

(so that $R'' = 66$ and $r'' = 10$). Then the posterior mean of S is

$$\mathsf{E}S = 25 \times 64/8 - 1 = 199$$

and hence that of N is $R + \mathsf{E}S$, that is, $41 + 199 = 240$. On the other hand, the same data with a reference prior Be(0, 0) for π (that is, $r' = R' = 0$) result in a posterior mean for S of

$$\mathsf{E}S = 25 \times 39/6 - 1 = 161.5$$

and hence that of N is $41 + 161.5 = 202.5$.

Either of these answers is notably different from the maximum likelihood answer that a classical statistician would be likely to quote, which is

$$\hat{S} = 41 \times 24/8 = 123$$

resulting in $\hat{N} = 41 + 123 = 164$. The conclusion is that an informative stopping rule can have a considerable impact on the conclusions, and (though this is scarcely surprising) that prior beliefs about the nuisance parameter π make a considerable difference.

7.4 The likelihood principle and reference priors

The case of Bernoulli trials and its general implications

Care should be taken when using reference priors as a representation of prior ignorance. We have already seen in Section 2.4 on "Dominant likelihoods" that the improper densities which often arise as reference priors should be regarded as approximations, reflecting the fact that our prior beliefs about an unknown parameter (or some function of it) are more or less uniform over a wide range. A different point to be aware of is that some ways of arriving at such priors, such as Jeffreys' rule, depend on the experiment that is to be performed, and so on *intentions*. (The same objection applies, of course, to arguments based on data-translated likelihoods). Consequently, an analysis using such a prior is not in accordance with the likelihood principle.

To make this clearer, let us consider a sequence of independent trials, each of which results in success with probability π or failure with probability $1-\pi$ (that is, a sequence of Bernoulli trials). If we look at the number of successes x in a fixed number n of trials, so that

$$p(x\,|\,\pi) \propto \binom{n}{x}\pi^x(1-\pi)^{n-x} \qquad (x=0, 1, \ldots, n)$$

then, as was shown in Section 3.3, Jeffreys' rule results in an arc-sine distribution

$$p(\pi) \propto \pi^{-\frac{1}{2}}(1-\pi)^{-\frac{1}{2}}$$

for the prior.

Now suppose that we decide to observe the number of failures y before the mth success. Evidently there will be m successes and y failures, and the probability of any particular sequence with that number of successes and failures is $\pi^m(1-\pi)^y$. The number of such sequences is $\binom{m+y-1}{y}$, because the y failures and $m-1$ of the successes can occur in any order, but the sequence must conclude with a success. It follows that

$$p(y\,|\,\pi) = \binom{m+y-1}{y}\pi^m(1-\pi)^y$$

that is, that $y \sim \mathrm{NB}(m, \pi)$ has a negative binomial distribution (see the Appendix). For such a distribution

$$L(\pi\,|\,y) = m \log \pi + y \log (1-\pi) + \text{constant}$$

so that

$$\partial^2 L/\partial\pi^2 = -m/\pi^2 - y/(1-\pi)^2.$$

Because $\mathsf{E}y = m(1-\pi)/\pi$ it follows that

$$I(\pi\,|\,y) = m/\pi^2 + m/\pi(1-\pi) = m/\pi^2(1-\pi)$$

so that Jeffreys' rule implies that we should take a prior

$$p(\pi) \propto \pi^{-1}(1-\pi)^{-\frac{1}{2}}$$

that is, $\pi \sim \mathrm{Be}(0, \frac{1}{2})$ instead of $\pi \sim \mathrm{Be}(\frac{1}{2}, \frac{1}{2})$.

Conclusion

Consequently, on being told that an experiment resulted in, say, 10 successes and 10 failures, Jeffreys' rule does not allow us to decide which prior to use until we know whether the experimental design involved a fixed number of trials, or waiting until a fixed number of successes, or some other method. This clearly violates the likelihood principle (cf. Lindley 1971, Section 12.4); in so far as they appear to include Jeffreys' work it is hard to see how Berger and Wolpert (1984, Section 4.1.2) came to the conclusion that "use of noninformative priors, purposely not involving subjective prior opinions . . . is consistent with the LP [Likelihood Principle]". Some further difficulties inherent in the notion of a uniform reference prior are discussed by Hill (1980) and by Berger and Wolpert (1984).

However, it has been argued that a reference prior should express ignorance *relative to* the information which can be supplied by a particular experiment; see, for example, Box and Tiao (1973, Section 1.3). In any case, provided they are used critically, reference priors can be very useful, and, of course, if there is a reasonable amount of detail, the precise form of the prior adopted will not make a great deal of difference.

7.5 Bayesian decision theory

The elements of game theory

Only a very brief account of this important topic is included here; readers who want to know more should begin by consulting Berger (1985) and Ferguson (1967).

The elements of decision theory are very similar to those of the mathematical theory of games as developed by von Neumann and Morgenstern (1944), although for statistical purposes one of the players is nature (in some sense) rather than another player. Only those aspects of the theory of games which are strictly necessary are given here; an entertaining popular account, is given by Williams (1954). A two-person zero-sum game (Θ, A, L) has the following three basic elements:

(1) a non-empty set Θ of possible states of nature θ, sometimes called the *parameter space*;
(2) a non-empty set A of *actions* available to the statistician;
(3) a *loss function* L, which defines the loss $L(\theta, a)$ which a statistician suffers if he takes action a when the true state of nature is θ (this loss being expressed as a real number).

A *statistical decision problem* or a *statistical game* is a game (Θ, A, L) coupled with an experiment whose result x lies in a *sample space* $\mathscr{X}$ and is randomly distributed with a density $p(x|\theta)$ which depends on the state $\theta \in \Theta$ "chosen" by nature. The data x can be, and usually are, more than one-dimensional.

Now suppose that on the basis of the result x of the experiment, the statistician chooses an action $d(x) \in A$, resulting in a random loss $L(\theta, d(x))$. Taking expectations over possible outcomes x of the experiment, we obtain a *risk function*

$$\begin{aligned} R(\theta, d) &= \mathsf{E}\, L(\theta, d(x)) \\ &= \int L(\theta, d(x)) p(x|\theta)\, dx \end{aligned}$$

which depends on the true state of nature and the form of the function d by which the action to be taken once the result of the experiment is known is determined. It is possible that this expectation may not exist, or may be infinite, but we shall exclude such cases and define a (nonrandomized) *decision rule* or a *decision function* as any function d for which $R(\theta, d)$ exists and is finite for all $\theta \in \Theta$.

As usual, we suppose that we have prior beliefs about θ which can be expressed in terms of a prior density $p(\theta)$. The *Bayes risk* $r(d)$ of the decision rule d can then be defined as the expectation of $R(\theta, d)$ over all possible values of θ, that is,

$$\begin{aligned} r(d) &= \mathsf{E}\, R(\theta, d) \\ &= \int R(\theta, d) p(\theta)\, d\theta. \end{aligned}$$

It seems sensible to minimize one's losses, and accordingly a *Bayes decision rule* d is defined as one which minimizes the Bayes risk $r(d)$. Now

$$\begin{aligned} r(d) &= \int R(\theta, d) p(\theta)\, d\theta \\ &= \int\int L(\theta, d(x)) p(x|\theta) p(\theta)\, dx\, d\theta \\ &= \int\int L(\theta, d(x)) p(x, \theta)\, dx\, d\theta \\ &= \int \{ \int L(\theta, d(x)) p(\theta|x)\, d\theta \} p(x)\, dx. \end{aligned}$$

It follows that if the posterior expected loss of an action a is defined by

$$\rho(a, x) = \int L(\theta, a) p(\theta|x)\, d\theta$$

then the Bayes risk is minimized if the decision rule d is so chosen that $\rho(d(x), x)$ is a minimum for all x (technically, for those who know measure theory, it need only be a minimum for almost all x).

Raiffa and Schlaifer (1961, Sections 1.2 and 1.3) refer to the overall minimization of $r(d)$ as the *normal form* of Bayesian analysis and to the minimization of $\rho(d(x), x)$ for all x as the *extensive form*; the remark above shows that the two are equivalent.

When a number of possible prior distributions are under consideration, one sometimes finds that the term Bayes rule as such is restricted to rules resulting from *proper* priors, while those resulting from *im*proper priors are called *generalized* Bayes rules. Further extensions are mentioned by Ferguson (1967).

Point estimators resulting from quadratic loss

We now apply Bayesian decision theory to find *point estimators* of a parameter θ, that is, to find a single number from the data which in some way best represents a parameter under study. A Bayes decision rule in such a problem is referred to as a *Bayes estimator*. In such problems it is easiest to work with *quadratic loss*, that is, with a *squared-error* loss function

$$L(\theta, a) = (\theta - a)^2.$$

In this case $\rho(a, x)$ is the *mean square error*, that is,

$$\begin{aligned}\rho(a, x) &= \int [\theta - a]^2 p(\theta \mid x)\,\mathrm{d}\theta \\ &= \int [\{\theta - \mathsf{E}(\theta \mid x)\} + \{\mathsf{E}(\theta \mid x) - a\}]^2 p(\theta \mid x)\,\mathrm{d}\theta \\ &= \int \{\theta - \mathsf{E}(\theta \mid x)\}^2 p(\theta \mid x)\,\mathrm{d}\theta \\ &\quad + 2\{\mathsf{E}(\theta \mid x) - a\} \int \{\theta - \mathsf{E}(\theta \mid x)\}\, p(\theta \mid x)\,\mathrm{d}\theta \\ &\quad + \{\mathsf{E}(\theta \mid x) - a\}^2.\end{aligned}$$

The second term clearly vanishes, so that

$$\rho(a, x) = \mathcal{V}(\theta \mid x) + \{\mathsf{E}(\theta \mid x) - a\}^2$$

which is a minimum when $a = \mathsf{E}(\theta \mid x)$, so that a Bayes estimator $d(x)$ is the posterior mean of θ, and in this case $\rho(d(x), x)$ is the posterior variance of θ.

Particular cases of quadratic loss

As a particular case, if we have a single observation $x \sim \mathrm{N}(\theta, \varphi)$ where φ is known and our prior for θ is $\mathrm{N}(\theta_0, \varphi_0)$, so that our posterior is $\mathrm{N}(\theta_1, \varphi_1)$ (cf. Section 2.2 on "Normal prior and likelihood"), then an estimate of θ that minimizes quadratic loss is the posterior mean θ_1, and if that estimate is used the mean square error is the posterior variance φ.

For another example, suppose that $x \sim \mathrm{P}(\lambda)$, that is, that x has a Poisson distribution of mean λ, and that our prior density for λ is $p(\lambda)$.

We first note that the predictive density of x is

$$p(x) = \int p(x, \lambda)\,\mathrm{d}\lambda = \int p(x\,|\,\lambda)p(\lambda)\,\mathrm{d}\lambda$$
$$= \int (x!)^{-1}\lambda^x \exp(-\lambda)p(\lambda)\,\mathrm{d}\lambda.$$

To avoid ambiguity in what follows, $p_{\tilde{x}}(x)$ is used for this predictive distribution, so that $p_{\tilde{x}}(z)$ just denotes $\int (z!)^{-1}\lambda^z \exp(-\lambda)p(\lambda)\,\mathrm{d}\lambda$. Then as

$$p(\lambda\,|\,x) = p(\lambda)p(x\,|\,\lambda)/p_{\tilde{x}}(x)$$
$$= (x!)^{-1}\lambda^x \exp(-\lambda)p(\lambda)/p_{\tilde{x}}(x)$$

it follows that the posterior mean is

$$\mathsf{E}(\lambda\,|\,x) = \int \lambda(x!)^{-1}\lambda^x \exp(-\lambda)p(\lambda)\,\mathrm{d}\lambda/p_{\tilde{x}}(x)$$
$$= (x+1)\int \{(x+1)!\}^{-1}\lambda^{x+1} \exp(-\lambda)p(\lambda)\,\mathrm{d}\lambda/p_{\tilde{x}}(x)$$
$$= (x+1)p_{\tilde{x}}(x+1)/p_{\tilde{x}}(x).$$

We shall return to this example in Section 7.7 in connection with empirical Bayes methods.

We note incidentally that if the prior is $\lambda \sim S_0^{-1}\chi_\nu^2$, then the posterior is $(S_0+2)^{-1}\chi_{\nu+2x}^2$ (as shown in Section 3.4 on "The Poisson distribution"), so that in this particular case

$$\mathsf{E}(\lambda\,|\,x) = (\nu+2x)/(S_0+2).$$

Weighted quadratic loss

However, we must avoid concluding that the solution to all problems of point estimation from a Bayesian point of view is simply to quote the posterior mean—the answer depends on the loss function. Thus if we take as loss function a *weighted quadratic loss*, that is,

$$L(\theta, a) = w(\theta)(\theta - a)^2$$

then

$$\rho(a, x) = \int w(\theta)[\theta - a]^2 p(\theta\,|\,x)\,\mathrm{d}x.$$

If we now define

$$\mathsf{E}^w(\theta\,|\,x) = \frac{\mathsf{E}(w(\theta)\theta\,|\,x)}{\mathsf{E}(w(\theta)\,|\,x)} = \frac{\int w(\theta)\theta p(\theta\,|\,x)\,\mathrm{d}\theta}{\int w(\theta)p(\theta\,|\,x)\,\mathrm{d}\theta}$$

then similar calculations to those above show that

$$\rho(a, x) = \mathsf{E}[w(\theta)\{\theta - a\}^2\,|\,x] + [\mathsf{E}^w(\theta\,|\,x) - a]^2$$

and hence that a Bayes decision results if

$$d(x) = \mathsf{E}^w(\theta\,|\,x)$$

that is, $d(x)$ is a weighted posterior mean of θ.

Absolute error loss

A further answer results if we take as loss function the absolute error

$$L(\theta, a) = |\theta - a|$$

in which case

$$\rho(a, x) = \int |\theta - a| p(\theta | x) \mathrm{d}\theta$$

is sometimes referred to as the *mean absolute deviation* or M.A.D. In this case any median $m(x)$ of the posterior distribution given x, that is, any value such that

$$\mathrm{P}(\theta \leqslant m(x) | x) \geqslant \tfrac{1}{2} \quad \text{and} \quad \mathrm{P}(\theta \geqslant m(x) | x) \geqslant \tfrac{1}{2},$$

is a Bayes rule. To show this let us suppose that $d(x)$ is any other rule and, for definiteness, that $d(x) > m(x)$ for some particular x [the proof is similar if $d(x) < m(x)$]. Then

$$L(\theta, m(x)) - L(\theta, d(x)) = \begin{cases} m(x) - d(x) & \text{if } \theta \leqslant m(x) \\ 2\theta - (m(x) + d(x)) & \text{if } m(x) < \theta < d(x) \\ d(x) - m(x) & \text{if } \theta \geqslant d(x) \end{cases}$$

while for $m(x) < \theta < d(x)$

$$2\theta - (m(x) + d(x)) < \theta - m(x) < d(x) - m(x)$$

so that

$$L(\theta, m(x)) - L(\theta, d(x)) \leqslant \begin{cases} m(x) - d(x) & \text{if } \theta \leqslant m(x) \\ d(x) - m(x) & \text{if } \theta > m(x) \end{cases}$$

and hence on taking expectations over θ

$$\begin{aligned} \rho(m(x), x) - \rho(d(x), x) &\leqslant \{m(x) - d(x)\} \mathrm{P}(\theta \leqslant m(x) | x) \\ &\quad + \{d(x) - m(x)\} \mathrm{P}(\theta > m(x) | x) \\ &\leqslant \{d(x) - m(x)\} \{-\mathrm{P}(\theta \leqslant m(x) | x) \\ &\quad + 1 - \mathrm{P}(\theta \leqslant m(x) | x)\} \\ &\leqslant 0 \end{aligned}$$

from which it follows that taking the posterior median is indeed the appropriate Bayes rule for this loss function. More generally, a loss function $L(\theta, a)$ which is $K_0 |\theta - a|$ if $\theta \geqslant a$ but $K_1 |\theta - a|$ if $\theta < a$ results in a Bayes estimator which is a $K_0/(K_0 + K_1)$ fractile of the posterior distribution.

Zero–one loss

Yet another answer results from the loss function

$$L(\theta, a) = \begin{cases} 0 & \text{if } |\theta - a| \leqslant \varepsilon \\ 1 & \text{if } |\theta - a| > \varepsilon \end{cases}$$

which results in

$$\rho(a, x) = \mathrm{P}(|\theta - a| > \varepsilon \,|\, x) = 1 - \mathrm{P}(|\theta - a| \leqslant \varepsilon \,|\, x).$$

Consequently, if a modal interval of length 2ε is defined as an interval

$$(\mathrm{mod}^{\varepsilon}(x) - \varepsilon, \quad \mathrm{mod}^{\varepsilon}(x) + \varepsilon)$$

which has highest probability for given ε, then the midpoint $\mathrm{mod}^{\varepsilon}(x)$ of this interval is a Bayes estimate for this loss function. If ε is fairly small, this value is clearly very close to the posterior mode of the distribution, which in its turn will be close to the maximum likelihood estimator if the prior is reasonably smooth.

Thus all three of mean, median and mode of the posterior distribution can arise as Bayes estimators for suitable loss functions (namely, quadratic error, absolute error, and zero–one error respectively).

General discussion of point estimation

Some Bayesian statisticians pour scorn on the whole idea of point estimators; see, for example, Box and Tiao (1973, Section A5.6). There are certainly doubtful points about the preceding analysis. It is difficult to be convinced in any particular case that a particular loss function represents real economic penalties in a particular case, and in many scientific contexts it is difficult to give any meaning at all to the notion of the loss suffered by making a wrong point estimate. Certainly the same loss function will not be valid in all cases. Moreover, even with quadratic loss, which is often treated as a norm, there are problems which do not admit of an easy solution. If, for example, $x_1, x_2, \ldots, x_n \sim \mathrm{N}(1/\theta, 1)$, then it would seem reasonable to estimate θ by $1/\bar{x}$, and yet the mean square error $\mathsf{E}(1/\bar{x} - \theta)^2$ is infinite. Of course, such decision functions are excluded by requiring that the risk function should be finite, but this is clearly a case of what Good (1965, Section 6.2) referred to as "adhockery".

Even though there are cases (for example, where the posterior distribution is bimodal) in which there is no sensible point estimator, I think there are cases where it is reasonable to ask for such a thing, though I have considerable doubts as to whether the ideas of decision theory add much to the appeal that quantities such as the posterior mean, median or mode have in themselves.

7.6 Decision theory and hypothesis testing

Relationship between decision theory and classical hypothesis testing

It is possible to reformulate hypothesis testing in the language of decision theory. If we want to test H_0: $\theta \in \Theta_0$ versus H_1: $\theta \in \Theta_1$, we have two actions open to us, namely,

$$a_0\text{: accept } H_0 \qquad \text{and} \qquad a_1\text{: accept } H_1.$$

As before, we write π_0 and π_1 for the prior probabilities of H_0 and H_1 and p_0 and p_1 for their posterior probabilities and

$$B = \frac{(p_0/p_1)}{(\pi_0/\pi_1)}$$

for the Bayes factor. We also need the notation

$$\rho_0(\theta) = p(\theta)/\pi_0 \qquad \text{for } \theta \in \Theta_0$$
$$\rho_1(\theta) = p(\theta)/\pi_1 \qquad \text{for } \theta \in \Theta_1$$

where $p(\theta)$ is the prior density function.

Now let us suppose that there is a loss function $L(\theta, a)$ defined by

$a \backslash \theta$	$\theta \in \Theta_0$	$\theta \in \Theta_1$
a_0	0	1
a_1	1	0

so that the use of a decision rule $d(x)$ results in a posterior expected loss function

$$\rho(a_0, x) = p_1$$
$$\rho(a_1, x) = p_0$$

so that a decision $d(x)$ which minimizes the posterior expected loss is just a decision to accept the hypothesis with the greater posterior probability, which is the way of choosing between hypotheses suggested when hypothesis testing was first introduced.

More generally, if there is a "$0 - K_i$" loss function, that is,

$a \backslash \theta$	$\theta \in \Theta_0$	$\theta \in \Theta_1$
a_0	0	K_0
a_1	K_1	0

then the posterior expected losses of the two actions are

$$\rho(a_0, x) = K_0 p_1,$$
$$\rho(a_1, x) = K_1 p_0$$

so that a Bayes decision rule results in rejecting the null hypothesis, that is, in taking action a_1, if and only if $K_1 p_0 < K_0 p_1$, that is,

$$B = \frac{(p_0/p_1)}{(\pi_0/\pi_1)} < \frac{(K_0/K_1)}{(\pi_0/\pi_1)}.$$

In the terminology of classical statistics, this corresponds to the use of a rejection region

$$R = \{x;\ B < K_0\pi_1/K_1\pi_1\}.$$

When hypothesis testing was first introduced in Section 4.1, we noted that in the case where $\Theta_0 = \{\theta_0\}$ and $\Theta_1 = \{\theta_1\}$ are simple hypotheses, then Bayes theorem implies that

$$B = \frac{p(x\,|\,\theta_0)}{p(x\,|\,\theta_1)}$$

so that the rejection region takes the form

$$R = \{x;\ p(x\,|\,\theta_0)/p(x\,|\,\theta_1) < K_0\pi_1/K_1\pi_0\}$$

which is the likelihood ratio test prescribed by Neyman–Pearson theory. A difference is that in the Neyman–Pearson theory, the "critical value" of the rejection region is determined by fixing the size α, that is, the probability that x lies in the rejection region R if the null hypothesis is true, whereas in a decision theoretic approach it is fixed in terms of the loss function and the prior probabilities of the hypotheses.

Composite hypotheses

If the hypotheses are composite, then, again as in Section 3.2,

$$B = \frac{(p_0/p_1)}{(\pi_0/\pi_1)} = \frac{\int_{\theta\in\Theta_0} p(x\,|\,\theta)\,\rho_0(\theta)\,d\theta}{\int_{\theta\in\Theta_1} p(x\,|\,\theta)\,\rho_1(\theta)\,d\theta}$$

so that there is still a rejection region that can be interpreted in a similar manner. However, it should be noted that classical statisticians faced with similar problems are more inclined to work in terms of a likelihood ratio

$$\frac{\max_{\theta\in\Theta_0} p(x\,|\,\theta)}{\max_{\theta\in\Theta_1} p(x\,|\,\theta)}$$

(cf. Lehmann, 1959; 1986, Section 1.71). In fact it is possible to express quite a lot of the ideas of classical statistics in a language involving loss functions.

It may be noted that it is easy to extend the above discussion about dichotomies (situations where a choice has to be made between two hypotheses) to deal with trichotomies or polytomies, although some theories of statistical inference find choices between more than two hypotheses difficult to deal with.

7.7 Empirical Bayes methods

Von Mises' example

Only a very brief idea about empirical Bayes methods will be given here; readers who wish to learn more about them, should consult Maritz (1970). One of the reasons for the brief treatment here is that, despite their name, very few empirical Bayes procedures are, in fact, Bayesian.

The problems we will consider in this section are concerned with a sequence x_i of observations such that the distribution of the ith observation x_i depends on a parameter θ_i, typically in such a way that $p(x_i|\theta_i)$ has the same functional form for all i. The parameters θ_i are themselves supposed to be a random sample from some (unknown) distribution, and it is this unknown distribution that plays the role of a prior distribution and so accounts for the use of the name of Bayes. There is a clear contrast with the situation in the rest of the book, where the prior distribution represents our prior beliefs, and so by definition it cannot be unknown. Further, the prior distribution in empirical Bayes methods is usually given a frequency interpretation, by contrast with the situation arising in true Bayesian methods.

One of the earliest examples of an empirical Bayes procedure was due to von Mises (1942). He supposed that in examining the quality of a batch of water for possible contamination by certain bacteria, $m = 5$ samples of a given volume were taken, and he was interested in determining the probability θ that a sample contains at least one bacterium. Evidently, the probability of x positive result in the five samples is

$$p(x|\theta) = \binom{5}{x}\theta^x(1-\theta)^{5-x}$$

for a given value of θ. If the same procedure is to be used with a number of batches of different quality, then the predictive distribution (denoted $p_{\tilde{x}}(x)$ to avoid ambiguity) is

$$p_{\tilde{x}}(x) = \int \binom{5}{x}\theta^x(1-\theta)^{5-x}p(\theta)\,d\theta$$

where the density $p(\theta)$ represents the variation of the quality θ of batches. [If $p(\theta)$ comes from the beta family, and there is no particular reason why it should, then $p_{\tilde{x}}(x)$ is a beta–binomial distribution, as mentioned at the end of Section 3.1 on "The binomial distribution"]. In his example, von Mises wished to estimate the density function $p(\theta)$ on the basis of $n = 3420$ observations.

The Poisson case

Instead of considering the binomial distribution further, we shall consider a problem to do with the Poisson distribution. Let us suppose that we have observations $x_i \sim \mathrm{P}(\lambda_i)$ where the λ_i have a distribution with a density $p(\lambda)$, and that we have available n past observations, among which $f_n(x)$ were equal to x for $x = 0, 1, 2, \ldots$ Thus $f_n(x)$ is an empirical frequency and $f_n(x)/n$ is an estimate of the predictive density $p_{\tilde{x}}(x)$. As x has a Poisson distribution for given λ

$$\begin{aligned} p_{\tilde{x}}(x) &= \int p(x, \lambda)\,\mathrm{d}\lambda = \int p(x|\lambda)p(\lambda)\,\mathrm{d}\lambda \\ &= \int (x!)^{-1}\lambda^x \exp(-\lambda)p(\lambda)\,\mathrm{d}\lambda. \end{aligned}$$

Now let us suppose that, with this past data available, a new observation ξ is made, and we want to say something about the corresponding value of λ. In Section 7.5 on "Bayesian decision theory" we saw that the posterior mean of λ is

$$\mathsf{E}(\lambda|\xi) = \frac{(\xi+1)p_{\tilde{x}}(\xi+1)}{p_{\tilde{x}}(\xi)}.$$

To use this formula, we need to know the prior $p(\lambda)$ or at least to know $p_{\tilde{x}}(\xi)$ and $p_{\tilde{x}}(\xi+1)$, which we do not know. However, it is clear that a reasonable estimate of $p_{\tilde{x}}(\xi)$ is $(f_n(\xi)+1)/(n+1)$, after allowing for the latest observation. Similarly, a reasonable estimate for $p_{\tilde{x}}(\xi+1)$ is $f_n(\xi+1)/(n+1)$. It follows that a possible point estimate for the current value of λ, corresponding to the value $\mathsf{E}(\lambda|x)$ resulting from a quadratic loss function, is

$$\delta_n = \frac{(\xi+1)\,f_n(\xi+1)}{f_n(\xi)+1}.$$

This formula could be used in a case like that investigated by von Mises if the number m of samples taken from each batch were fairly large and the probability θ that a sample contained at least one bacterium were fairly small, so that the Poisson approximation to the binomial could be used.

This method can easily be adapted to any case where the posterior mean $\mathsf{E}(\theta \mid \xi)$ of the parameter of the interest takes the form

$$\frac{c(\xi)p_{\tilde{x}}(\xi+1)}{p_{\tilde{x}}(\xi)}$$

and there are quite a number of such cases (Maritz, 1970, Section 1.3).

Going back to the Poisson case, if it were known that the underlying distribution $p(\lambda)$ were of the form $S_0{}^{-1}\chi_\nu{}^2$ for some S_0 and ν, then it is known (cf. Section 7.5) that

$$\mathsf{E}(\lambda \mid \xi) = (\nu + 2\xi)/(S_o + 2).$$

In this case, we could use $x_1, x_2, \ldots, x_n$ to estimate S_0 and ν in some way, by, say, $\hat{S}_0$ and $\hat{\nu}$, giving an alternative point estimate for the current value of λ

$$(\hat{\nu} + 2\xi)/(\hat{S}_0 + 2).$$

The advantage of an estimate like this is that because, considered as a function of ξ, it is smoother than δ_n, it could be expected to do better. This is analogous with the situation in regression analysis, where a fitted regression line can be expected to give a better estimate of the mean of the dependent variable y at a particular value of the independent variable x than would be obtained by concentrating on values of y obtained at that single value of x. On the other hand, the method just described does depend on assuming a particular form for the prior, which is probably not justifiable. There are, however, other methods of producing a "smoother" estimate.

Empirical Bayes methods can also be used for testing whether a parameter θ lies in one or another of a number of sets, that is, for hypothesis testing and its generalizations.

Exercises on Chapter 7

1. Show that in any experiment E in which there is a possible value y for the random variable $\tilde{x}$ such that $p_{\tilde{x}}(y \mid \theta) = 0$, then if z is *any* other possible value of $\tilde{x}$, the statistic $t = t(x)$ defined by

$$t(x) = \begin{cases} z & \text{if } x = y \\ x & \text{if } x \neq y \end{cases}$$

is sufficient for θ given x. Hence show that if $\tilde{x}$ is a continuous random variable, then a naïve application of the weak sufficiency principle as defined in Section 7.1 would result in $\text{Ev}\{E, y, \theta\} = \text{Ev}\{E, z, \theta\}$ for *any* two possible values y and z of $\tilde{x}$.

2. Suppose that the density function $p(x \mid \theta)$ is defined as follows for $x = 1, 2, 3, \ldots$ and $\theta = 1, 2, 3, \ldots$. If θ is even, then

$$p(x \mid \theta) = \begin{cases} \frac{1}{3} & \text{if } x = \theta/2, 2\theta \text{ or } 2\theta + 1 \\ 0 & \text{otherwise;} \end{cases}$$

if θ is odd but $\theta \neq 1$, then

$$p(x \mid \theta) = \begin{cases} \frac{1}{3} & \text{if } x = (\theta - 1)/2, 2\theta \text{ or } 2\theta + 1 \\ 0 & \text{otherwise;} \end{cases}$$

while if $\theta = 1$ then

$$p(x \mid \theta) = \begin{cases} \frac{1}{3} & \text{if } x = \theta, 2\theta \text{ or } 2\theta + 1 \\ 0 & \text{otherwise;} \end{cases}$$

Show that, for any x the data intuitively give equal support to the three possible values of θ compatible with that observation, and hence that on likelihood grounds any of the three would be a suitable estimate. Consider, therefore, the three possible estimators d_1, d_2 and d_3 corresponding to the smallest, middle and largest possible θ. Show that

$$p(d_2 = \theta) = \begin{cases} \frac{1}{3} & \text{when } \theta \text{ is even} \\ 0 & \text{otherwise,} \end{cases}$$

$$p(d_3 = \theta) = \begin{cases} \frac{1}{3} & \text{when } \theta \text{ is odd but } \theta \neq 1 \\ 0 & \text{otherwise,} \end{cases}$$

but that

$$p(d_1 = \theta) = \begin{cases} 1 & \text{when } \theta = 1 \\ \frac{2}{3} & \text{otherwise.} \end{cases}$$

Does this apparent discrepancy cause problems for a Bayesian analysis? (Due to G. Monette and D. A. S. Fraser.)

3. A drunken soldier, starting at an intersection O in a city which has square blocks, staggers around a random path trailing a taut string. Eventually he stops at an intersection (after walking at least one block) and buries a treasure. Let θ denote the path of the string from O to the treasure. Letting N, S, E and W stand for a path segment one block long in the indicated direction, so that θ can be expressed as a sequence

of such letters, say $\theta = NNESWSWW$. (Note that NS, SN, EW and WE cannot appear as the taut string would be rewound). After burying the treasure, the soldier walks one block further in a random direction (still keeping the string taut). Let X denote this augmented path, so that X is one of θN, θS, θE and θW, each with probability $\frac{1}{4}$. You observe X and are then to find the treasure. Show that if a reference prior $p(\theta) \propto 1$ is used for all possible paths θ, then all four possible values of θ given X are equally likely. Note, however, that intuition would suggest that θ is three times as likely to extend the path as to backtrack, suggesting that one particular value of θ is more likely than the others after X is observed. (Due to M. Stone).

4. Suppose that $x_1, x_2, \ldots$ is a sequential sample from an $N(\theta, 1)$ distribution and it is desired to test H_0: $\theta = \theta_0$ versus H_1: $\theta \neq \theta_0$. The experimenter reports that he used a proper stopping rule and obtained the data $3, -1, 2, 1$.
   (a) What could a frequentist conclude?
   (b) What could a Bayesian conclude?

5. Let $x_1, x_2, \ldots$ be a sequential sample from a Poisson distribution $P(\lambda)$. Suppose that the stopping rule is to stop sampling at time $n \geqslant 2$ with probability

$$\sum_{i=1}^{n-1} x_i \Big/ \sum_{i=1}^{n} x_i$$

for $n = 2, 3, \ldots$ (define $0/0 = 1$). Suppose that the first five observations are 3, 1, 2, 5, 7 and that sampling then stops. Find the likelihood function for λ (Berger, 1985).

6. Show that the mean of the beta–Pascal distribution

$$p(S\,|\,R, r, s) = \binom{S}{s} \frac{B(r''+s,\, R''-r''+S-s)}{B(r''-1,\, R''-r'')}$$

is given by the formula in Section 7.3, namely,

$$\mathsf{E}S = (s+1)\left(\frac{R''-2}{r''-2}\right) - 1.$$

7. Suppose that you intend to observe the number x of successes in n Bernoulli trials *and* the number y of failures before the nth success *after* the first n trials, so that $x \sim B(n, \pi)$ and $y \sim NB(n, \pi)$. Find the

likelihood function $L(\pi|x, y)$ and deduce the reference prior that Jeffreys' rule would suggest for this case.

8. Suppose that you want to estimate the parameter π of a binomial distribution $B(n, \pi)$. Show that if the loss function is

$$L(\theta, a) = (\theta - a)^2/\{\theta(1-\theta)\}$$

then the Bayes rule corresponding to a uniform prior for π is given by $d(x) = x/n$ for any x such that $0<x<n$, that is, the maximum likelihood estimator. Is $d(x) = x/n$ a Bayes rule if $x = 0$ or $x = n$?

9. Let $x \sim B(n, \pi)$ and $y \sim B(n, \rho)$ have independent binomial distributions of the same index but possibly different parameters. Find the Bayes rule corresponding to the loss

$$L((\pi, \rho), a) = (\pi - \rho - a)^2$$

when the priors for π and ρ are independent uniform distributions.

10. Suppose that your prior for the proportion π of defective items supplied by a manufacturer is given by the beta distribution Be(2, 12), and that you then observe that none of a random sample of size 6 is defective. Find the posterior distribution and use it to carry out a test of the hypothesis H_0: $\pi<0.1$ using

(a) a "$0-1$" loss function, and
(b) the loss function

$a\backslash\theta$	$\theta\in\Theta_0$	$\theta\in\Theta_1$
a_0	0	1
a_1	2	0

11. A child is given an intelligence test. We assume that the test result x is $N(\theta, 100)$ where θ is the true intelligence quotient of the child, as measured by the test (in other words, if the child took a large number of similar tests, the average score would be θ). Assume also that, in the population as a whole, θ is distributed according to an $N(100, 225)$ distribution. If it is desired, on the basis of the intelligence quotient, to decide whether to put the child into a slow, average or fast group for reading, the actions available are:

a_1: Put in slow group, that is, decide $\theta\in\Theta_1 = (0, 90)$
a_1: Put in average group, that is, decide $\theta\in\Theta_2 = (90, 100)$
a_1: Put in fast group, that is, decide $\theta\in\Theta_3 = (100, \infty)$.

A loss function $L(\theta, a)$ of the following form might be deemed appropriate:

$a \backslash \theta$	$\theta \in \Theta_1$	$\theta \in \Theta_2$	$\theta \in \Theta_3$
a_1	0	$\theta - 90$	$2(\theta - 90)$
a_2	$90 - \theta$	0	$\theta - 110$
a_3	$2(110 - \theta)$	$110 - \theta$	0

Assume that we observe that the test result $x = 115$. By using tables of the normal distribution and the fact that if $\varphi(t)$ is the density function of the standard normal distribution, then $\int t\varphi(t)\,\mathrm{d}t = -\varphi(t)$, find the appropriate action to take on the basis of this observation. [See Berger (1985, Sections 4.2–4.4)].

12. In Section 7.7, a point estimator δ_n for the current value λ of the parameter of a Poisson distribution was found. Adapt the argument to deal with the case where the underlying distribution is negative binomial, that is,

$$p(x \mid \pi) = \binom{n+x-1}{x}(1-\pi)^n \pi^x.$$

Appendix

Common Statistical Distributions

Some facts are given below about various common statistical distributions. In the case of continuous distributions, the (probability) density (function) $p(x)$ equals the derivative of the distribution function $F(x) = \mathrm{P}(X \leqslant x)$. In the case of discrete distributions, the (probability) density (function) $p(x)$ equals the probability that the random variable X takes the value x.

The mean is defined as

$$\mathsf{E}X = \int xp(x)\,\mathrm{d}x \quad \text{or} \quad \sum xp(x)$$

depending on whether the random variable is discrete or continuous. The variance is defined as

$$\mathcal{V}X = \int (x - \mathsf{E}X)^2 p(x)\,\mathrm{d}x \quad \text{or} \quad \sum (x - \mathsf{E}X)^2 p(x)$$

depending on whether the random variable is discrete or continuous. A mode is any value for which $p(x)$ is a maximum; most common distributions have only one mode and so are called unimodal. A median is any value m such that both

$$\mathrm{P}(X \leqslant m) \geqslant \tfrac{1}{2} \quad \text{and} \quad \mathrm{P}(X \geqslant m) \geqslant \tfrac{1}{2}.$$

In the case of most continuous distributions there is a unique median m and

$$F(x) = \mathrm{P}(X \leqslant x) = \tfrac{1}{2}.$$

There is a well-known empirical relationship that

$$\text{mean} - \text{mode} \cong 3(\text{mean} - \text{median})$$

or equivalently

$$\text{median} \cong (2\ \text{mean} + \text{mode})/3.$$

Some theoretical grounds for this relationship based on Gram–Charlier or Edgeworth expansions can be found in Kendall, Stuart and Ord (1987, Section 2.11).

Further material can be found in Rothschild and Logothetis (1986) or Johnson and Kotz (1969–1972).

A.1 Normal distribution

X is normal with mean θ and variance φ, denoted

$$X \sim \mathrm{N}(\theta, \varphi)$$

if it has density

$$p(X) = \frac{1}{\sqrt{(2\pi\varphi)}} \exp\left(-\tfrac{1}{2}(X-\theta)^2/\varphi\right) \qquad (-\infty < X < \infty).$$

The mean and variance are

$$\mathsf{E}X = \theta$$
$$\mathcal{V}X = \varphi.$$

Because the distribution is symmetrical and unimodal, the median and mode both equal the mean, that is,

$$\text{median}\ (X) = \theta$$
$$\text{mode}\ (X) = \theta.$$

If $\theta = 0$ and $\varphi = 1$, that is, $X \sim \mathrm{N}(0, 1)$, X is said to have a standard normal distribution.

A.2 Chi-squared distribution

X has a chi-squared distribution on ν degrees of freedom, denoted

$$X \sim \chi_\nu^2$$

if it has the same distribution as

$$Z_1^2 + Z_2^2 + \ldots + Z_\nu^2$$

where $Z_1, Z_2, \ldots, Z_\nu$ are independent standard normal variates, or equivalently if it has density

$$p(X) = \frac{1}{2^{\nu/2}\Gamma(\nu/2)} X^{\nu/2-1} \exp\left(-\tfrac{1}{2}X\right) \qquad (0 < X < \infty).$$

If $Y = X/S$ where S is a constant, then Y is a chi-squared variate on ν degrees of freedom divided by S, denoted

$$Y \sim S^{-1}\chi_\nu^2$$

and it has density

$$p(Y) = \frac{S^{\nu/2}}{2^{\nu/2}\Gamma(\nu/2)} Y^{\nu/2-1} \exp(-\tfrac{1}{2}SY) \qquad (0<Y<\infty).$$

The mean and variance are

$$\mathsf{E}Y = \nu/S$$
$$\mathcal{V}Y = 2\nu/S^2$$

The mode is

$$\text{mode}(Y) = (\nu-2)/S \qquad (\text{provided } \nu \geq 2)$$

and the approximate relationship between mean, mode and median implies that the median is approximately

$$\text{median}(Y) = (\nu - (\tfrac{2}{3}))/S$$

at least for reasonably large ν, say $\nu \geq 5$.

A.3 Normal approximation to chi-squared

If $X \sim \chi_\nu^2$ then for large ν we have that approximately

$$\surd(2X) - \surd(2\nu-1) \sim \mathrm{N}(0, 1)$$

has a standard normal distribution.

A.4 Gamma distribution

X has a (one-parameter) gamma distribution with parameter α, denoted

$$X \sim \mathrm{G}(\alpha)$$

if it has density

$$p(X) = \frac{1}{\Gamma(\alpha)} X^{\alpha-1} \exp(-X) \qquad (0<X<\infty).$$

This is simply an alternative name for the distribution we refer to as

$$\tfrac{1}{2}\chi_{2\alpha}{}^2.$$

If $Y = \beta X$, then Y has a two-parameter gamma distribution with parameters α and β denoted

$$Y \sim \mathrm{G}(\alpha, \beta)$$

and it has density

$$p(Y) = \frac{1}{\beta^{\alpha}\Gamma(\alpha)} Y^{\alpha-1} \exp(-Y/\beta) \qquad (0<Y<\infty).$$

This is simply an alternative name for the distribution we refer to as

$$\tfrac{1}{2}\beta\chi_{2\alpha}^{\;2}.$$

If $\beta = 1$ we recover the one-parameter gamma distribution; if $\alpha = 1$ so that the density is

$$p(Y) = \beta^{-1} \exp(-Y/\beta)$$

we obtain another special case sometimes called the (negative) exponential distribution and denoted

$$Y \sim \mathrm{E}(\beta).$$

The distribution function of any variable with a gamma distribution is easily found in terms of the incomplete gamma function

$$\gamma(\alpha, x) = \int_0^x \xi^{\alpha-1} \exp(-\xi)\, \mathrm{d}\xi$$

or in terms of Karl Pearson's incomplete gamma function

$$\mathrm{I}(u, p) = \frac{1}{\Gamma(p+1)} \int_0^{u\sqrt{(p+1)}} t^p \exp(-t)\, \mathrm{d}t.$$

Extensive tables can be found in Pearson (1922, 1934).

A.5 Inverse chi-squared distribution

X has an inverse chi-squared distribution on ν degrees of freedom, denoted

$$X \sim \chi_{\nu}^{-2}$$

if $1/X \sim \chi_{\nu}^2$, or equivalently if it has density

$$p(X) = \frac{1}{2^{\nu/2}\Gamma(\nu/2)} X^{-\nu/2-1} \exp(-\tfrac{1}{2}X^{-1}) \qquad (0<X<\infty).$$

If $Y = SX$, so that $1/Y \sim S^{-1}\chi_{\nu}^{\;2}$, then Y is S times an inverse chi-squared distribution on ν degrees of freedom, denoted

$$Y \sim S\chi_{\nu}^{-2}$$

and it has density

$$p(Y)=\frac{S^{\nu/2}}{2^{\nu/2}\Gamma(\nu/2)}\,Y^{-\nu/2-1}\exp\left(-\tfrac{1}{2}S/Y\right) \qquad (0<Y<\infty).$$

The mean and variance are

$$\mathsf{E}Y=S/(\nu-2) \qquad \text{(provided } \nu>2)$$

$$\mathcal{V}\,Y=\frac{2S^2}{(\nu-2)^2(\nu-4)} \qquad \text{(provided } \nu>4).$$

The mode is

$$\text{mode}\,(Y)=S/(\nu+2)$$

and the median is in the range

$$S/(\nu-(\tfrac{1}{2}))<\text{median}\,(Y)<S/(\nu-(\tfrac{2}{3}))$$

provided $\nu\geqslant 1$, with the upper limit approached closely when $\nu>5$ [see Novick and Jackson (1974, Section 7.3)].

A.6 Inverse chi distribution

X has an inverse chi distribution on ν degrees of freedom, denoted

$$X\sim\chi_\nu^{-1}$$

if $1/X^2\sim\chi_\nu^{\,2}$, or equivalently if it has density

$$p(X)=\frac{1}{2^{\nu/2-1}\Gamma(\nu/2)}\,X^{-\nu-1}\exp\left(-\tfrac{1}{2}X^{-2}\right) \qquad (0<X<\infty).$$

If $Y=S^{\frac{1}{2}}X$ so that $1/Y^2\sim S^{-1}\chi_\nu^{\,2}$, then Y is $S^{\frac{1}{2}}$ times an inverse chi distribution on ν degrees of freedom, denoted

$$Y\sim S^{\frac{1}{2}}\chi_\nu^{-1}$$

and it has density

$$p(Y)=\frac{S^{\nu/2}}{2^{\nu/2-1}\Gamma(\nu/2)}\,Y^{-\nu-1}\exp\left(-\tfrac{1}{2}S/Y^2\right) \qquad (0<Y<\infty).$$

The mean and variance do not greatly simplify. They are

$$\mathsf{E}Y=\frac{S^{\frac{1}{2}}\Gamma(\nu-1)/2}{\sqrt{2}\,\Gamma(\nu/2)}$$

$$\mathcal{V}\,Y=S/(\nu-2)-(\mathsf{E}Y)^2$$

but very good approximations, at least if $\nu \geqslant 5$, are

$$\mathsf{E}Y = S^{\frac{1}{2}}/\sqrt{(\nu - (\tfrac{3}{2}))}$$

$$\mathscr{V}Y = \frac{S}{2(\nu-2)(\nu-(\tfrac{5}{3}))}$$

[see Novick and Jackson (1974, Section 7.3)]. The mode is exactly

$$\text{mode}\,(Y) = S^{\frac{1}{2}}/\sqrt{(\nu+1)}$$

and a good approximation to the median at least if $\nu \geqslant 4$ is (*ibid.*)

$$\text{median}\,(Y) = S^{\frac{1}{2}}/\sqrt{(\nu - (\tfrac{2}{3}))}.$$

A.7 Log chi-squared distribution

X has a log chi-squared distribution on ν degrees of freedom, denoted

$$X \sim \log \chi_\nu^{\,2}$$

if $X = \log W$ where $W \sim \chi_\nu^{\,2}$, or equivalently if X has density

$$p(X) = \frac{1}{2^{\nu/2}\Gamma(\nu/2)} \exp\{\tfrac{1}{2}\nu X - \tfrac{1}{2}\exp(X)\} \qquad (-\infty < X < \infty)$$

(note that unlike $\chi_\nu^{\,2}$ itself this is a distribution over the whole line).

Because the logarithm of an $S^{-1}\chi_\nu^{\,2}$ variable differs from a log chi-squared variable simply by an additive constant, it is not necessary to consider such variables in any detail.

By considering the tth moment of a $\chi_\nu^{\,2}$ variable, it is easily shown that the moment generating functions of a log chi-squared variable is

$$2^t\Gamma(t + (\nu/2))/\Gamma(\nu/2).$$

Writing
$$\psi(z) = \frac{\mathrm{d}}{\mathrm{d}z} \log \Gamma(z) = \Gamma'(z)/\Gamma(z)$$

it follows that the mean and variance are

$$\mathsf{E}X = \log 2 + \psi(\nu/2)$$
$$\mathscr{V}X = \psi'(\nu/2)$$

or (using Stirling's approximation and its derivatives) approximately

$$\mathsf{E}X = \log \nu - \nu^{-1} \cong \log(\nu - 1)$$
$$\mathscr{V}X = 2/\nu.$$

The mode is

$$\text{mode}\,(X) = \log \nu.$$

A.8 Student's t distribution

X has a Student's t distribution on ν degrees of freedom, denoted

$$X \sim t_\nu$$

if it has the same distribution as

$$\frac{Z}{\sqrt{(W/\nu)}}$$

where $Z \sim N(0, 1)$ and $W \sim \chi_\nu^2$ are independent, or equivalently if X has density

$$p(X) = \frac{\Gamma((\nu+1)/2)}{\sqrt{(\pi\nu)}\Gamma(\nu/2)}\left(1+\frac{X^2}{\nu}\right)^{-(\nu+1)/2}$$

$$= B(\nu/2, \tfrac{1}{2})^{-1}\left(1+\frac{X^2}{\nu}\right)^{-(\nu+1)/2}$$

It follows that if $X_1, X_2, \ldots, X_n$ are independently $N(\mu, \sigma^2)$ and

$$\bar{X} = \sum X_i/n$$
$$S = \sum (X_i - \bar{X})^2$$
$$s^2 = S/(n-1)$$

then

$$\frac{(\bar{X}-\mu)}{s/\sqrt{n}} \sim t_{n-1}.$$

The mean and variance are

$$\mathsf{E}X = 0$$
$$\mathcal{V}X = \nu/(\nu-2).$$

Because the distribution is symmetrical and unimodal, the median and mode both equal the mean, that is,

$$\text{median}\,(X) = 0$$
$$\text{mode}\,(X) = 0.$$

As $\nu \to \infty$ the distribution approaches the standard normal form.

It may be noted that Student's t distribution on one degree of freedom is the standard Cauchy distribution C(0, 1).

A.9 Normal/chi-squared distribution

The ordered pair (X, Y) has a normal/chi-squared distribution if

$$Y \sim S\chi_\nu^{-2}$$

for some S and ν and, conditional on Y,

$$X \sim \mathrm{N}(\mu, Y/n)$$

for some μ and n. An equivalent condition is that the joint density function (for $-\infty < X < \infty$ and $0 < Y < \infty$) is

$$p(X, Y) = \frac{1}{\sqrt{(2\pi Y/n)}} \exp\left(-\tfrac{1}{2}(X-\mu)^2/(Y/n)\right) \times$$

$$\frac{S^{\nu/2}}{2^{\nu/2}\Gamma(\nu/2)} Y^{\nu/2-1} \exp(-S/Y)$$

$$= \frac{\sqrt{n}S^{\nu/2}}{\sqrt{\pi}2^{(\nu+1)/2}\Gamma(\nu/2)} Y^{-(\nu+1)/2-1} \times$$

$$\exp\left(-\tfrac{1}{2}\{S + n(X-\mu)^2\}/Y\right)$$

$$\propto Y^{-(\nu+1)/2-1} \exp\left(-\tfrac{1}{2}Q/Y\right)$$

where

$$Q = S + n(X-\mu)^2$$
$$= nX^2 - 2(n\mu)X + (n\mu^2 + S).$$

If we define

$$s^2 = S/\nu$$

then the marginal distribution of X is given by the fact that

$$\frac{X-\mu}{s/\sqrt{n}} \sim \mathrm{t}_\nu.$$

The marginal distribution of Y is of course

$$Y \sim S\chi_\nu^{-2}.$$

Approximate methods of constructing two-dimensional highest density regions for this distribution are described in Box and Tiao (1973, Section 2.4).

A.10 Beta distribution

X has a beta distribution with parameters α and β, denoted

$$X \sim \mathrm{Be}(\alpha, \beta)$$

if it has density

$$p(X) = \frac{1}{\mathrm{B}(\alpha, \beta)} X^{\alpha-1}(1-X)^{\beta-1} \qquad (0<X<1)$$

where the beta function $\mathrm{B}(\alpha, \beta)$ is defined by

$$\mathrm{B}(\alpha, \beta) = \frac{\Gamma(\alpha)\Gamma(\beta)}{\Gamma(\alpha+\beta)}.$$

The mean and variance are

$$\mathsf{E}X = \alpha/(\alpha+\beta)$$

$$\mathscr{V}X = \frac{\alpha\beta}{(\alpha+\beta)^2(\alpha+\beta+1)}.$$

The mode is

$$\text{mode}\,(X) = (\alpha-1)/(\alpha+\beta-2)$$

and the approximate relationship between mean, mode and median can be used to find an approximate median.

The distribution function of any variable with a beta distribution is easily found in terms of the incomplete beta function

$$\mathrm{I}_x(\alpha, \beta) = \int_0^x \frac{1}{\mathrm{B}(\alpha, \beta)} \xi^{\alpha-1}(1-\xi)^{\beta-1}\,\mathrm{d}\xi.$$

Extensive tables can be found in Pearson (1934, 1968) or in Pearson and Hartley (1954, 1958, 1966, Table 17).

A.11 Binomial distribution

X has a binomial distribution of index n and parameter π, denoted

$$X \sim \mathrm{B}(n, \pi)$$

if it has a discrete distribution with density

$$p(X) = \binom{n}{X} \pi X(1-\pi)^{n-X} \qquad (X = 0, 1, 2, \ldots, n).$$

The mean and variance are

$$\mathsf{E}X = n\pi$$
$$\mathcal{V}X = n\pi(1-\pi).$$

Because

$$\frac{p(X+1)}{p(X)} = \frac{(n-X)\pi}{(X+1)(1-\pi)}$$

we see that $p(X+1) > p(X)$ if and only if

$$X < (n+1)\pi - 1$$

and hence that a mode occurs at

$$\text{mode}\,(X) = [(n+1)\pi]$$

the square brackets denoting "integer part of", and this mode is unique unless $(n+1)\pi$ is an integer.

Integration by parts shows that the distribution function is expressible in terms of the incomplete beta function, namely,

$$\sum_{\xi=1}^{x} \binom{n}{\xi} \pi^{\xi}(1-\pi)^{n-\xi} = \mathrm{I}_{1-\pi}(n-x, x+1) = 1 - \mathrm{I}_{\pi}(x+1, n-x)$$

see, for example, Kendall, Stuart and Ord (1987, Section 5.7).

A.12 Poisson distribution

X has a Poisson distribution of mean λ, denoted

$$X \sim \mathrm{P}(\lambda)$$

if it has a discrete distribution with density

$$p(X) = \frac{\lambda^X}{X!} \exp(-\lambda) \qquad (X = 0, 1, 2, \ldots).$$

The mean and variance are

$$\mathsf{E}X = \lambda$$
$$\mathcal{V}X = \lambda.$$

Because

$$p(X+1)/p(X) = \lambda/(X+1)$$

we see that $p(X+1)>p(X)$ if and only if

$$X<\lambda-1$$

and hence that a mode occurs at

$$\text{mode}(X)=[\lambda]$$

the square brackets denoting "integer part of", and this mode is unique unless λ is an integer.

Integration by parts shows that the distribution function is expressible in terms of the incomplete gamma function, namely,

$$\sum_{\xi=0}^{x}\frac{\lambda^{\xi}}{\xi!}\exp(-\lambda)=1-\frac{\gamma(x+1,\lambda)}{\Gamma(x+1)}$$

see, for example, Kendall, Stuart and Ord (1987, Section 5.9).

The Poisson distribution often occurs as the limit of the binomial as

$$n\to\infty,\quad \pi\to 0,\quad n\pi\to\lambda.$$

A.13 Negative binomial distribution

X has a negative binomial distribution of index n and parameter π, denoted

$$X\sim \text{NB}(n,\pi)$$

if it has a discrete distribution with density

$$p(X)=\binom{n+X-1}{X}\pi^{n}(1-\pi)^{x}\qquad (x=0,1,2,\ldots).$$

Because

$$(1+z)^{-n}=\sum_{x=0}^{\infty}\binom{n+x-1}{x}(-1)^{x}z^{x}$$

we sometimes use the notation

$$\binom{-n}{x}=\binom{n+x-1}{x}(-1)^{x}.$$

The mean and variance are

$$\mathsf{E}X=n(1-\pi)/\pi$$
$$\mathcal{V}X=n(1-\pi)/\pi^{2}.$$

Because

$$\frac{p(X+1)}{p(X)} = \frac{(n+X)}{(X+1)}(1-\pi)$$

we see that $p(X+1) > p(X)$ if and only if

$$X < \{n(1-\pi)-1\}/\pi$$

and hence that a mode occurs at

$$\text{mode}\,(X) = [(n-1)(1-\pi)/\pi]$$

the square brackets denoting "integer part of", and this mode is unique unless $(n-1)(1-\pi)/\pi$ is an integer.

It can be shown that the distribution function can be found in terms of that of the binomial distribution, or equivalently in terms of the incomplete beta function; for details see Johnson and Kotz (1969–1972, Chapter 5, Section 6). Just as the Poisson distribution can arise as a limit of the binomial distribution, so it can as a limit of the negative binomial, but in this case as

$$n \to \infty, \quad 1-\pi \to 0, \quad n(1-\pi) \to \lambda.$$

A.14 Hypergeometric distribution

X has a hypergeometric distribution of population size N, index n and parameter π, denoted

$$X \sim \mathrm{H}(N, n, \pi)$$

if it has a discrete distribution with density

$$p(X) = \frac{\binom{N\pi}{X}\binom{N(1-\pi)}{n-X}}{\binom{N}{n}} \qquad (X = 0, 1, 2, \ldots, n).$$

The mean and variance are

$$\mathsf{E}X = n\pi$$
$$\mathscr{V}X = n\pi(1-\pi)(N-n)/(N-1).$$

Because

$$\frac{p(X+1)}{p(X)} = \frac{(n-X)(N\pi-X)}{(X+1)[N(1-\pi)-n+X+1]}$$

we see that $p(X+1)>p(X)$ if and only if

$$X<(n+1)\pi-1+(n+1)(1-2\pi)/(N+2)$$

and hence that if, as is usually the case, N is fairly large, if and only if

$$X<(n+1)\pi-1.$$

Hence the mode occurs very close to the binomial value

$$\text{mode}\,(X)=[(n+1)\pi].$$

As $N\to\infty$ this distribution approaches the binomial distribution $\mathrm{B}(n,\pi)$.

Tables of it can be found in Lieberman and Owen (1961).

A.15 Uniform distribution

X has a uniform distribution on the interval (a, b) denoted

$$X\sim\mathrm{U}(a,b)$$

if it has density

$$p(X)=(b-a)^{-1}\mathrm{I}_{(a,b)}(X)$$

where

$$\mathrm{I}_{(a,b)}(X)=\begin{cases}1 & (X\in(a,b))\\ 0 & (X\notin(a,b))\end{cases}$$

is the *indicator function* of the set (a, b). The mean and variance are

$$\mathsf{E}X=\tfrac{1}{2}(a+b)$$
$$\mathcal{V}X=(b-a)^2/12.$$

There is no unique mode, but the distribution is symmetrical and hence

$$\text{median}\,(X)=\tfrac{1}{2}(a+b).$$

Sometimes we have occasion to refer to a discrete version; Y has a discrete uniform distribution on the interval $[a, b]$ denoted

$$Y\sim\mathrm{UD}(a,b)$$

if it has a discrete distribution with density

$$p(Y)=(b-a+1)^{-1}\qquad (Y=a,a+1,\ldots,b).$$

The mean and variance are

$$\mathsf{E}Y=\tfrac{1}{2}(a+b)$$
$$\mathcal{V}Y=(b-a)(b-a+2)/12$$

using formulae for the sum and sum of squares of the first n natural numbers [the variance is best found by noting that the variance of $\mathrm{UD}(a, b)$ equals that of $\mathrm{UD}(1, n)$ where $n = b - a + 1$]. Again, there is no unique mode, but the distribution is symmetrical and hence

$$\text{median } (Y) = \tfrac{1}{2}(a+b).$$

A.16 Pareto distribution

X has a Pareto distribution with parameters ξ and γ, denoted

$$X \sim \mathrm{Pa}(\xi, \gamma)$$

if it has density

$$p(X) = \gamma\xi^{\gamma}X^{-\gamma-1}\mathrm{I}_{(\xi,\infty)}(X)$$

where

$$\mathrm{I}_{(\xi,\infty)}(X) = \begin{cases} 1 & (X > \xi) \\ 0 & (\text{otherwise}). \end{cases}$$

The mean and variance are

$$\mathsf{E}X = \gamma\xi/(\gamma-1) \qquad (\text{provided } \gamma > 1)$$

$$\mathscr{V}X = \frac{\gamma\xi^2}{(\gamma-1)^2(\gamma-2)} \qquad (\text{provided } \gamma > 2).$$

The distribution function is

$$F(x) = [1 - (\xi/x)^{\gamma}]\,\mathrm{I}_{(\xi,\infty)}(x)$$

and in particular the median is

$$\text{median } (X) = 2^{1/\gamma}\xi.$$

The mode, of course, occurs at

$$\text{mode } (X) = \xi.$$

The ordered pair (Y, Z) has a bilateral bivariate Pareto distribution with parameters ξ, η and γ, denoted

$$(Y, Z) \sim \text{Pabb } (\xi, \eta, \gamma)$$

if it has joint density function

$$p(Y, Z) = \gamma(\gamma+1)(\xi-\eta)^{\gamma}(Y-Z)^{-\gamma-2}\mathrm{I}_{(\xi,\infty)}(Y)\mathrm{I}_{(-\infty,\eta)}(Z).$$

The means and variances are

$$\mathsf{E}Y = (\gamma\xi - \eta)/(\gamma - 1)$$
$$\mathsf{E}Z = (\gamma\eta - \xi)/(\gamma - 1)$$
$$\mathcal{V}Y = \mathcal{V}Z = \frac{\gamma(\xi-\eta)^2}{(\gamma-1)^2(\gamma-2)}$$

and the correlation between Y and Z is

$$\rho(Y, Z) = -1/\gamma.$$

It is also sometimes useful that

$$\mathsf{E}(Y-Z) = (\gamma+1)(\xi-\eta)/(\gamma-1)$$
$$\mathcal{V}(Y-Z) = \frac{2(\gamma+1)(\xi-\eta)^2}{(\gamma-1)^2(\gamma-2)}.$$

The marginal distribution function of Y is

$$F(y) = \left[1 - \left(\frac{\xi-\eta}{y-\eta}\right)^{\gamma}\right] \mathrm{I}_{(\xi,\infty)}(y)$$

and in particular the median is

$$\text{median}\,(Y) = \eta + 2^{1/\gamma}(\xi-\eta).$$

The distribution function of Z is similar, and in particular the median is

$$\text{median}\,(Z) = \xi - 2^{1/\gamma}(\xi-\eta).$$

The modes, of course, occur at

$$\text{mode}\,(Y) = \xi$$
$$\text{mode}\,(Z) = \eta.$$

The distribution is discussed in DeGroot (1970, Sections 4.11, 5.7, and 9.7).

A.17 Circular normal distribution

X has a circular normal or von Mises' distribution with mean direction μ and concentration parameter κ, denoted

$$X \sim \mathrm{M}(\mu, \kappa)$$

if it has density

$$p(X) = (2\pi \mathrm{I}_0(\kappa))^{-1} \exp\,(\kappa \cos(X-\mu))$$

where X is any angle, so $0 \leqslant X < 2\pi$, and $I_0(\kappa)$ is a constant called the modified Bessel function of the first kind and order zero. It turns out that

$$I_0(\kappa) = \sum_{r=0}^{\infty} \frac{1}{(r!)^2} (\tfrac{1}{2}\kappa)^{2r}$$

and that asymptotically for large X

$$I_0(\kappa) \sim \frac{1}{\sqrt{(2\pi\kappa)}} \exp(\kappa).$$

For large κ we have approximately

$$X \sim N(\mu, 1/\kappa)$$

while for small κ we have approximately

$$p(X) = (2\pi)^{-1}\{1 + \tfrac{1}{2}\kappa \cos(X - \mu)\}$$

which density is sometimes referred to as a cardioid distribution. The circular normal distribution is discussed by Mardia (1972) and Batschelet (1981).

One point related to this distribution arises in a Bayesian context in connection with the reference prior

$$p(\mu, \kappa) \propto 1$$

when we have observations $X_1, X_2, \ldots, X_n$ such that

$$\begin{aligned} c &= n^{-1} \textstyle\sum \cos X_i \\ s &= n^{-1} \textstyle\sum \sin X_i \\ \rho &= \sqrt{(c^2 + s^2)} \\ \hat{\mu} &= \tan^{-1}(s/c). \end{aligned}$$

The only sensible estimator of μ on the basis of the posterior distribution is $\hat{\mu}$.

The mode of the posterior distribution of κ is approximately $\hat{\kappa}$, where

$$\hat{\kappa} = \rho\left(\frac{2 - \rho^2}{1 - \rho^2}\right) \qquad (\rho < 2/3)$$

$$\hat{\kappa} = \frac{\rho + 1}{4\rho(1 - \rho)} \qquad (\rho > 2/3)$$

(both of which are approximately 1.87 when $\rho = 2/3$), according to Schmitt (1969, Section 10.2). Because of the skewness of the distribution of κ, its posterior mean is greater than its posterior mode.

A.18 Behrens' distribution

X is said to have Behrens' (or Behrens–Fisher or Fisher–Behrens) distribution with degrees of freedom ν_1 and ν_2 and angle φ, denoted

$$X \sim \mathrm{BF}(\nu_1, \nu_2, \varphi)$$

if X has the same distribution as

$$T_2 \cos \varphi - T_1 \sin \varphi$$

where T_1 and T_2 are independent and

$$T_1 \sim \mathrm{t}_{\nu_1} \text{ and } T_2 \sim \mathrm{t}_{\nu_2}.$$

Equivalently, X has density

$$p(X) = k \int_{-\infty}^{\infty} g(z)\,\mathrm{d}z$$

where

$$g(z) = \left[1 + \frac{(z \cos \varphi - X \sin \varphi)^2}{\nu_1}\right]^{-(\nu_1+1)/2} \left[1 + \frac{(z \sin \varphi - X \cos \varphi)^2}{\nu_2}\right]^{-(\nu_2+1)/2}$$

over the whole real line, where

$$k = \mathrm{B}(\tfrac{1}{2}\nu_1, \tfrac{1}{2})\mathrm{B}(\tfrac{1}{2}\nu_2, \tfrac{1}{2})\sqrt{(\nu_1\nu_2)}.$$

This distribution usually arises as the posterior distribution of

$$\theta_2 - \theta_1$$

when we have samples of size $n_1 = \nu_1 + 1$ from $\mathrm{N}(\theta_1, \sigma_1^2)$ and of size $n_2 = \nu_2 + 1$ from $\mathrm{N}(\theta_2, \sigma_2^2)$ and neither σ_1^2 nor σ_1^2 is known, and conventional priors are adopted. In this case, in a fairly obvious notation

$$T_1 = \frac{\theta_1 - \bar{X}_1}{s_1/\sqrt{n_1}} \quad \text{and} \quad T_2 = \frac{\theta_2 - \bar{X}_2}{s_2/\sqrt{n_2}}$$

$$\tan \varphi = \frac{s_1/\sqrt{n_1}}{s_2/\sqrt{n_2}}.$$

An approximation to this distribution due to Patil (1965) is as follows.

Define

$$f_1 = \left(\frac{\nu_2}{\nu_2 - 2}\right) \cos^2 \varphi + \left(\frac{\nu_1}{\nu_1 - 2}\right) \sin^2 \varphi$$

$$f_2 = \frac{\nu_2^2}{(\nu_2-2)^2(\nu_2-4)}\cos^4\varphi + \frac{\nu_1^2}{(\nu_1-2)^2(\nu_1-4)}\sin^4\varphi$$
$$a = \sqrt{\{f_1(b-2)/b\}}$$
$$b = 4 + (f_1^2/f_2)$$
$$s^2 = (s_1^2/n_1) + (s_2^2/n_2)$$
$$T = \frac{(\theta_2-\theta_1)-(\bar{X}_2-\bar{X}_1)}{s}$$

Then, approximately,

$$T/a \sim \mathrm{t}_b.$$

Obviously b is usually not an integer, and consequently this approximation requires interpolation in the t-tables.

Clearly Behrens' distribution has mean and variance

$$\mathsf{E}X = 0$$

$$\mathcal{V}X = \frac{\nu_2\cos^2\varphi}{\nu_2-2} + \frac{\nu_1\sin^2\theta}{\nu_1-2}$$

using the mean and variance of t distributions and the independence of T_1 and T_2. The distribution is symmetrical and unimodal and hence the mean, mode and median are all equal, so

$$\text{mode }(X) = 0$$
$$\text{median }(X) = 0.$$

A.19 Snedecor's F distribution

X has an F distribution on ν_1 and ν_2 degrees of freedom, denoted

$$X \sim \mathrm{F}_{\nu_1,\nu_2}$$

if X has the same distribution as

$$\frac{W_1/\nu_1}{W_2/\nu_2}$$

where W_1 and W_2 are independent and

$$W_1 \sim \chi_{\nu_1}^2 \quad \text{and} \quad W_2 \sim \chi_{\nu_2}^2.$$

Equivalently, X has density

$$p(X) = \frac{\nu_1^{(\nu_1/2)}\nu_2^{(\nu_2/2)}}{\mathrm{B}(\frac{1}{2}\nu_1, \frac{1}{2}\nu_2)} \frac{X^{\nu_1/2-1}}{(\nu_2+\nu_1X)^{(\nu_1+\nu_2)/2}}.$$

The mean and variance are

$$\mathsf{E}X = \nu_2/(\nu_2 - 2)$$

$$\mathscr{V}X = \frac{2\nu_2{}^2(\nu_1 + \nu_2 - 2)}{\nu_1(\nu_2 - 2)^2(\nu_2 - 4)}.$$

The mode is

$$\text{mode }(X) = \frac{\nu_2}{\nu_2 + 1}\,\frac{\nu_1 - 2}{\nu_1}.$$

If $X \sim \mathrm{F}_{\nu_1, \nu_2}$, then

$$\frac{\nu_1 X}{\nu_2 + \nu_1 X} \sim \mathrm{Be}(\tfrac{1}{2}\nu_1, \tfrac{1}{2}\nu_2).$$

Conversely, if $Y \sim \mathrm{B}(\delta_1, \delta_2)$ then

$$X = \frac{\delta_2 Y}{\delta_1(1 - Y)} \sim \mathrm{F}_{2\delta_1, 2\delta_2}.$$

A.20 Fisher's z distribution

X has a z distribution on ν_1 and ν_2 degrees of freedom, denoted

$$X \sim \mathrm{z}_{\nu_1, \nu_2}$$

if $Y = \exp(2X) \sim \mathrm{F}_{\nu_1, \nu_2}$, or equivalently if it has the density

$$p(X) = 2\frac{\nu_1^{\nu_1/2}\nu_2^{\nu_2/2}}{\mathrm{B}(\nu_1/2, \nu_2/2)}\,\frac{\mathrm{e}^{\nu_1 X}}{(\nu_2 + \nu_1 \mathrm{e}^{2X})(\nu_1 + \nu_2)/2}.$$

Another definition is that if $Y \sim \mathrm{F}_{\nu_1, \nu_2}$, then

$$X = \tfrac{1}{2}\log Y \sim \mathrm{z}_{\nu_1, \nu_2}.$$

The mean and variance are easily deduced from those of the log chi-squared distribution; they are

$$\mathsf{E}X = \tfrac{1}{2}\log(\nu_2/\nu_1) + \tfrac{1}{2}\psi(\nu_1/2) - \tfrac{1}{2}\psi(\nu_2/2)$$
$$\mathscr{V}X = \tfrac{1}{4}\psi'(\nu_1/2) + \tfrac{1}{4}\psi'(\nu_2/2)$$

or approximately

$$\mathsf{E}X = \tfrac{1}{2}(\nu_2^{-1} - \nu_1^{-1}) \cong \tfrac{1}{2}\log[(1 - \nu_1^{-1})/(1 - \nu_2^{-1})]$$
$$\mathscr{V}X = \tfrac{1}{2}(\nu_1^{-1} + \nu_2^{-1}).$$

The mode is zero.

Unless ν_1 and ν_2 are very small, the distribution of z is approximately normal. The z distribution was introduced by Fisher (1924).

A.21 Cauchy distribution

X has a Cauchy distribution with location parameter μ and scale parameter σ^2, denoted

$$X \sim \mathrm{C}(\mu, \sigma^2)$$

if

$$p(X) = \frac{1}{\pi} \frac{\sigma}{\sigma^2 + (X-\mu)^2}.$$

Because the relevant integral is not absolutely convergent, this distribution does not have a finite mean, nor, *a fortiori*, a finite variance. However, it is symmetrical about μ, and hence

$$\text{median } (X) = \mu$$
$$\text{mode } (X) = \mu.$$

The distribution function is

$$F(x) = \tfrac{1}{2} + \pi^{-1} \tan^{-1} \{(x-\mu)/\sigma\}$$

so that when $x = \mu - \sigma$ then $F(x) = 1/4$ and when $x = \mu + \sigma$ then $F(x) = 3/4$. Thus $\mu - \sigma$ and $\mu + \sigma$ are, respectively, the lower and upper quartiles and hence σ may be thought of as the *semi-interquartile range*. Note that for a normal $\mathrm{N}(\mu, \sigma^2)$ distribution the semi-interquartile range is $0.67449\,\sigma$ rather than σ.

It may be noted that the $\mathrm{C}(0, 1)$ distribution is also Student's t distribution on 1 degree of freedom.

A.22 Probability one beta variable is greater than another

Suppose π and ρ have independent beta distributions

$$\pi \sim \mathrm{Be}(\alpha, \beta), \qquad \rho \sim \mathrm{Be}(\gamma, \delta).$$

Then

$$\mathrm{P}(\pi < \rho) = \sum_{\kappa = \max(\gamma-\beta, 0)}^{\gamma-1} \frac{\binom{\gamma+\delta-1}{\kappa}\binom{\alpha+\beta+1}{\alpha+\gamma-1-\kappa}}{\binom{\alpha+\beta+\gamma+\delta-2}{\alpha+\gamma-1}}$$

[see Altham (1969)]. For an expression (albeit a complicated one) for

$$P(\pi/\rho \leqslant c)$$

see Weisberg (1972).

When α, β, γ and δ are large we can approximate the beta variates by normal variates of the same means and variances and hence approximate the distribution of $\pi-\rho$ by a normal distribution.

A.23 Distribution of the correlation coefficient

If the prior density of the correlation coefficient is ρ, then its posterior density, given n pairs of observation (X_i, Y_i) with sample correlation coefficient r, is given by

$$\begin{aligned} p(\rho|\boldsymbol{X},\boldsymbol{Y}) &\propto p(\rho)(1-\rho^2)^{(n-1)/2}\int_0^\infty \omega^{-1}(\omega+\omega^{-1}-2\rho r)^{-(n-1)}\,\mathrm{d}\omega \\ &\propto p(\rho)(1-\rho^2)^{(n-1)/2}\int_0^\infty (\cosh t-\rho r)^{-(n-1)}\,\mathrm{d}t \\ &\propto p(\rho)(1-\rho^2)^{(n-1)/2}\int_0^\infty (\cosh t+\cos\theta)^{-(n-1)}\,\mathrm{d}t \end{aligned}$$

on writing $-\rho r=\cos\theta$. It can also be shown that

$$p(\rho|\boldsymbol{X},\boldsymbol{Y}) \propto p(\rho)(1-\rho^2)^{(n-1)/2}\left(\frac{\partial}{\sin\theta\,\partial\theta}\right)^{n-2}\left(\frac{\theta}{\sin\theta}\right).$$

When $r=0$ the density simplifies to

$$p(\rho|\boldsymbol{X},\boldsymbol{Y}) \propto p(\rho)(1-\rho^2)^{(n-1)/2}$$

and so if the prior is of the form

$$p(\rho) \propto (1-\rho^2)^{(\nu_0-3)/2}$$

it can be shown that

$$(\nu_0+n-2)^{\frac{1}{2}}\frac{\rho}{(1-\rho^2)^{\frac{1}{2}}} \sim \mathrm{t}_{\nu_0+n-2}$$

has a Student's t distribution on ν_0+n-2 degrees of freedom.

Going back to the general case, it can be shown that

$$\begin{aligned} p(\rho|\boldsymbol{X},\boldsymbol{Y}) &\propto p(\rho)\frac{(1-\rho^2)^{(n-1)/2}}{(1-\rho r)^{n-(3/2)}} \\ &\quad\times\int_0^1 \frac{(1-u)^{n-2}}{(2u)^{\frac{1}{2}}}\,[1-\tfrac{1}{2}(1+\rho r)u]^{-\frac{1}{2}}\,\mathrm{d}u. \end{aligned}$$

Expanding the term in square brackets as a power series in u we can express the last integral as a sum of beta functions. Taking only the first term we have as an approximation

$$p(\rho \mid \boldsymbol{X}, \boldsymbol{Y}) \propto p(\rho) \frac{(1-\rho^2)^{(n-1)/2}}{(1-\rho r)^{n-(3/2)}}.$$

On writing

$$\rho = \tanh \zeta, \qquad r = \tanh z$$

it can be shown that

$$p(\zeta \mid \boldsymbol{X}, \boldsymbol{Y}) \propto \frac{p(\zeta)}{\cosh^{5/2}(\zeta)\cosh^{n-(3/2)}(\zeta - z)}$$

and hence that for large n

$$\zeta \propto \mathrm{N}(z, 1/n)$$

approximately [whatever $p(\rho)$ is]. A better approximation is

$$\zeta \sim \mathrm{N}(z - (\tfrac{5}{2})n, \{n - (\tfrac{3}{2}) + (\tfrac{5}{2})(1 - r^2)\}^{-1}).$$

Table A.1 Percentage points of the Behrens–Fisher distribution

		$\psi = 0°$						$\psi = 15°$			
	ν_2	75%	90%	95%	97.5%		ν_2	75%	90%	95%	97.5%
	6	0.72	1.44	1.94	2.45		6	0.72	1.45	1.95	2.45
$\nu_1 = 6$	8	0.71	1.40	1.86	2.31	$\nu_1 = 6$	8	0.72	1.41	1.87	2.32
	12	0.70	1.36	1.78	2.18		12	0.71	1.37	1.80	2.19
	24	0.68	1.32	1.71	2.06		24	0.69	1.34	1.73	2.09
	∞	0.67	1.28	1.65	1.96		∞	0.68	1.30	1.67	2.00

		$\psi = 0°$						$\psi = 15°$			
	ν_2	75%	90%	95%	97.5%		ν_2	75%	90%	95%	97.5%
	6	0.72	1.44	1.94	2.45		6	0.72	1.44	1.94	2.44
$\nu_1 = 8$	8	0.71	1.40	1.86	2.31	$\nu_1 = 8$	8	0.71	1.40	1.86	2.31
	12	0.70	1.36	1.78	2.18		12	0.70	1.37	1.79	2.18
	24	0.68	1.32	1.71	2.06		24	0.69	1.33	1.72	2.08
	∞	0.67	1.28	1.65	1.96		∞	0.68	1.30	1.66	1.98

		$\psi = 0°$						$\psi = 15°$			
	ν_2	75%	90%	95%	97.5%		ν_2	75%	90%	95%	97.5%
	6	0.72	1.44	1.94	2.45		6	0.72	1.44	1.94	2.43
$\nu_1 = 12$	8	0.71	1.40	1.86	2.31	$\nu_1 = 12$	8	0.71	1.40	1.86	2.30
	12	0.70	1.36	1.78	2.18		12	0.70	1.36	1.78	2.18
	24	0.68	1.32	1.71	2.06		24	0.69	1.32	1.72	2.07
	∞	0.67	1.28	1.65	1.96		∞	0.68	1.29	1.66	1.98

		$\psi = 0°$						$\psi = 15°$			
	ν_2	75%	90%	95%	97.5%		ν_2	75%	90%	95%	97.5%
	6	0.72	1.44	1.94	2.45		6	0.72	1.43	1.93	2.43
$\nu_1 = 24$	8	0.71	1.40	1.86	2.31	$\nu_1 = 24$	8	0.71	1.39	1.85	2.29
	12	0.70	1.36	1.78	2.18		12	0.70	1.36	1.78	2.17
	24	0.68	1.32	1.71	2.06		24	0.69	1.32	1.71	2.06
	∞	0.67	1.28	1.65	1.96		∞	0.68	1.29	1.65	1.97

		$\psi = 0°$						$\psi = 15°$			
	ν_2	75%	90%	95%	97.5%		ν_2	75%	90%	95%	97.5%
	6	0.72	1.44	1.94	2.45		6	0.71	1.43	1.93	2.42
$\nu_1 = \infty$	8	0.71	1.40	1.86	2.31	$\nu_1 = \infty$	8	0.70	1.39	1.85	2.29
	12	0.70	1.36	1.78	2.18		12	0.70	1.35	1.77	2.16
	24	0.68	1.32	1.71	2.06		24	0.69	1.32	1.71	2.06
	∞	0.67	1.28	1.65	1.96		∞	0.67	1.28	1.65	1.96

(*continued*)

Table A.1—continued

		$\psi = 30°$						$\psi = 45°$			
	ν_2	75%	90%	95%	97.5%		ν_2	75%	90%	95%	97.5%
	6	0.74	1.47	1.96	2.45		6	0.75	1.48	1.97	2.45
$\nu_1 = 6$	8	0.73	1.44	1.90	2.34	$\nu_1 = 6$	8	0.74	1.45	1.93	2.37
	12	0.72	1.40	1.84	2.24		12	0.72	1.42	1.88	2.32
	24	0.71	1.37	1.79	2.16		24	0.71	1.39	1.84	2.27
	∞	0.69	1.34	1.74	2.10		∞	0.70	1.37	1.81	2.23

		$\psi = 30°$						$\psi = 45°$			
	ν_2	75%	90%	95%	97.5%		ν_2	75%	90%	95%	97.5%
	6	0.73	1.45	1.94	2.41		6	0.74	1.45	1.93	2.37
$\nu_1 = 8$	8	0.73	1.42	1.87	2.30	$\nu_1 = 8$	8	0.73	1.43	1.88	2.30
	12	0.72	1.39	1.81	2.20		12	0.72	1.40	1.84	2.23
	24	0.70	1.36	1.76	2.12		24	0.71	1.38	1.80	2.18
	∞	0.69	1.32	1.71	2.05		∞	0.70	1.35	1.77	2.14

		$\psi = 30°$						$\psi = 45°$			
	ν_2	75%	90%	95%	97.5%		ν_2	75%	90%	95%	97.5%
	6	0.72	1.43	1.91	2.39		6	0.72	1.42	1.88	2.32
$\nu_1 = 12$	8	0.72	1.40	1.85	2.27	$\nu_1 = 12$	8	0.72	1.40	1.84	2.23
	12	0.71	1.37	1.79	2.17		12	0.71	1.38	1.79	2.17
	24	0.70	1.34	1.73	2.09		24	0.70	1.35	1.75	2.11
	∞	0.69	1.31	1.69	2.01		∞	0.69	1.33	1.72	2.07

		$\psi = 30°$						$\psi = 45°$			
	ν_2	75%	90%	95%	97.5%		ν_2	75%	90%	95%	97.5%
	6	0.71	1.42	1.89	2.36		6	0.71	1.39	1.84	2.27
$\nu_1 = 24$	8	0.71	1.39	1.83	2.25	$\nu_1 = 24$	8	0.71	1.38	1.80	2.18
	12	0.70	1.36	1.77	2.15		12	0.70	1.35	1.75	2.11
	24	0.69	1.33	1.71	2.06		24	0.69	1.33	1.71	2.06
	∞	0.68	1.30	1.66	1.99		∞	0.68	1.30	1.68	2.01

		$\psi = 30°$						$\psi = 45°$			
	ν_2	75%	90%	95%	97.5%		ν_2	75%	90%	95%	97.5%
	6	0.70	1.40	1.88	2.34		6	0.70	1.37	1.81	2.23
$\nu_1 = \infty$	8	0.70	1.37	1.81	2.23	$\nu_1 = \infty$	8	0.70	1.35	1.77	2.14
	12	0.69	1.34	1.75	2.13		12	0.69	1.33	1.72	2.07
	24	0.69	1.31	1.70	2.04		24	0.68	1.30	1.68	2.01
	∞	0.67	1.28	1.65	1.96		∞	0.67	1.28	1.65	1.96

Table A.2 Highest density regions for the chi-squared distribution

ν	50%		60%		67%		70%		75%	
3	0.259	2.543	0.170	3.061	0.120	3.486	0.099	3.731	0.070	4.155
4	0.871	3.836	0.684	4.411	0.565	4.876	0.506	5.143	0.420	5.603
5	1.576	5.097	1.315	5.730	1.141	6.238	1.052	6.527	0.918	7.023
6	2.327	6.330	2.004	7.016	1.783	7.563	1.669	7.874	1.493	8.404
7	3.107	7.540	2.729	8.276	2.467	8.860	2.331	9.190	2.118	9.753
8	3.907	8.732	3.480	9.515	3.181	10.133	3.025	10.482	2.779	11.075
9	4.724	9.911	4.251	10.737	3.917	11.387	3.742	11.754	3.465	12.376
10	5.552	11.079	5.037	11.946	4.671	12.626	4.479	13.009	4.174	13.658
11	6.391	12.238	5.836	13.143	5.440	13.851	5.231	14.250	4.899	14.925
12	7.238	13.388	6.645	14.330	6.221	15.066	5.997	15.480	5.639	16.179
13	8.093	14.532	7.464	15.508	7.013	16.271	6.774	16.699	6.392	17.422
14	8.954	15.669	8.290	16.679	7.814	17.467	7.560	17.909	7.155	18.654
15	9.821	16.801	9.124	17.844	8.622	18.656	8.355	19.111	7.928	19.878
16	10.692	17.929	9.963	19.003	9.438	19.838	9.158	20.306	8.709	21.094
17	11.568	19.052	10.809	20.156	10.260	21.014	9.968	21.494	9.497	22.303
18	12.448	20.171	11.659	21.304	11.088	22.184	10.783	22.677	10.293	23.505
19	13.331	21.287	12.514	22.448	11.921	23.349	11.605	23.854	11.095	24.701
20	14.218	22.399	13.373	23.589	12.759	24.510	12.431	25.026	11.902	25.892
21	15.108	23.509	14.235	24.725	13.602	25.667	13.262	26.194	12.715	27.078
22	16.001	24.615	15.102	25.858	14.448	26.820	14.098	27.357	13.532	28.259
23	16.897	25.719	15.971	26.987	15.298	27.969	14.937	28.517	14.354	29.436
24	17.794	26.821	16.844	28.114	16.152	29.114	15.780	29.672	15.180	30.609
25	18.695	27.921	17.719	29.238	17.008	30.257	16.627	30.825	16.010	31.778
26	19.597	29.018	18.597	30.360	17.868	31.396	17.477	31.974	16.843	32.944
27	20.501	30.113	19.478	31.479	18.731	32.533	18.329	33.121	17.680	34.106
28	21.407	31.207	20.361	32.595	19.596	33.667	19.185	34.264	18.520	35.265
29	22.315	32.299	21.246	33.710	20.464	34.798	20.044	35.405	19.363	36.421
30	23.225	33.389	22.133	34.822	21.335	35.927	20.905	36.543	20.209	37.575
35	27.795	38.818	26.597	40.357	25.718	41.541	25.245	42.201	24.477	43.304
40	32.396	44.216	31.099	45.853	30.146	47.112	29.632	47.812	28.797	48.981
45	37.023	49.588	35.633	51.318	34.611	52.646	34.059	53.383	33.161	54.616
50	41.670	54.940	40.194	56.757	39.106	58.150	38.518	58.924	37.560	60.216
55	46.336	60.275	44.776	62.174	43.626	63.629	43.004	64.437	41.990	65.784
60	51.017	65.593	49.378	67.572	48.169	69.086	47.514	69.926	46.446	71.328

(continued)

Table A.2—continued

ν	80%		90%		95%		99%		99.5%	
3	0.046	4.672	0.012	6.260	0.003	7.817	0.000	11.346	0.000	12.840
4	0.335	6.161	0.168	7.864	0.085	9.530	0.017	13.287	0.009	14.860
5	0.779	7.622	0.476	9.434	0.296	11.191	0.101	15.128	0.064	16.771
6	1.308	9.042	0.883	10.958	0.607	12.802	0.264	16.903	0.186	18.612
7	1.891	10.427	1.355	12.442	0.989	14.369	0.496	18.619	0.372	20.390
8	2.513	11.784	1.875	13.892	1.425	15.897	0.786	20.295	0.614	22.116
9	3.165	13.117	2.431	15.314	1.903	17.393	1.122	21.931	0.904	23.802
10	3.841	14.430	3.017	16.711	2.414	18.860	1.498	23.532	1.233	25.450
11	4.535	15.727	3.628	18.087	2.953	20.305	1.906	25.108	1.596	27.073
12	5.246	17.009	4.258	19.447	3.516	21.729	2.344	26.654	1.991	28.659
13	5.970	18.279	4.906	20.789	4.099	23.135	2.807	28.176	2.410	30.231
14	6.707	19.537	5.570	22.119	4.700	24.525	3.291	29.685	2.853	31.777
15	7.454	20.786	6.246	23.437	5.317	25.901	3.795	31.171	3.317	33.305
16	8.210	22.026	6.935	24.743	5.948	27.263	4.315	32.644	3.797	34.821
17	8.975	23.258	7.634	26.039	6.591	28.614	4.853	34.099	4.296	36.315
18	9.747	24.483	8.343	27.325	7.245	29.955	5.404	35.539	4.811	37.788
19	10.527	25.701	9.060	28.604	7.910	31.285	5.968	36.972	5.339	39.253
20	11.312	26.913	9.786	29.876	8.584	32.608	6.545	38.388	5.879	40.711
21	12.104	28.120	10.519	31.140	9.267	33.921	7.132	39.796	6.430	42.160
22	12.900	29.322	11.259	32.398	9.958	35.227	7.730	41.194	6.995	43.585
23	13.702	30.519	12.005	33.649	10.656	36.526	8.337	42.583	7.566	45.016
24	14.508	31.711	12.756	34.896	11.362	37.817	8.951	43.969	8.152	46.421
25	15.319	32.899	13.514	36.136	12.073	39.103	9.574	45.344	8.742	47.832
26	16.134	34.083	14.277	37.372	12.791	40.384	10.206	46.708	9.341	49.232
27	16.952	35.264	15.044	38.603	13.515	41.657	10.847	48.062	9.949	50.621
28	17.774	36.441	15.815	39.830	14.243	42.927	11.491	49.419	10.566	52.000
29	18.599	37.615	16.591	41.052	14.977	44.191	12.143	50.764	11.186	53.381
30	19.427	38.786	17.372	42.271	15.715	45.452	12.804	52.099	11.815	54.752
35	23.611	44.598	21.327	48.311	19.473	51.687	16.179	58.716	15.051	61.504
40	27.855	50.352	25.357	54.276	23.319	57.836	19.668	65.223	18.408	68.143
45	32.146	56.059	29.449	60.182	27.238	63.913	23.257	71.624	21.871	74.672
50	36.478	61.726	33.591	66.037	31.217	69.931	26.919	77.962	25.416	81.127
55	40.842	67.360	37.777	71.849	35.249	75.896	30.648	84.230	29.036	87.501
60	45.236	72.965	41.999	77.625	39.323	81.821	34.436	90.440	32.717	93.818

Table A.3 Highest density regions for the inverse chi-squared distribution

ν	50%		60%		67%		70%		75%	
3	0.106	0.446	0.093	0.553	0.085	0.653	0.082	0.716	0.076	0.837
4	0.098	0.320	0.087	0.380	0.080	0.435	0.077	0.469	0.072	0.532
5	0.089	0.249	0.081	0.289	0.075	0.324	0.072	0.346	0.068	0.385
6	0.082	0.204	0.075	0.233	0.070	0.258	0.067	0.273	0.064	0.299
7	0.075	0.173	0.069	0.195	0.065	0.213	0.063	0.224	0.060	0.244
8	0.070	0.150	0.064	0.167	0.061	0.182	0.059	0.190	0.056	0.206
9	0.065	0.133	0.060	0.147	0.057	0.158	0.055	0.165	0.053	0.177
10	0.061	0.119	0.056	0.130	0.054	0.140	0.052	0.146	0.050	0.156
11	0.057	0.107	0.053	0.117	0.051	0.125	0.049	0.130	0.047	0.138
12	0.054	0.098	0.050	0.106	0.048	0.113	0.047	0.118	0.045	0.125
13	0.051	0.090	0.048	0.097	0.045	0.104	0.044	0.107	0.042	0.113
14	0.048	0.083	0.045	0.090	0.043	0.095	0.042	0.098	0.041	0.104
15	0.046	0.078	0.043	0.083	0.041	0.088	0.040	0.091	0.039	0.096
16	0.044	0.072	0.041	0.078	0.039	0.082	0.038	0.084	0.037	0.089
17	0.042	0.068	0.039	0.073	0.038	0.077	0.037	0.079	0.036	0.083
18	0.040	0.064	0.038	0.068	0.036	0.072	0.035	0.074	0.034	0.077
19	0.038	0.061	0.036	0.065	0.035	0.068	0.034	0.069	0.033	0.073
20	0.037	0.057	0.035	0.061	0.033	0.064	0.033	0.066	0.032	0.068
21	0.035	0.055	0.033	0.058	0.032	0.061	0.032	0.062	0.031	0.065
22	0.034	0.052	0.032	0.055	0.031	0.058	0.031	0.059	0.030	0.061
23	0.033	0.050	0.031	0.053	0.030	0.055	0.030	0.056	0.029	0.058
24	0.032	0.047	0.030	0.050	0.029	0.052	0.029	0.054	0.028	0.056
25	0.031	0.046	0.029	0.048	0.028	0.050	0.028	0.051	0.027	0.053
26	0.030	0.044	0.028	0.046	0.027	0.048	0.027	0.049	0.026	0.051
27	0.029	0.042	0.027	0.044	0.027	0.046	0.026	0.047	0.025	0.049
28	0.028	0.040	0.027	0.043	0.026	0.044	0.025	0.045	0.025	0.047
29	0.027	0.039	0.026	0.041	0.025	0.043	0.025	0.043	0.024	0.045
30	0.026	0.038	0.025	0.039	0.024	0.041	0.024	0.042	0.023	0.043
35	0.023	0.032	0.022	0.034	0.021	0.035	0.021	0.035	0.021	0.036
40	0.020	0.028	0.020	0.029	0.019	0.030	0.019	0.031	0.018	0.031
45	0.018	0.025	0.018	0.026	0.017	0.026	0.017	0.027	0.017	0.028
50	0.017	0.022	0.016	0.023	0.016	0.024	0.016	0.024	0.015	0.025
55	0.015	0.020	0.015	0.021	0.015	0.021	0.014	0.022	0.014	0.022
60	0.014	0.018	0.014	0.019	0.014	0.019	0.013	0.020	0.013	0.020

Table A.3—continued

ν	80%		90%		95%		99%		99.5%	
3	0.070	1.005	0.057	1.718	0.048	2.847	0.036	8.711	0.033	13.946
4	0.067	0.616	0.055	0.947	0.047	1.412	0.036	3.370	0.033	4.829
5	0.063	0.436	0.053	0.627	0.046	0.878	0.036	1.807	0.033	2.433
6	0.060	0.334	0.050	0.460	0.044	0.616	0.035	1.150	0.032	1.482
7	0.056	0.270	0.048	0.359	0.042	0.466	0.033	0.810	0.031	1.013
8	0.053	0.225	0.045	0.292	0.040	0.370	0.032	0.610	0.030	0.746
9	0.050	0.193	0.043	0.245	0.038	0.305	0.031	0.481	0.029	0.578
10	0.047	0.168	0.041	0.210	0.037	0.257	0.030	0.393	0.028	0.466
11	0.045	0.149	0.039	0.184	0.035	0.222	0.029	0.330	0.027	0.386
12	0.043	0.134	0.037	0.163	0.034	0.195	0.028	0.282	0.026	0.327
13	0.041	0.121	0.036	0.146	0.032	0.173	0.027	0.245	0.025	0.282
14	0.039	0.110	0.034	0.132	0.031	0.155	0.026	0.216	0.024	0.247
15	0.037	0.102	0.033	0.120	0.030	0.140	0.025	0.193	0.024	0.219
16	0.036	0.094	0.032	0.111	0.029	0.128	0.024	0.174	0.023	0.196
17	0.034	0.087	0.031	0.102	0.028	0.118	0.024	0.158	0.022	0.177
18	0.033	0.082	0.029	0.095	0.027	0.109	0.023	0.144	0.022	0.161
19	0.032	0.076	0.028	0.089	0.026	0.101	0.022	0.132	0.021	0.147
20	0.031	0.072	0.028	0.083	0.025	0.094	0.022	0.122	0.020	0.136
21	0.029	0.068	0.027	0.078	0.025	0.088	0.021	0.114	0.020	0.126
22	0.029	0.064	0.026	0.074	0.024	0.083	0.020	0.106	0.019	0.117
23	0.028	0.061	0.025	0.070	0.023	0.078	0.020	0.099	0.019	0.109
24	0.027	0.058	0.024	0.066	0.022	0.074	0.019	0.093	0.018	0.102
25	0.026	0.055	0.024	0.063	0.022	0.070	0.019	0.088	0.018	0.096
26	0.025	0.053	0.023	0.060	0.021	0.067	0.018	0.083	0.018	0.091
27	0.024	0.051	0.022	0.057	0.021	0.063	0.018	0.079	0.017	0.086
28	0.024	0.049	0.022	0.055	0.020	0.061	0.018	0.075	0.017	0.081
29	0.023	0.047	0.021	0.052	0.020	0.058	0.017	0.071	0.016	0.077
30	0.023	0.045	0.021	0.050	0.019	0.055	0.017	0.068	0.016	0.073
35	0.020	0.038	0.018	0.042	0.017	0.046	0.015	0.055	0.015	0.059
40	0.018	0.032	0.017	0.036	0.016	0.039	0.014	0.046	0.013	0.049
45	0.016	0.028	0.015	0.031	0.014	0.034	0.013	0.039	0.012	0.042
50	0.015	0.025	0.014	0.027	0.013	0.030	0.012	0.034	0.011	0.036
55	0.014	0.023	0.013	0.025	0.012	0.026	0.011	0.030	0.011	0.032
60	0.013	0.021	0.012	0.022	0.011	0.024	0.010	0.027	0.010	0.029

(continued)

Table A.4 Values of chi-squared corresponding to HDRs for log chi-squared

ν	50%		60%		67%		70%		75%	
3	1.576	5.097	1.315	5.730	1.141	6.238	1.052	6.527	0.918	7.023
4	2.327	6.330	2.004	7.016	1.783	7.563	1.669	7.874	1.493	8.404
5	3.107	7.540	2.729	8.276	2.467	8.860	2.331	9.190	2.118	9.753
6	3.907	8.732	3.480	9.515	3.181	10.133	3.025	10.482	2.779	11.075
7	4.724	9.911	4.251	10.737	3.917	11.387	3.742	11.754	3.465	12.376
8	5.552	11.079	5.037	11.946	4.671	12.626	4.479	13.009	4.174	13.658
9	6.391	12.238	5.836	13.143	5.440	13.851	5.231	14.250	4.899	14.925
10	7.238	13.388	6.645	14.330	6.221	15.066	5.997	15.480	5.639	16.179
11	8.093	14.532	7.464	15.508	7.013	16.271	6.774	16.699	6.392	17.422
12	8.954	15.669	8.290	16.679	7.814	17.467	7.560	17.909	7.155	18.654
13	9.821	16.801	9.124	17.844	8.622	18.656	8.355	19.111	7.928	19.878
14	10.692	17.929	9.963	19.003	9.438	19.838	9.158	20.306	8.709	21.094
15	11.568	19.052	10.809	20.156	10.260	21.014	9.968	21.494	9.497	22.303
16	12.448	20.171	11.659	21.304	11.088	22.184	10.783	22.677	10.293	23.505
17	13.331	21.287	12.514	22.448	11.921	23.349	11.605	23.854	11.095	24.701
18	14.218	22.399	13.373	23.589	12.759	24.510	12.431	25.026	11.902	25.892
19	15.108	23.509	14.235	24.725	13.602	25.667	13.262	26.194	12.715	27.078
20	16.001	24.615	15.102	25.858	14.448	26.820	14.098	27.357	13.532	28.259
21	16.897	25.719	15.971	26.987	15.298	27.969	14.937	28.517	14.354	29.436
22	17.794	26.821	16.844	28.114	16.152	29.114	15.780	29.672	15.180	30.609
23	18.695	27.921	17.719	29.238	17.008	30.257	16.627	30.825	16.010	31.778
24	19.597	29.018	18.597	30.360	17.868	31.396	17.477	31.974	16.843	32.944
25	20.501	30.113	19.478	31.479	18.731	32.533	18.329	33.121	17.680	34.106
26	21.407	31.207	20.361	32.595	19.596	33.667	19.185	34.264	18.520	35.265
27	22.315	32.299	21.246	33.710	20.464	34.798	20.044	35.405	19.363	36.421
28	23.225	33.389	22.133	34.822	21.335	35.927	20.905	36.543	20.209	37.575
29	24.136	34.478	23.022	35.933	22.207	37.054	21.769	37.679	21.058	38.725
30	25.048	35.565	23.913	37.042	23.082	38.179	22.635	38.812	21.909	39.873
35	29.632	40.980	28.393	42.560	27.485	43.774	26.995	44.450	26.199	45.581
40	34.244	46.368	32.909	48.042	31.929	49.329	31.399	50.044	30.538	51.240
45	38.879	51.731	37.455	53.496	36.406	54.851	35.839	55.603	34.917	56.860
50	43.534	57.076	42.024	58.926	40.911	60.345	40.309	61.132	39.329	62.447
55	48.206	62.404	46.615	64.335	45.440	65.815	44.805	66.635	43.769	68.005
60	52.893	67.717	51.223	69.726	49.991	71.263	49.323	72.117	48.234	73.539

(continued)

Table A.4—continued

ν	80%		90%		95%		99%		99.5%	
3	0.779	7.622	0.476	9.434	0.296	11.191	0.101	15.128	0.064	16.771
4	1.308	9.042	0.883	10.958	0.607	12.802	0.264	16.903	0.186	18.612
5	1.891	10.427	1.355	12.442	0.989	14.369	0.496	18.619	0.372	20.390
6	2.513	11.784	1.875	13.892	1.425	15.897	0.786	20.295	0.614	22.116
7	3.165	13.117	2.431	15.314	1.903	17.393	1.122	21.931	0.904	23.802
8	3.841	14.430	3.017	16.711	2.414	18.860	1.498	23.532	1.233	25.450
9	4.535	15.727	3.628	18.087	2.953	20.305	1.906	25.108	1.596	27.073
10	5.246	17.009	4.258	19.447	3.516	21.729	2.344	26.654	1.991	28.659
11	5.970	18.279	4.906	20.789	4.099	23.135	2.807	28.176	2.410	30.231
12	6.707	19.537	5.570	22.119	4.700	24.525	3.291	29.685	2.853	31.777
13	7.454	20.786	6.246	23.437	5.317	25.901	3.795	31.171	3.317	33.305
14	8.210	22.026	6.935	24.743	5.948	27.263	4.315	32.644	3.797	34.821
15	8.975	23.258	7.634	26.039	6.591	28.614	4.853	34.099	4.296	36.315
16	9.747	24.483	8.343	27.325	7.245	29.955	5.404	35.539	4.811	37.788
17	10.527	25.701	9.060	28.604	7.910	31.285	5.968	36.972	5.339	39.253
18	11.312	26.913	9.786	29.876	8.584	32.608	6.545	38.388	5.879	40.711
19	12.104	28.120	10.519	31.140	9.267	33.921	7.132	39.796	6.430	42.160
20	12.900	29.322	11.259	32.398	9.958	35.227	7.730	41.194	6.995	43.585
21	13.702	30.519	12.005	33.649	10.656	36.526	8.337	42.583	7.566	45.016
22	14.508	31.711	12.756	34.896	11.362	37.817	8.951	43.969	8.152	46.421
23	15.319	32.899	13.514	36.136	12.073	39.103	9.574	45.344	8.742	47.832
24	16.134	34.083	14.277	37.372	12.791	40.384	10.206	46.708	9.341	49.232
25	16.952	35.264	15.044	38.603	13.515	41.657	10.847	48.062	9.949	50.621
26	17.774	36.441	15.815	39.830	14.243	42.927	11.491	49.419	10.566	52.000
27	18.599	37.615	16.591	41.052	14.977	44.191	12.143	50.764	11.186	53.381
28	19.427	38.786	17.372	42.271	15.715	45.452	12.804	52.099	11.815	54.752
29	20.259	39.953	18.156	43.486	16.459	46.706	13.467	53.436	12.451	56.111
30	21.093	41.119	18.944	44.696	17.206	47.958	14.138	54.761	13.091	57.473
35	25.303	46.906	22.931	50.705	21.002	54.156	17.563	61.330	16.384	64.165
40	29.566	52.640	26.987	56.645	24.879	60.275	21.094	67.792	19.782	70.766
45	33.874	58.330	31.100	62.530	28.823	66.326	24.711	74.172	23.277	77.269
50	38.220	63.983	35.260	68.366	32.824	72.324	28.401	80.480	26.857	83.681
55	42.597	69.605	39.461	74.164	36.873	78.272	32.158	86.717	30.499	90.039
60	47.001	75.200	43.698	79.926	40.965	84.178	35.966	92.908	34.207	96.324

Table A.5 Values of F corresponding to HDRs for log F

($\nu_1 = 3$)

ν_2	50%		67%		75%		90%		95%		99%	
3	0.42	2.36	0.29	3.48	0.22	4.47	0.11	9.28	0.06	15.44	0.02	47.45

($\nu_1 = 4$)

ν_2	50%		67%		75%		90%		95%		99%	
3	0.46	2.24	0.33	3.26	0.26	4.16	0.14	8.48	0.09	14.00	0.04	42.61
4	0.48	2.06	0.35	2.86	0.28	3.52	0.16	6.39	0.10	9.60	0.04	23.15

($\nu_1 = 5$)

ν_2	50%		67%		75%		90%		95%		99%	
3	0.48	2.17	0.35	3.13	0.29	3.98	0.17	8.02	0.12	13.17	0.05	39.74
4	0.51	2.00	0.38	2.74	0.31	3.35	0.19	6.00	0.13	8.97	0.06	21.45
5	0.53	1.89	0.40	2.52	0.33	3.02	0.20	5.05	0.14	7.15	0.07	14.94

($\nu_1 = 6$)

ν_2	50%		67%		75%		90%		95%		99%	
3	0.50	2.13	0.37	3.05	0.31	3.86	0.19	7.72	0.13	12.63	0.07	37.93
4	0.53	1.95	0.40	2.65	0.34	3.24	0.21	5.75	0.15	8.56	0.08	20.34
5	0.55	1.85	0.42	2.44	0.35	2.91	0.22	4.82	0.16	6.79	0.08	14.10
6	0.56	1.78	0.43	2.30	0.37	2.71	0.23	4.28	0.17	5.82	0.09	11.07

($\nu_1 = 7$)

ν_2	50%		67%		75%		90%		95%		99%	
3	0.51	2.10	0.38	2.99	0.32	3.77	0.20	7.50	0.15	12.25	0.08	36.66
4	0.54	1.92	0.42	2.60	0.35	3.16	0.22	5.57	0.17	8.27	0.09	19.55
5	0.56	1.82	0.44	2.38	0.37	2.84	0.24	4.66	0.18	6.54	0.10	13.50
6	0.58	1.75	0.45	2.25	0.39	2.64	0.25	4.13	0.19	5.59	0.11	10.57
7	0.59	1.70	0.46	2.15	0.40	2.50	0.26	3.79	0.20	4.99	0.11	8.89

(continued)

Note: If $\nu_2 > \nu_1$ then an interval corresponding to a *P*% HDR for log F is given by $(1/\bar{F}, 1/\underline{F})$ where $(\underline{F}, \bar{F})$ is the appropriate interval with ν_1 and ν_2 interchanged.

Table A.5—continued

($\nu_1 = 8$)

ν_2	50%		67%		75%		90%		95%		99%	
3	0.52	2.07	0.39	2.94	0.33	3.71	0.21	7.35	0.16	11.97	0.09	35.69
4	0.55	1.89	0.43	2.55	0.37	3.10	0.24	5.44	0.18	8.05	0.10	18.97
5	0.57	1.79	0.45	2.34	0.39	2.78	0.26	4.54	0.20	6.35	0.11	13.05
6	0.59	1.72	0.47	2.20	0.40	2.58	0.27	4.01	0.21	5.42	0.12	10.20
7	0.60	1.68	0.48	2.11	0.42	2.44	0.28	3.68	0.22	4.83	0.13	8.55
8	0.61	1.64	0.49	2.04	0.43	2.34	0.29	3.44	0.23	4.43	0.13	7.50

($\nu_1 = 9$)

ν_2	50%		67%		75%		90%		95%		99%	
3	0.53	2.05	0.40	2.91	0.34	3.66	0.22	7.22	0.17	11.75	0.10	34.99
4	0.56	1.88	0.44	2.52	0.38	3.05	0.25	5.34	0.19	7.88	0.11	18.52
5	0.58	1.77	0.46	2.30	0.40	2.73	0.27	4.44	0.21	6.20	0.13	12.71
6	0.60	1.70	0.48	2.17	0.42	2.53	0.28	3.92	0.22	5.28	0.13	9.91
7	0.61	1.66	0.49	2.07	0.43	2.40	0.30	3.59	0.23	4.70	0.14	8.29
8	0.62	1.62	0.50	2.01	0.44	2.30	0.31	3.35	0.24	4.31	0.15	7.26
9	0.63	1.59	0.51	1.95	0.45	2.22	0.31	3.18	0.25	4.03	0.15	6.54

($\nu_1 = 10$)

ν_2	50%		67%		75%		90%		95%		99%	
3	0.53	2.04	0.41	2.88	0.35	3.62	0.23	7.13	0.18	11.58	0.11	34.41
4	0.57	1.86	0.45	2.49	0.38	3.01	0.26	5.25	0.20	7.75	0.12	18.17
5	0.59	1.76	0.47	2.28	0.41	2.69	0.28	4.37	0.22	6.08	0.14	12.44
6	0.61	1.69	0.49	2.14	0.43	2.50	0.30	3.85	0.23	5.17	0.15	9.67
7	0.62	1.64	0.50	2.05	0.44	2.36	0.31	3.52	0.25	4.60	0.15	8.09
8	0.63	1.60	0.51	1.98	0.45	2.26	0.32	3.28	0.25	4.21	0.16	7.07
9	0.64	1.57	0.52	1.92	0.46	2.19	0.33	3.11	0.26	3.93	0.17	6.36
10	0.64	1.55	0.53	1.88	0.47	2.13	0.34	2.98	0.27	3.72	0.17	5.85

($\nu_1 = 11$)

ν_2	50%		67%		75%		90%		95%		99%	
3	0.54	2.03	0.42	2.85	0.36	3.58	0.24	7.05	0.18	11.44	0.11	33.97
4	0.57	1.85	0.45	2.47	0.39	2.98	0.27	5.19	0.21	7.64	0.13	17.89
5	0.59	1.74	0.48	2.25	0.42	2.66	0.29	4.30	0.23	5.99	0.14	12.22
6	0.61	1.67	0.50	2.12	0.44	2.47	0.31	3.79	0.24	5.09	0.16	9.48
7	0.63	1.63	0.51	2.02	0.45	2.33	0.32	3.46	0.26	4.52	0.16	7.92
8	0.64	1.59	0.52	1.95	0.46	2.23	0.33	3.23	0.27	4.13	0.17	6.91
9	0.65	1.56	0.53	1.90	0.47	2.16	0.34	3.06	0.27	3.85	0.18	6.21
10	0.65	1.54	0.54	1.86	0.48	2.10	0.35	2.92	0.28	3.64	0.18	5.70
11	0.66	1.52	0.55	1.82	0.49	2.05	0.35	2.82	0.29	3.47	0.19	5.32

(*continued*)

Table A.5—continued

($\nu_1 = 12$)

ν_2	50%		67%		75%		90%		95%		99%	
3	0.54	2.02	0.42	2.84	0.36	3.56	0.24	6.99	0.19	11.33	0.12	33.58
4	0.58	1.84	0.46	2.45	0.40	2.96	0.27	5.13	0.22	7.55	0.14	17.64
5	0.60	1.73	0.48	2.24	0.42	2.64	0.30	4.25	0.24	5.91	0.15	12.03
6	0.62	1.66	0.50	2.10	0.44	2.44	0.32	3.74	0.25	5.01	0.16	9.33
7	0.63	1.61	0.52	2.01	0.46	2.30	0.33	3.41	0.27	4.45	0.17	7.77
8	0.64	1.58	0.53	1.94	0.47	2.21	0.34	3.18	0.28	4.07	0.18	6.78
9	0.65	1.55	0.54	1.88	0.48	2.13	0.35	3.01	0.29	3.79	0.19	6.09
10	0.66	1.53	0.55	1.84	0.49	2.07	0.36	2.88	0.29	3.58	0.19	5.59
11	0.67	1.51	0.56	1.80	0.50	2.02	0.37	2.77	0.30	3.41	0.20	5.20
12	0.67	1.49	0.56	1.78	0.50	1.98	0.37	2.69	0.31	3.28	0.20	4.91

($\nu_1 = 13$)

ν_2	50%		67%		75%		90%		95%		99%	
3	0.54	2.01	0.43	2.82	0.37	3.54	0.25	6.93	0.20	11.23	0.12	33.26
4	0.58	1.83	0.46	2.43	0.40	2.94	0.28	5.08	0.22	7.48	0.14	17.45
5	0.60	1.72	0.49	2.22	0.43	2.62	0.30	4.21	0.24	5.84	0.16	11.88
6	0.62	1.65	0.51	2.08	0.45	2.42	0.32	3.70	0.26	4.95	0.17	9.20
7	0.64	1.60	0.53	1.99	0.47	2.28	0.34	3.37	0.27	4.39	0.18	7.66
8	0.65	1.57	0.54	1.92	0.48	2.18	0.35	3.14	0.29	4.01	0.19	6.67
9	0.66	1.54	0.55	1.87	0.49	2.11	0.36	2.97	0.29	3.73	0.20	5.99
10	0.67	1.51	0.56	1.82	0.50	2.05	0.37	2.84	0.30	3.52	0.20	5.49
11	0.67	1.50	0.56	1.79	0.51	2.00	0.38	2.73	0.31	3.36	0.21	5.11
12	0.68	1.48	0.57	1.76	0.51	1.96	0.38	2.65	0.32	3.22	0.21	4.81
13	0.68	1.47	0.58	1.73	0.52	1.93	0.39	2.58	0.32	3.12	0.22	4.57

($\nu_1 = 14$)

ν_2	50%		67%		75%		90%		95%		99%	
3	0.55	2.00	0.43	2.81	0.37	3.52	0.25	6.88	0.20	11.15	0.13	32.99
4	0.58	1.82	0.47	2.42	0.41	2.92	0.29	5.04	0.23	7.41	0.15	17.28
5	0.61	1.71	0.49	2.21	0.44	2.60	0.31	4.17	0.25	5.79	0.16	11.75
6	0.63	1.65	0.51	2.07	0.46	2.40	0.33	3.67	0.27	4.90	0.18	9.09
7	0.64	1.60	0.53	1.98	0.47	2.27	0.34	3.34	0.28	4.34	0.19	7.55
8	0.65	1.56	0.54	1.91	0.49	2.17	0.36	3.11	0.29	3.96	0.20	6.57
9	0.66	1.53	0.55	1.85	0.50	2.09	0.37	2.94	0.30	3.68	0.21	5.90
10	0.67	1.51	0.56	1.81	0.51	2.03	0.38	2.80	0.31	3.47	0.21	5.40
11	0.68	1.49	0.57	1.77	0.51	1.98	0.38	2.70	0.32	3.31	0.22	5.03
12	0.68	1.47	0.58	1.74	0.52	1.94	0.39	2.61	0.33	3.18	0.22	4.73
13	0.69	1.46	0.58	1.72	0.53	1.91	0.40	2.54	0.33	3.07	0.23	4.49
14	0.69	1.44	0.59	1.70	0.53	1.88	0.40	2.48	0.34	2.98	0.23	4.30

(*continued*)

Table A.5—continued

($\nu_1 = 15$)

ν_2	50%		67%		75%		90%		95%		99%	
3	0.55	1.99	0.43	2.79	0.37	3.50	0.26	6.84	0.20	11.08	0.13	32.77
4	0.59	1.81	0.47	2.41	0.41	2.90	0.29	5.01	0.23	7.35	0.15	17.12
5	0.61	1.71	0.50	2.19	0.44	2.58	0.32	4.14	0.26	5.74	0.17	11.63
6	0.63	1.64	0.52	2.06	0.46	2.39	0.34	3.64	0.27	4.85	0.18	8.98
7	0.65	1.59	0.54	1.96	0.48	2.25	0.35	3.31	0.29	4.30	0.20	7.47
8	0.66	1.55	0.55	1.89	0.49	2.15	0.36	3.08	0.30	3.92	0.21	6.49
9	0.67	1.52	0.56	1.84	0.50	2.07	0.37	2.91	0.31	3.64	0.21	5.82
10	0.67	1.50	0.57	1.80	0.51	2.01	0.38	2.78	0.32	3.43	0.22	5.33
11	0.68	1.48	0.58	1.76	0.52	1.97	0.39	2.67	0.33	3.27	0.23	4.96
12	0.69	1.46	0.58	1.73	0.53	1.93	0.40	2.59	0.33	3.14	0.23	4.66
13	0.69	1.45	0.59	1.71	0.53	1.89	0.41	2.51	0.34	3.03	0.24	4.43
14	0.70	1.44	0.60	1.68	0.54	1.86	0.41	2.45	0.34	2.94	0.24	4.23
15	0.70	1.43	0.60	1.67	0.54	1.84	0.42	2.40	0.35	2.86	0.25	4.07

($\nu_1 = 16$)

ν_2	50%		67%		75%		90%		95%		99%	
3	0.55	1.99	0.44	2.78	0.38	3.48	0.26	6.81	0.21	11.02	0.13	32.54
4	0.59	1.81	0.47	2.40	0.42	2.89	0.29	4.98	0.24	7.31	0.16	17.01
5	0.61	1.70	0.50	2.18	0.44	2.57	0.32	4.11	0.26	5.69	0.18	11.53
6	0.63	1.63	0.52	2.05	0.47	2.37	0.34	3.61	0.28	4.81	0.19	8.90
7	0.65	1.58	0.54	1.95	0.48	2.24	0.36	3.28	0.29	4.26	0.20	7.39
8	0.66	1.54	0.55	1.88	0.50	2.14	0.37	3.05	0.31	3.88	0.21	6.42
9	0.67	1.52	0.56	1.83	0.51	2.06	0.38	2.88	0.32	3.61	0.22	5.75
10	0.68	1.49	0.57	1.78	0.52	2.00	0.39	2.75	0.33	3.40	0.23	5.26
11	0.69	1.47	0.58	1.75	0.53	1.95	0.40	2.65	0.33	3.23	0.23	4.89
12	0.69	1.46	0.59	1.72	0.53	1.91	0.41	2.56	0.34	3.10	0.24	4.60
13	0.70	1.44	0.60	1.69	0.54	1.88	0.41	2.49	0.35	3.00	0.25	4.37
14	0.70	1.43	0.60	1.67	0.55	1.85	0.42	2.43	0.35	2.90	0.25	4.17
15	0.71	1.42	0.61	1.65	0.55	1.82	0.42	2.38	0.36	2.83	0.25	4.01
16	0.71	1.41	0.61	1.64	0.56	1.80	0.43	2.33	0.36	2.76	0.26	3.87

($\nu_1 = 17$)

ν_2	50%		67%		75%		90%		95%		99%	
3	0.55	1.98	0.44	2.78	0.38	3.47	0.26	6.78	0.21	10.96	0.14	32.38
4	0.59	1.80	0.48	2.39	0.42	2.88	0.30	4.95	0.24	7.26	0.16	16.89
5	0.62	1.70	0.51	2.18	0.45	2.56	0.32	4.09	0.27	5.66	0.18	11.44
6	0.64	1.63	0.53	2.04	0.47	2.36	0.34	3.59	0.28	4.78	0.19	8.83
7	0.65	1.58	0.54	1.94	0.49	2.22	0.36	3.26	0.30	4.22	0.21	7.32
8	0.66	1.54	0.56	1.87	0.50	2.12	0.38	3.03	0.31	3.85	0.22	6.36
9	0.67	1.51	0.57	1.82	0.51	2.05	0.39	2.86	0.32	3.57	0.23	5.69
10	0.68	1.49	0.58	1.77	0.52	1.99	0.40	2.73	0.33	3.37	0.23	5.20
11	0.69	1.47	0.59	1.74	0.53	1.94	0.41	2.62	0.34	3.20	0.24	4.83
12	0.70	1.45	0.59	1.71	0.54	1.90	0.41	2.54	0.35	3.07	0.25	4.55
13	0.70	1.44	0.60	1.68	0.55	1.86	0.42	2.47	0.35	2.96	0.25	4.31
14	0.71	1.42	0.61	1.66	0.55	1.83	0.43	2.41	0.36	2.87	0.26	4.12
15	0.71	1.41	0.61	1.64	0.56	1.81	0.43	2.36	0.37	2.80	0.26	3.96
16	0.71	1.40	0.62	1.63	0.56	1.79	0.44	2.31	0.37	2.73	0.27	3.82
17	0.72	1.39	0.62	1.61	0.57	1.77	0.44	2.27	0.37	2.67	0.27	3.71

(*continued*)

Table A.5—continued

($\nu_1 = 18$)

ν_2	50%		67%		75%		90%		95%		99%	
3	0.56	1.98	0.44	2.77	0.38	3.46	0.27	6.75	0.21	10.92	0.14	32.23
4	0.59	1.80	0.48	2.38	0.42	2.87	0.30	4.93	0.24	7.22	0.16	16.78
5	0.62	1.69	0.51	2.17	0.45	2.55	0.33	4.07	0.27	5.62	0.18	11.37
6	0.64	1.62	0.53	2.03	0.47	2.35	0.35	3.56	0.29	4.75	0.20	8.76
7	0.65	1.57	0.55	1.93	0.49	2.21	0.37	3.24	0.30	4.19	0.21	7.26
8	0.67	1.53	0.56	1.86	0.51	2.11	0.38	3.01	0.32	3.82	0.22	6.30
9	0.68	1.50	0.57	1.81	0.52	2.04	0.39	2.84	0.33	3.55	0.23	5.64
10	0.69	1.48	0.58	1.77	0.53	1.98	0.40	2.71	0.34	3.34	0.24	5.15
11	0.69	1.46	0.59	1.73	0.54	1.93	0.41	2.60	0.35	3.18	0.25	4.79
12	0.70	1.44	0.60	1.70	0.54	1.89	0.42	2.52	0.35	3.04	0.25	4.50
13	0.70	1.43	0.61	1.68	0.55	1.85	0.43	2.45	0.36	2.94	0.26	4.26
14	0.71	1.42	0.61	1.65	0.56	1.82	0.43	2.39	0.37	2.85	0.26	4.07
15	0.71	1.41	0.62	1.63	0.56	1.80	0.44	2.33	0.37	2.77	0.27	3.91
16	0.72	1.40	0.62	1.62	0.57	1.78	0.44	2.29	0.38	2.70	0.27	3.78
17	0.72	1.39	0.62	1.60	0.57	1.76	0.45	2.25	0.38	2.65	0.28	3.66
18	0.72	1.38	0.63	1.59	0.58	1.74	0.45	2.22	0.39	2.60	0.28	3.56

($\nu_1 = 19$)

ν_2	50%		67%		75%		90%		95%		99%	
3	0.56	1.97	0.44	2.76	0.38	3.45	0.27	6.73	0.22	10.87	0.14	32.09
4	0.60	1.79	0.48	2.37	0.42	2.86	0.30	4.91	0.25	7.19	0.17	16.70
5	0.62	1.69	0.51	2.16	0.45	2.54	0.33	4.05	0.27	5.59	0.19	11.30
6	0.64	1.62	0.53	2.02	0.48	2.34	0.35	3.55	0.29	4.72	0.20	8.70
7	0.66	1.57	0.55	1.93	0.49	2.20	0.37	3.22	0.31	4.17	0.22	7.21
8	0.67	1.53	0.57	1.86	0.51	2.10	0.38	2.99	0.32	3.79	0.23	6.25
9	0.68	1.50	0.58	1.80	0.52	2.03	0.40	2.82	0.33	3.52	0.24	5.59
10	0.69	1.48	0.59	1.76	0.53	1.97	0.41	2.69	0.34	3.31	0.25	5.11
11	0.70	1.46	0.60	1.72	0.54	1.92	0.42	2.58	0.35	3.15	0.25	4.74
12	0.70	1.44	0.60	1.69	0.55	1.88	0.42	2.50	0.36	3.02	0.26	4.45
13	0.71	1.42	0.61	1.67	0.56	1.84	0.43	2.43	0.37	2.91	0.27	4.22
14	0.71	1.41	0.62	1.65	0.56	1.81	0.44	2.37	0.37	2.82	0.27	4.03
15	0.72	1.40	0.62	1.63	0.57	1.79	0.44	2.32	0.38	2.74	0.28	3.87
16	0.72	1.39	0.63	1.61	0.57	1.77	0.45	2.27	0.38	2.68	0.28	3.74
17	0.72	1.38	0.63	1.60	0.58	1.75	0.45	2.23	0.39	2.62	0.28	3.62
18	0.73	1.38	0.63	1.58	0.58	1.73	0.46	2.20	0.39	2.57	0.29	3.52
19	0.73	1.37	0.64	1.57	0.58	1.71	0.46	2.17	0.40	2.53	0.29	3.43

(*continued*)

Table A.5—continued

($\nu_1 = 20$)

ν_2	50%		67%		75%		90%		95%		99%	
3	0.56	1.97	0.44	2.75	0.39	3.44	0.27	6.70	0.22	10.83	0.15	31.95
4	0.60	1.79	0.48	2.37	0.43	2.85	0.31	4.89	0.25	7.16	0.17	16.63
5	0.62	1.68	0.51	2.15	0.46	2.53	0.33	4.03	0.28	5.57	0.19	11.24
6	0.64	1.61	0.54	2.02	0.48	2.33	0.36	3.53	0.30	4.69	0.21	8.64
7	0.66	1.56	0.55	1.92	0.50	2.19	0.37	3.20	0.31	4.14	0.22	7.16
8	0.67	1.52	0.57	1.85	0.51	2.09	0.39	2.97	0.33	3.77	0.23	6.20
9	0.68	1.49	0.58	1.79	0.52	2.02	0.40	2.80	0.34	3.50	0.24	5.55
10	0.69	1.47	0.59	1.75	0.54	1.96	0.41	2.67	0.35	3.29	0.25	5.07
11	0.70	1.45	0.60	1.71	0.54	1.91	0.42	2.57	0.36	3.13	0.26	4.70
12	0.70	1.43	0.61	1.68	0.55	1.87	0.43	2.48	0.37	3.00	0.27	4.41
13	0.71	1.42	0.61	1.66	0.56	1.83	0.44	2.41	0.37	2.89	0.27	4.18
14	0.72	1.41	0.62	1.64	0.57	1.80	0.44	2.35	0.38	2.80	0.28	3.99
15	0.72	1.40	0.62	1.62	0.57	1.78	0.45	2.30	0.38	2.72	0.28	3.83
16	0.72	1.39	0.63	1.60	0.58	1.76	0.45	2.26	0.39	2.66	0.29	3.70
17	0.73	1.38	0.63	1.59	0.58	1.74	0.46	2.22	0.39	2.60	0.29	3.58
18	0.73	1.37	0.64	1.57	0.58	1.72	0.46	2.18	0.40	2.55	0.29	3.48
19	0.73	1.36	0.64	1.56	0.59	1.70	0.47	2.15	0.40	2.50	0.30	3.40
20	0.74	1.36	0.64	1.55	0.59	1.69	0.47	2.12	0.41	2.46	0.30	3.32

($\nu_1 = 21$)

ν_2	50%		67%		75%		90%		95%		99%	
3	0.56	1.97	0.45	2.75	0.39	3.43	0.27	6.68	0.22	10.80	0.15	31.85
4	0.60	1.79	0.49	2.36	0.43	2.84	0.31	4.87	0.25	7.13	0.17	16.55
5	0.63	1.68	0.52	2.15	0.46	2.52	0.34	4.01	0.28	5.54	0.19	11.18
6	0.65	1.61	0.54	2.01	0.48	2.32	0.36	3.51	0.30	4.67	0.21	8.59
7	0.66	1.56	0.56	1.91	0.50	2.19	0.38	3.19	0.32	4.12	0.22	7.11
8	0.67	1.52	0.57	1.84	0.52	2.09	0.39	2.96	0.33	3.75	0.24	6.16
9	0.68	1.49	0.58	1.79	0.53	2.01	0.41	2.79	0.34	3.48	0.25	5.51
10	0.69	1.47	0.59	1.74	0.54	1.95	0.42	2.66	0.35	3.27	0.26	5.03
11	0.70	1.45	0.60	1.71	0.55	1.90	0.43	2.55	0.36	3.11	0.26	4.66
12	0.71	1.43	0.61	1.68	0.56	1.86	0.43	2.47	0.37	2.98	0.27	4.38
13	0.71	1.42	0.62	1.65	0.56	1.82	0.44	2.40	0.38	2.87	0.28	4.15
14	0.72	1.40	0.62	1.63	0.57	1.79	0.45	2.34	0.38	2.78	0.28	3.96
15	0.72	1.39	0.63	1.61	0.58	1.77	0.45	2.28	0.39	2.70	0.29	3.80
16	0.73	1.38	0.63	1.60	0.58	1.75	0.46	2.24	0.39	2.64	0.29	3.67
17	0.73	1.37	0.64	1.58	0.59	1.73	0.46	2.20	0.40	2.58	0.30	3.55
18	0.73	1.37	0.64	1.57	0.59	1.71	0.47	2.17	0.40	2.53	0.30	3.45
19	0.74	1.36	0.64	1.56	0.59	1.69	0.47	2.14	0.41	2.48	0.30	3.36
20	0.74	1.35	0.65	1.55	0.60	1.68	0.48	2.11	0.41	2.44	0.31	3.29
21	0.74	1.35	0.65	1.54	0.60	1.67	0.48	2.08	0.42	2.41	0.31	3.22

(*continued*)

Table A.5—continued

($\nu_1 = 22$)

ν_2	50%		67%		75%		90%		95%		99%	
3	0.56	1.96	0.45	2.74	0.39	3.43	0.27	6.67	0.22	10.77	0.15	31.75
4	0.60	1.78	0.49	2.36	0.43	2.83	0.31	4.86	0.26	7.11	0.18	16.48
5	0.63	1.68	0.52	2.14	0.46	2.51	0.34	4.00	0.28	5.52	0.20	11.12
6	0.65	1.61	0.54	2.00	0.48	2.31	0.36	3.50	0.30	4.65	0.21	8.55
7	0.66	1.56	0.56	1.91	0.50	2.18	0.38	3.17	0.32	4.10	0.23	7.07
8	0.68	1.52	0.57	1.84	0.52	2.08	0.40	2.94	0.33	3.73	0.24	6.13
9	0.69	1.49	0.59	1.78	0.53	2.00	0.41	2.77	0.35	3.46	0.25	5.47
10	0.70	1.46	0.60	1.74	0.54	1.94	0.42	2.64	0.36	3.25	0.26	4.99
11	0.70	1.44	0.61	1.70	0.55	1.89	0.43	2.54	0.37	3.09	0.27	4.63
12	0.71	1.43	0.61	1.67	0.56	1.85	0.44	2.45	0.37	2.96	0.28	4.35
13	0.72	1.41	0.62	1.65	0.57	1.82	0.45	2.38	0.38	2.85	0.28	4.12
14	0.72	1.40	0.63	1.62	0.57	1.79	0.45	2.32	0.39	2.76	0.29	3.93
15	0.73	1.39	0.63	1.61	0.58	1.76	0.46	2.27	0.39	2.68	0.29	3.77
16	0.73	1.38	0.64	1.59	0.58	1.74	0.46	2.23	0.40	2.62	0.30	3.64
17	0.73	1.37	0.64	1.57	0.59	1.72	0.47	2.19	0.40	2.56	0.30	3.52
18	0.74	1.36	0.64	1.56	0.59	1.70	0.47	2.15	0.41	2.51	0.31	3.42
19	0.74	1.36	0.65	1.55	0.60	1.69	0.48	2.12	0.41	2.47	0.31	3.33
20	0.74	1.35	0.65	1.54	0.60	1.67	0.48	2.09	0.42	2.43	0.31	3.26
21	0.74	1.34	0.65	1.53	0.60	1.66	0.49	2.07	0.42	2.39	0.32	3.19
22	0.75	1.34	0.66	1.52	0.61	1.65	0.49	2.05	0.42	2.36	0.32	3.12

($\nu_1 = 23$)

ν_2	50%		67%		75%		90%		95%		99%	
3	0.56	1.96	0.45	2.74	0.39	3.42	0.28	6.65	0.22	10.74	0.15	31.66
4	0.60	1.78	0.49	2.35	0.43	2.82	0.31	4.84	0.26	7.08	0.18	16.42
5	0.63	1.67	0.52	2.14	0.46	2.51	0.34	3.98	0.28	5.50	0.20	11.09
6	0.65	1.60	0.54	2.00	0.49	2.31	0.37	3.49	0.31	4.63	0.22	8.51
7	0.66	1.55	0.56	1.90	0.51	2.17	0.38	3.16	0.32	4.08	0.23	7.04
8	0.68	1.51	0.58	1.83	0.52	2.07	0.40	2.93	0.34	3.71	0.24	6.09
9	0.69	1.48	0.59	1.78	0.53	1.99	0.41	2.76	0.35	3.44	0.25	5.44
10	0.70	1.46	0.60	1.73	0.54	1.93	0.42	2.63	0.36	3.23	0.26	4.96
11	0.71	1.44	0.61	1.70	0.55	1.88	0.43	2.53	0.37	3.07	0.27	4.60
12	0.71	1.42	0.62	1.67	0.56	1.84	0.44	2.44	0.38	2.94	0.28	4.32
13	0.72	1.41	0.62	1.64	0.57	1.81	0.45	2.37	0.39	2.83	0.29	4.09
14	0.72	1.40	0.63	1.62	0.58	1.78	0.46	2.31	0.39	2.74	0.29	3.90
15	0.73	1.38	0.63	1.60	0.58	1.75	0.46	2.26	0.40	2.67	0.30	3.74
16	0.73	1.37	0.64	1.58	0.59	1.73	0.47	2.21	0.40	2.60	0.30	3.61
17	0.74	1.37	0.64	1.57	0.59	1.71	0.47	2.17	0.41	2.54	0.31	3.49
18	0.74	1.36	0.65	1.56	0.60	1.69	0.48	2.14	0.41	2.49	0.31	3.39
19	0.74	1.35	0.65	1.54	0.60	1.68	0.48	2.11	0.42	2.45	0.32	3.30
20	0.74	1.35	0.66	1.53	0.60	1.66	0.49	2.08	0.42	2.41	0.32	3.23
21	0.75	1.34	0.66	1.52	0.61	1.65	0.49	2.06	0.43	2.37	0.32	3.16
22	0.75	1.33	0.66	1.51	0.61	1.64	0.49	2.03	0.43	2.34	0.33	3.10
23	0.75	1.33	0.66	1.51	0.61	1.63	0.50	2.01	0.43	2.31	0.33	3.04

(continued)

Table A.5—continued
($\nu_1 = 24$)

ν_2	50%		67%		75%		90%		95%		99%	
3	0.56	1.96	0.45	2.73	0.39	3.41	0.28	6.64	0.23	10.71	0.15	31.55
4	0.60	1.78	0.49	2.35	0.43	2.82	0.32	4.83	0.26	7.06	0.18	16.36
5	0.63	1.67	0.52	2.13	0.47	2.50	0.35	3.97	0.29	5.48	0.20	11.04
6	0.65	1.60	0.54	1.99	0.49	2.30	0.37	3.47	0.31	4.61	0.22	8.48
7	0.67	1.55	0.56	1.90	0.51	2.17	0.39	3.15	0.33	4.07	0.23	7.01
8	0.68	1.51	0.58	1.83	0.52	2.06	0.40	2.92	0.34	3.69	0.25	6.06
9	0.69	1.48	0.59	1.77	0.54	1.99	0.42	2.75	0.35	3.42	0.26	5.41
10	0.70	1.46	0.60	1.73	0.55	1.93	0.43	2.62	0.36	3.22	0.27	4.94
11	0.71	1.44	0.61	1.69	0.56	1.88	0.44	2.52	0.37	3.06	0.28	4.57
12	0.71	1.42	0.62	1.66	0.57	1.84	0.45	2.43	0.38	2.93	0.28	4.29
13	0.72	1.40	0.63	1.64	0.57	1.80	0.45	2.36	0.39	2.82	0.29	4.06
14	0.73	1.39	0.63	1.61	0.58	1.77	0.46	2.30	0.40	2.73	0.30	3.87
15	0.73	1.38	0.64	1.59	0.59	1.75	0.47	2.25	0.40	2.65	0.30	3.71
16	0.73	1.37	0.64	1.58	0.59	1.72	0.47	2.20	0.41	2.59	0.31	3.58
17	0.74	1.36	0.65	1.56	0.60	1.70	0.48	2.16	0.41	2.53	0.31	3.47
18	0.74	1.36	0.65	1.55	0.60	1.69	0.48	2.13	0.42	2.48	0.32	3.37
19	0.74	1.35	0.65	1.54	0.60	1.67	0.49	2.10	0.42	2.43	0.32	3.28
20	0.75	1.34	0.66	1.53	0.61	1.66	0.49	2.07	0.43	2.39	0.32	3.20
21	0.75	1.34	0.66	1.52	0.61	1.64	0.49	2.05	0.43	2.36	0.33	3.13
22	0.75	1.33	0.66	1.51	0.62	1.63	0.50	2.02	0.43	2.33	0.33	3.07
23	0.75	1.33	0.67	1.50	0.62	1.62	0.50	2.00	0.44	2.30	0.33	3.02
24	0.76	1.32	0.67	1.49	0.62	1.61	0.50	1.98	0.44	2.27	0.34	2.97

($\nu_1 = 25$)

ν_2	50%		67%		75%		90%		95%		99%	
3	0.56	1.96	0.45	2.73	0.39	3.41	0.28	6.62	0.23	10.69	0.16	31.48
4	0.60	1.78	0.49	2.34	0.44	2.81	0.32	4.82	0.26	7.04	0.18	16.32
5	0.63	1.67	0.52	2.13	0.47	2.50	0.35	3.96	0.29	5.46	0.20	11.00
6	0.65	1.60	0.55	1.99	0.49	2.30	0.37	3.46	0.31	4.60	0.22	8.44
7	0.67	1.55	0.57	1.89	0.51	2.16	0.39	3.14	0.33	4.05	0.24	6.97
8	0.68	1.51	0.58	1.82	0.53	2.06	0.40	2.91	0.34	3.68	0.25	6.03
9	0.69	1.48	0.59	1.77	0.54	1.98	0.42	2.74	0.36	3.41	0.26	5.38
10	0.70	1.45	0.60	1.72	0.55	1.92	0.43	2.61	0.37	3.20	0.27	4.91
11	0.71	1.43	0.61	1.69	0.56	1.87	0.44	2.50	0.38	3.04	0.28	4.55
12	0.72	1.42	0.62	1.66	0.57	1.83	0.45	2.42	0.39	2.91	0.29	4.26
13	0.72	1.40	0.63	1.63	0.58	1.80	0.46	2.35	0.39	2.80	0.30	4.03
14	0.73	1.39	0.63	1.61	0.58	1.77	0.46	2.29	0.40	2.71	0.30	3.85
15	0.73	1.38	0.64	1.59	0.59	1.74	0.47	2.24	0.41	2.64	0.31	3.69
16	0.74	1.37	0.65	1.57	0.59	1.72	0.48	2.19	0.41	2.57	0.31	3.56
17	0.74	1.36	0.65	1.56	0.60	1.70	0.48	2.15	0.42	2.51	0.32	3.44
18	0.74	1.35	0.65	1.54	0.60	1.68	0.49	2.12	0.42	2.46	0.32	3.34
19	0.75	1.34	0.66	1.53	0.61	1.66	0.49	2.09	0.43	2.42	0.33	3.26
20	0.75	1.34	0.66	1.52	0.61	1.65	0.49	2.06	0.43	2.38	0.33	3.18
21	0.75	1.33	0.66	1.51	0.62	1.64	0.50	2.03	0.44	2.34	0.33	3.11
22	0.75	1.33	0.67	1.50	0.62	1.62	0.50	2.01	0.44	2.31	0.34	3.05
23	0.76	1.32	0.67	1.49	0.62	1.61	0.51	1.99	0.44	2.28	0.34	2.99
24	0.76	1.32	0.67	1.49	0.62	1.60	0.51	1.97	0.45	2.25	0.34	2.94
25	0.76	1.31	0.68	1.48	0.63	1.59	0.51	1.96	0.45	2.23	0.35	2.90

(*continued*)

Table A.5—continued
($\nu_1 = 26$)

ν_2	50%		67%		75%		90%		95%		99%	
3	0.57	1.96	0.45	2.72	0.40	3.40	0.28	6.61	0.23	10.67	0.16	31.39
4	0.60	1.77	0.49	2.34	0.44	2.81	0.32	4.81	0.26	7.03	0.18	16.27
5	0.63	1.67	0.52	2.12	0.47	2.49	0.35	3.95	0.29	5.45	0.21	10.96
6	0.65	1.60	0.55	1.99	0.49	2.29	0.37	3.45	0.31	4.58	0.22	8.41
7	0.67	1.54	0.57	1.89	0.51	2.15	0.39	3.13	0.33	4.04	0.24	6.94
8	0.68	1.51	0.58	1.82	0.53	2.05	0.41	2.90	0.35	3.67	0.25	6.01
9	0.69	1.48	0.59	1.76	0.54	1.98	0.42	2.73	0.36	3.39	0.26	5.36
10	0.70	1.45	0.61	1.72	0.55	1.92	0.43	2.60	0.37	3.19	0.27	4.88
11	0.71	1.43	0.61	1.68	0.56	1.87	0.44	2.49	0.38	3.03	0.28	4.52
12	0.72	1.41	0.62	1.65	0.57	1.82	0.45	2.41	0.39	2.90	0.29	4.24
13	0.72	1.40	0.63	1.63	0.58	1.79	0.46	2.34	0.40	2.79	0.30	4.01
14	0.73	1.39	0.64	1.60	0.59	1.76	0.47	2.28	0.40	2.70	0.31	3.82
15	0.73	1.38	0.64	1.59	0.59	1.73	0.47	2.23	0.41	2.62	0.31	3.67
16	0.74	1.37	0.65	1.57	0.60	1.71	0.48	2.18	0.42	2.56	0.32	3.53
17	0.74	1.36	0.65	1.55	0.60	1.69	0.48	2.14	0.42	2.50	0.32	3.42
18	0.75	1.35	0.66	1.54	0.61	1.67	0.49	2.11	0.43	2.45	0.33	3.32
19	0.75	1.34	0.66	1.53	0.61	1.66	0.49	2.08	0.43	2.41	0.33	3.23
20	0.75	1.34	0.66	1.52	0.61	1.64	0.50	2.05	0.44	2.37	0.33	3.16
21	0.75	1.33	0.67	1.51	0.62	1.63	0.50	2.02	0.44	2.33	0.34	3.09
22	0.76	1.32	0.67	1.50	0.62	1.62	0.51	2.00	0.44	2.30	0.34	3.03
23	0.76	1.32	0.67	1.49	0.62	1.61	0.51	1.98	0.45	2.27	0.34	2.97
24	0.76	1.31	0.68	1.48	0.63	1.60	0.51	1.96	0.45	2.24	0.35	2.92
25	0.76	1.31	0.68	1.48	0.63	1.59	0.52	1.95	0.45	2.22	0.35	2.88
26	0.77	1.31	0.68	1.47	0.63	1.58	0.52	1.93	0.46	2.19	0.35	2.84

(continued)

Table A.5—continued
($\nu_1 = 27$)

ν_2	50%		67%		75%		90%		95%		99%	
3	0.57	1.95	0.45	2.72	0.40	3.40	0.28	6.60	0.23	10.65	0.16	31.35
4	0.61	1.77	0.50	2.34	0.44	2.80	0.32	4.80	0.27	7.01	0.19	16.23
5	0.63	1.66	0.53	2.12	0.47	2.49	0.35	3.94	0.29	5.43	0.21	10.93
6	0.65	1.59	0.55	1.98	0.49	2.29	0.37	3.44	0.32	4.57	0.23	8.38
7	0.67	1.54	0.57	1.89	0.51	2.15	0.39	3.12	0.33	4.02	0.24	6.92
8	0.68	1.50	0.58	1.81	0.53	2.05	0.41	2.89	0.35	3.65	0.26	5.98
9	0.69	1.47	0.60	1.76	0.54	1.97	0.42	2.72	0.36	3.38	0.27	5.33
10	0.70	1.45	0.61	1.71	0.55	1.91	0.44	2.59	0.37	3.18	0.28	4.86
11	0.71	1.43	0.62	1.68	0.56	1.86	0.45	2.49	0.38	3.02	0.29	4.50
12	0.72	1.41	0.63	1.65	0.57	1.82	0.45	2.40	0.39	2.89	0.30	4.22
13	0.73	1.40	0.63	1.62	0.58	1.78	0.46	2.33	0.40	2.78	0.30	3.99
14	0.73	1.38	0.64	1.60	0.59	1.75	0.47	2.27	0.41	2.69	0.31	3.80
15	0.74	1.37	0.64	1.58	0.59	1.73	0.48	2.22	0.41	2.61	0.32	3.65
16	0.74	1.36	0.65	1.56	0.60	1.71	0.48	2.17	0.42	2.55	0.32	3.52
17	0.74	1.35	0.65	1.55	0.60	1.69	0.49	2.13	0.43	2.49	0.33	3.40
18	0.75	1.35	0.66	1.54	0.61	1.67	0.49	2.10	0.43	2.44	0.33	3.30
19	0.75	1.34	0.66	1.52	0.61	1.65	0.50	2.07	0.44	2.39	0.33	3.21
20	0.75	1.33	0.67	1.51	0.62	1.64	0.50	2.04	0.44	2.35	0.34	3.14
21	0.76	1.33	0.67	1.50	0.62	1.62	0.51	2.01	0.44	2.32	0.34	3.07
22	0.76	1.32	0.67	1.49	0.62	1.61	0.51	1.99	0.45	2.29	0.35	3.01
23	0.76	1.32	0.68	1.49	0.63	1.60	0.51	1.97	0.45	2.26	0.35	2.95
24	0.76	1.31	0.68	1.48	0.63	1.59	0.52	1.95	0.45	2.23	0.35	2.90
25	0.77	1.31	0.68	1.47	0.63	1.58	0.52	1.94	0.46	2.20	0.35	2.86
26	0.77	1.30	0.68	1.46	0.64	1.57	0.52	1.92	0.46	2.18	0.36	2.82
27	0.77	1.30	0.69	1.46	0.64	1.57	0.52	1.90	0.46	2.16	0.36	2.78

(continued)

Table A.5—continued
($\nu_1 = 28$)

ν_2	50%		67%		75%		90%		95%		99%	
3	0.57	1.95	0.45	2.72	0.40	3.39	0.28	6.59	0.23	10.63	0.16	31.30
4	0.61	1.77	0.50	2.33	0.44	2.80	0.32	4.79	0.27	6.99	0.19	16.20
5	0.63	1.66	0.53	2.12	0.47	2.48	0.35	3.93	0.29	5.42	0.21	10.90
6	0.66	1.59	0.55	1.98	0.50	2.28	0.38	3.44	0.32	4.55	0.23	8.35
7	0.67	1.54	0.57	1.88	0.52	2.14	0.40	3.11	0.34	4.01	0.25	6.89
8	0.69	1.50	0.59	1.81	0.53	2.04	0.41	2.88	0.35	3.64	0.26	5.96
9	0.70	1.47	0.60	1.76	0.55	1.97	0.43	2.71	0.36	3.37	0.27	5.31
10	0.71	1.45	0.61	1.71	0.56	1.91	0.44	2.58	0.38	3.17	0.28	4.84
11	0.71	1.43	0.62	1.67	0.57	1.86	0.45	2.48	0.39	3.00	0.29	4.48
12	0.72	1.41	0.63	1.64	0.58	1.81	0.46	2.39	0.40	2.87	0.30	4.20
13	0.73	1.39	0.63	1.62	0.58	1.78	0.47	2.32	0.40	2.77	0.31	3.97
14	0.73	1.38	0.64	1.60	0.59	1.75	0.47	2.26	0.41	2.68	0.31	3.78
15	0.74	1.37	0.65	1.58	0.60	1.72	0.48	2.21	0.42	2.60	0.32	3.63
16	0.74	1.36	0.65	1.56	0.60	1.70	0.49	2.16	0.42	2.53	0.32	3.50
17	0.75	1.35	0.66	1.54	0.61	1.68	0.49	2.12	0.43	2.48	0.33	3.38
18	0.75	1.34	0.66	1.53	0.61	1.66	0.50	2.09	0.43	2.43	0.33	3.28
19	0.75	1.34	0.67	1.52	0.62	1.65	0.50	2.06	0.44	2.38	0.34	3.20
20	0.76	1.33	0.67	1.51	0.62	1.63	0.51	2.03	0.44	2.34	0.34	3.12
21	0.76	1.32	0.67	1.50	0.62	1.62	0.51	2.01	0.45	2.31	0.35	3.05
22	0.76	1.32	0.68	1.49	0.63	1.61	0.51	1.98	0.45	2.27	0.35	2.99
23	0.76	1.31	0.68	1.48	0.63	1.60	0.52	1.96	0.45	2.24	0.35	2.93
24	0.77	1.31	0.68	1.47	0.63	1.59	0.52	1.94	0.46	2.22	0.36	2.88
25	0.77	1.30	0.68	1.47	0.64	1.58	0.52	1.93	0.46	2.19	0.36	2.84
26	0.77	1.30	0.69	1.46	0.64	1.57	0.53	1.91	0.46	2.17	0.36	2.80
27	0.77	1.30	0.69	1.45	0.64	1.56	0.53	1.90	0.47	2.15	0.36	2.76
28	0.77	1.29	0.69	1.45	0.64	1.55	0.53	1.88	0.47	2.13	0.37	2.72

(*continued*)

Table A.5—continued
($\nu_1 = 29$)

ν_2	50%		67%		75%		90%		95%		99%	
3	0.57	1.95	0.46	2.71	0.40	3.39	0.28	6.58	0.23	10.61	0.16	31.23
4	0.61	1.77	0.50	2.33	0.44	2.79	0.32	4.78	0.27	6.98	0.19	16.15
5	0.64	1.66	0.53	2.11	0.47	2.48	0.35	3.92	0.30	5.41	0.21	10.87
6	0.66	1.59	0.55	1.98	0.50	2.28	0.38	3.43	0.32	4.54	0.23	8.33
7	0.67	1.54	0.57	1.88	0.52	2.14	0.40	3.10	0.34	4.00	0.25	6.87
8	0.69	1.50	0.59	1.81	0.53	2.04	0.41	2.88	0.35	3.63	0.26	5.94
9	0.70	1.47	0.60	1.75	0.55	1.96	0.43	2.71	0.37	3.36	0.27	5.29
10	0.71	1.44	0.61	1.71	0.56	1.90	0.44	2.57	0.38	3.15	0.28	4.82
11	0.71	1.42	0.62	1.67	0.57	1.85	0.45	2.47	0.39	2.99	0.29	4.46
12	0.72	1.41	0.63	1.64	0.58	1.81	0.46	2.38	0.40	2.86	0.30	4.18
13	0.73	1.39	0.64	1.61	0.59	1.78	0.47	2.31	0.41	2.76	0.31	3.95
14	0.73	1.38	0.64	1.59	0.59	1.75	0.48	2.25	0.41	2.67	0.32	3.77
15	0.74	1.37	0.65	1.57	0.60	1.72	0.48	2.20	0.42	2.59	0.32	3.61
16	0.74	1.36	0.65	1.56	0.60	1.70	0.49	2.16	0.43	2.52	0.33	3.48
17	0.75	1.35	0.66	1.54	0.61	1.68	0.49	2.12	0.43	2.47	0.33	3.36
18	0.75	1.34	0.66	1.53	0.61	1.66	0.50	2.08	0.44	2.42	0.34	3.26
19	0.75	1.33	0.67	1.52	0.62	1.64	0.50	2.05	0.44	2.37	0.34	3.18
20	0.76	1.33	0.67	1.50	0.62	1.63	0.51	2.02	0.45	2.33	0.35	3.10
21	0.76	1.32	0.67	1.49	0.63	1.61	0.51	2.00	0.45	2.29	0.35	3.03
22	0.76	1.32	0.68	1.49	0.63	1.60	0.52	1.98	0.46	2.26	0.35	2.97
23	0.76	1.31	0.68	1.48	0.63	1.59	0.52	1.95	0.46	2.23	0.36	2.92
24	0.77	1.31	0.68	1.47	0.64	1.58	0.52	1.94	0.46	2.21	0.36	2.87
25	0.77	1.30	0.69	1.46	0.64	1.57	0.53	1.92	0.47	2.18	0.36	2.82
26	0.77	1.30	0.69	1.46	0.64	1.56	0.53	1.90	0.47	2.16	0.37	2.78
27	0.77	1.29	0.69	1.45	0.64	1.56	0.53	1.89	0.47	2.14	0.37	2.74
28	0.77	1.29	0.69	1.44	0.65	1.55	0.53	1.87	0.47	2.12	0.37	2.71
29	0.78	1.29	0.70	1.44	0.65	1.54	0.54	1.86	0.48	2.10	0.37	2.67

(*continued*)

Table A.5—continued
($\nu_1 = 30$)

ν_2	50%		67%		75%		90%		95%		99%	
3	0.57	1.95	0.46	2.71	0.40	3.38	0.29	6.57	0.23	10.60	0.16	31.19
4	0.61	1.77	0.50	2.33	0.44	2.79	0.33	4.77	0.27	6.97	0.19	16.12
5	0.64	1.66	0.53	2.11	0.47	2.47	0.36	3.92	0.30	5.39	0.21	10.84
6	0.66	1.59	0.55	1.97	0.50	2.27	0.38	3.42	0.32	4.53	0.23	8.31
7	0.67	1.54	0.57	1.88	0.52	2.14	0.40	3.10	0.34	3.99	0.25	6.85
8	0.69	1.50	0.59	1.80	0.53	2.04	0.42	2.87	0.36	3.62	0.26	5.92
9	0.70	1.47	0.60	1.75	0.55	1.96	0.43	2.70	0.37	3.35	0.28	5.27
10	0.71	1.44	0.61	1.70	0.56	1.90	0.44	2.57	0.38	3.14	0.29	4.80
11	0.72	1.42	0.62	1.67	0.57	1.85	0.45	2.46	0.39	2.98	0.30	4.44
12	0.72	1.40	0.63	1.64	0.58	1.81	0.46	2.38	0.40	2.85	0.30	4.16
13	0.73	1.39	0.64	1.61	0.59	1.77	0.47	2.31	0.41	2.75	0.31	3.94
14	0.74	1.38	0.64	1.59	0.59	1.74	0.48	2.24	0.42	2.66	0.32	3.75
15	0.74	1.37	0.65	1.57	0.60	1.71	0.49	2.19	0.42	2.58	0.33	3.59
16	0.74	1.36	0.66	1.55	0.61	1.69	0.49	2.15	0.43	2.51	0.33	3.46
17	0.75	1.35	0.66	1.54	0.61	1.67	0.50	2.11	0.44	2.46	0.34	3.35
18	0.75	1.34	0.67	1.52	0.62	1.65	0.50	2.07	0.44	2.41	0.34	3.25
19	0.76	1.33	0.67	1.51	0.62	1.64	0.51	2.04	0.45	2.36	0.35	3.16
20	0.76	1.32	0.67	1.50	0.63	1.62	0.51	2.02	0.45	2.32	0.35	3.09
21	0.76	1.32	0.68	1.49	0.63	1.61	0.52	1.99	0.45	2.28	0.35	3.02
22	0.76	1.31	0.68	1.48	0.63	1.60	0.52	1.97	0.46	2.25	0.36	2.96
23	0.77	1.31	0.68	1.47	0.64	1.59	0.52	1.95	0.46	2.22	0.36	2.90
24	0.77	1.30	0.69	1.47	0.64	1.58	0.53	1.93	0.47	2.20	0.36	2.85
25	0.77	1.30	0.69	1.46	0.64	1.57	0.53	1.91	0.47	2.17	0.37	2.81
26	0.77	1.30	0.69	1.45	0.64	1.56	0.53	1.89	0.47	2.15	0.37	2.76
27	0.77	1.29	0.69	1.45	0.65	1.55	0.54	1.88	0.47	2.13	0.37	2.73
28	0.78	1.29	0.70	1.44	0.65	1.54	0.54	1.87	0.48	2.11	0.38	2.69
29	0.78	1.29	0.70	1.43	0.65	1.54	0.54	1.85	0.48	2.09	0.38	2.66
30	0.78	1.28	0.70	1.43	0.65	1.53	0.54	1.84	0.48	2.07	0.38	2.63

(*continued*)

Table A.5—continued
($\nu_1 = 35$)

ν_2	50%		67%		75%		90%		95%		99%	
3	0.57	1.94	0.46	2.70	0.40	3.37	0.29	6.53	0.24	10.53	0.17	30.96
4	0.61	1.76	0.50	2.31	0.45	2.77	0.33	4.74	0.28	6.91	0.20	15.98
5	0.64	1.65	0.53	2.10	0.48	2.46	0.36	3.88	0.30	5.35	0.22	10.73
6	0.66	1.58	0.56	1.96	0.50	2.26	0.39	3.39	0.33	4.49	0.24	8.21
7	0.68	1.53	0.58	1.86	0.52	2.12	0.41	3.07	0.35	3.95	0.26	6.76
8	0.69	1.49	0.59	1.79	0.54	2.02	0.42	2.84	0.36	3.58	0.27	5.84
9	0.70	1.46	0.61	1.74	0.56	1.94	0.44	2.67	0.38	3.31	0.29	5.20
10	0.71	1.43	0.62	1.69	0.57	1.88	0.45	2.54	0.39	3.10	0.30	4.73
11	0.72	1.41	0.63	1.65	0.58	1.83	0.46	2.43	0.40	2.94	0.31	4.37
12	0.73	1.40	0.64	1.62	0.59	1.79	0.47	2.35	0.41	2.81	0.32	4.09
13	0.74	1.38	0.65	1.60	0.60	1.75	0.48	2.27	0.42	2.70	0.32	3.86
14	0.74	1.37	0.65	1.57	0.60	1.72	0.49	2.21	0.43	2.61	0.33	3.68
15	0.75	1.36	0.66	1.56	0.61	1.70	0.50	2.16	0.44	2.54	0.34	3.52
16	0.75	1.35	0.66	1.54	0.62	1.67	0.50	2.12	0.44	2.47	0.35	3.39
17	0.76	1.34	0.67	1.52	0.62	1.65	0.51	2.08	0.45	2.41	0.35	3.28
18	0.76	1.33	0.67	1.51	0.63	1.63	0.51	2.04	0.45	2.36	0.36	3.18
19	0.76	1.32	0.68	1.50	0.63	1.62	0.52	2.01	0.46	2.32	0.36	3.09
20	0.77	1.31	0.68	1.49	0.64	1.60	0.52	1.98	0.46	2.28	0.37	3.02
21	0.77	1.31	0.69	1.47	0.64	1.59	0.53	1.96	0.47	2.24	0.37	2.95
22	0.77	1.30	0.69	1.47	0.64	1.58	0.53	1.93	0.47	2.21	0.37	2.89
23	0.77	1.30	0.69	1.46	0.65	1.57	0.54	1.91	0.48	2.18	0.38	2.83
24	0.78	1.29	0.70	1.45	0.65	1.56	0.54	1.89	0.48	2.15	0.38	2.78
25	0.78	1.29	0.70	1.44	0.65	1.55	0.54	1.88	0.48	2.13	0.38	2.74
26	0.78	1.29	0.70	1.44	0.66	1.54	0.55	1.86	0.49	2.11	0.39	2.70
27	0.78	1.28	0.70	1.43	0.66	1.53	0.55	1.85	0.49	2.08	0.39	2.66
28	0.78	1.28	0.71	1.42	0.66	1.52	0.55	1.83	0.49	2.07	0.39	2.62
29	0.79	1.27	0.71	1.42	0.66	1.52	0.56	1.82	0.50	2.05	0.40	2.59
30	0.79	1.27	0.71	1.41	0.67	1.51	0.56	1.81	0.50	2.03	0.40	2.56
35	0.79	1.26	0.72	1.39	0.68	1.48	0.57	1.76	0.51	1.96	0.41	2.44

(continued)

Table A.5—continued
($\nu_1 = 40$)

ν_2	50%		67%		75%		90%		95%		99%	
3	0.57	1.94	0.46	2.69	0.41	3.35	0.29	6.50	0.24	10.47	0.17	30.78
4	0.61	1.75	0.51	2.31	0.45	2.76	0.33	4.71	0.28	6.87	0.20	15.87
5	0.64	1.65	0.54	2.09	0.48	2.44	0.37	3.86	0.31	5.31	0.23	10.65
6	0.66	1.58	0.56	1.95	0.51	2.25	0.39	3.37	0.33	4.45	0.25	8.14
7	0.68	1.52	0.58	1.85	0.53	2.11	0.41	3.04	0.35	3.91	0.27	6.70
8	0.70	1.48	0.60	1.78	0.55	2.01	0.43	2.81	0.37	3.54	0.28	5.77
9	0.71	1.45	0.61	1.73	0.56	1.93	0.45	2.64	0.39	3.27	0.29	5.13
10	0.72	1.43	0.62	1.68	0.57	1.87	0.46	2.51	0.40	3.07	0.31	4.67
11	0.73	1.41	0.63	1.64	0.58	1.82	0.47	2.41	0.41	2.91	0.32	4.31
12	0.73	1.39	0.64	1.61	0.59	1.77	0.48	2.32	0.42	2.78	0.33	4.04
13	0.74	1.37	0.65	1.59	0.60	1.74	0.49	2.25	0.43	2.67	0.33	3.81
14	0.75	1.36	0.66	1.56	0.61	1.71	0.50	2.19	0.44	2.58	0.34	3.63
15	0.75	1.35	0.67	1.54	0.62	1.68	0.51	2.14	0.45	2.51	0.35	3.47
16	0.76	1.34	0.67	1.53	0.62	1.66	0.51	2.09	0.45	2.44	0.36	3.34
17	0.76	1.33	0.68	1.51	0.63	1.64	0.52	2.05	0.46	2.38	0.36	3.23
18	0.76	1.32	0.68	1.50	0.63	1.62	0.52	2.02	0.47	2.33	0.37	3.13
19	0.77	1.31	0.69	1.48	0.64	1.60	0.53	1.99	0.47	2.29	0.37	3.04
20	0.77	1.31	0.69	1.47	0.64	1.59	0.53	1.96	0.48	2.25	0.38	2.97
21	0.77	1.30	0.69	1.46	0.65	1.58	0.54	1.93	0.48	2.21	0.38	2.90
22	0.78	1.30	0.70	1.45	0.65	1.56	0.54	1.91	0.48	2.18	0.39	2.84
23	0.78	1.29	0.70	1.44	0.66	1.55	0.55	1.89	0.49	2.15	0.39	2.78
24	0.78	1.29	0.70	1.44	0.66	1.54	0.55	1.87	0.49	2.12	0.40	2.73
25	0.78	1.28	0.71	1.43	0.66	1.53	0.56	1.85	0.50	2.10	0.40	2.69
26	0.79	1.28	0.71	1.42	0.66	1.52	0.56	1.84	0.50	2.07	0.40	2.64
27	0.79	1.27	0.71	1.42	0.67	1.51	0.56	1.82	0.50	2.05	0.41	2.61
28	0.79	1.27	0.71	1.41	0.67	1.51	0.56	1.81	0.51	2.03	0.41	2.57
29	0.79	1.27	0.72	1.40	0.67	1.50	0.57	1.79	0.51	2.01	0.41	2.54
30	0.79	1.26	0.72	1.40	0.67	1.49	0.57	1.78	0.51	2.00	0.41	2.51
35	0.80	1.25	0.73	1.38	0.68	1.46	0.58	1.73	0.52	1.93	0.43	2.39
40	0.81	1.24	0.73	1.36	0.69	1.44	0.59	1.69	0.53	1.88	0.44	2.30

(continued)

Table A.5—continued
($\nu_1 = 45$)

ν_2	50%		67%		75%		90%		95%		99%	
3	0.58	1.93	0.46	2.68	0.41	3.35	0.30	6.48	0.24	10.43	0.17	30.67
4	0.62	1.75	0.51	2.30	0.45	2.75	0.34	4.69	0.28	6.84	0.20	15.80
5	0.65	1.64	0.54	2.08	0.49	2.44	0.37	3.84	0.31	5.28	0.23	10.59
6	0.67	1.57	0.57	1.94	0.51	2.24	0.40	3.35	0.34	4.42	0.25	8.08
7	0.68	1.52	0.59	1.85	0.53	2.10	0.42	3.02	0.36	3.89	0.27	6.64
8	0.70	1.48	0.60	1.77	0.55	2.00	0.44	2.80	0.38	3.52	0.29	5.73
9	0.71	1.45	0.62	1.72	0.57	1.92	0.45	2.63	0.39	3.25	0.30	5.09
10	0.72	1.42	0.63	1.67	0.58	1.86	0.46	2.50	0.41	3.05	0.31	4.62
11	0.73	1.40	0.64	1.64	0.59	1.81	0.48	2.39	0.42	2.88	0.32	4.27
12	0.74	1.38	0.65	1.60	0.60	1.76	0.49	2.30	0.43	2.76	0.33	3.99
13	0.74	1.37	0.66	1.58	0.61	1.73	0.50	2.23	0.44	2.65	0.34	3.77
14	0.75	1.36	0.66	1.56	0.62	1.70	0.50	2.17	0.45	2.56	0.35	3.58
15	0.75	1.34	0.67	1.54	0.62	1.67	0.51	2.12	0.45	2.48	0.36	3.43
16	0.76	1.33	0.68	1.52	0.63	1.65	0.52	2.07	0.46	2.41	0.37	3.30
17	0.76	1.32	0.68	1.50	0.64	1.63	0.53	2.03	0.47	2.36	0.37	3.19
18	0.77	1.32	0.69	1.49	0.64	1.61	0.53	2.00	0.47	2.31	0.38	3.09
19	0.77	1.31	0.69	1.48	0.65	1.59	0.54	1.97	0.48	2.26	0.38	3.00
20	0.78	1.30	0.70	1.46	0.65	1.58	0.54	1.94	0.48	2.22	0.39	2.93
21	0.78	1.30	0.70	1.45	0.65	1.56	0.55	1.91	0.49	2.18	0.39	2.86
22	0.78	1.29	0.70	1.44	0.66	1.55	0.55	1.89	0.49	2.15	0.40	2.80
23	0.78	1.28	0.71	1.44	0.66	1.54	0.56	1.87	0.50	2.12	0.40	2.74
24	0.79	1.28	0.71	1.43	0.67	1.53	0.56	1.85	0.50	2.10	0.41	2.69
25	0.79	1.27	0.71	1.42	0.67	1.52	0.56	1.83	0.51	2.07	0.41	2.65
26	0.79	1.27	0.72	1.41	0.67	1.51	0.57	1.82	0.51	2.05	0.41	2.60
27	0.79	1.27	0.72	1.41	0.67	1.50	0.57	1.80	0.51	2.03	0.42	2.57
28	0.79	1.26	0.72	1.40	0.68	1.49	0.57	1.79	0.52	2.01	0.42	2.53
29	0.80	1.26	0.72	1.39	0.68	1.49	0.58	1.77	0.52	1.99	0.42	2.50
30	0.80	1.26	0.72	1.39	0.68	1.48	0.58	1.76	0.52	1.97	0.43	2.47
35	0.81	1.24	0.73	1.37	0.69	1.45	0.59	1.71	0.53	1.90	0.44	2.35
40	0.81	1.23	0.74	1.35	0.70	1.43	0.60	1.67	0.55	1.85	0.45	2.25
45	0.82	1.22	0.75	1.34	0.71	1.41	0.61	1.64	0.55	1.81	0.46	2.19

(*continued*)

Table A.5—continued
($\nu_1 = 50$)

ν_2	50%		67%		75%		90%		95%		99%	
3	0.58	1.93	0.47	2.68	0.41	3.34	0.30	6.46	0.25	10.40	0.18	30.57
4	0.62	1.75	0.51	2.29	0.46	2.75	0.34	4.67	0.29	6.82	0.21	15.72
5	0.65	1.64	0.54	2.08	0.49	2.43	0.37	3.83	0.32	5.26	0.23	10.53
6	0.67	1.57	0.57	1.94	0.52	2.23	0.40	3.33	0.34	4.40	0.26	8.04
7	0.69	1.52	0.59	1.84	0.54	2.09	0.42	3.01	0.36	3.87	0.28	6.61
8	0.70	1.48	0.61	1.77	0.55	1.99	0.44	2.78	0.38	3.50	0.29	5.69
9	0.71	1.44	0.62	1.71	0.57	1.91	0.46	2.61	0.40	3.23	0.31	5.05
10	0.72	1.42	0.63	1.67	0.58	1.85	0.47	2.48	0.41	3.03	0.32	4.59
11	0.73	1.40	0.64	1.63	0.59	1.80	0.48	2.38	0.42	2.87	0.33	4.24
12	0.74	1.38	0.65	1.60	0.60	1.76	0.49	2.29	0.43	2.74	0.34	3.96
13	0.75	1.36	0.66	1.57	0.61	1.72	0.50	2.22	0.44	2.63	0.35	3.74
14	0.75	1.35	0.67	1.55	0.62	1.69	0.51	2.16	0.45	2.54	0.36	3.55
15	0.76	1.34	0.67	1.53	0.63	1.66	0.52	2.10	0.46	2.46	0.37	3.40
16	0.76	1.33	0.68	1.51	0.63	1.64	0.53	2.06	0.47	2.39	0.37	3.27
17	0.77	1.32	0.69	1.49	0.64	1.62	0.53	2.02	0.47	2.34	0.38	3.15
18	0.77	1.31	0.69	1.48	0.65	1.60	0.54	1.98	0.48	2.29	0.39	3.05
19	0.77	1.30	0.70	1.47	0.65	1.58	0.54	1.95	0.49	2.24	0.39	2.97
20	0.78	1.30	0.70	1.46	0.66	1.57	0.55	1.92	0.49	2.20	0.40	2.89
21	0.78	1.29	0.70	1.45	0.66	1.55	0.55	1.90	0.50	2.16	0.40	2.82
22	0.78	1.28	0.71	1.44	0.66	1.54	0.56	1.87	0.50	2.13	0.41	2.76
23	0.79	1.28	0.71	1.43	0.67	1.53	0.56	1.85	0.51	2.10	0.41	2.71
24	0.79	1.27	0.71	1.42	0.67	1.52	0.57	1.83	0.51	2.07	0.42	2.66
25	0.79	1.27	0.72	1.41	0.67	1.51	0.57	1.82	0.51	2.05	0.42	2.61
26	0.79	1.27	0.72	1.40	0.68	1.50	0.57	1.80	0.52	2.03	0.42	2.57
27	0.80	1.26	0.72	1.40	0.68	1.49	0.58	1.78	0.52	2.00	0.43	2.53
28	0.80	1.26	0.73	1.39	0.68	1.48	0.58	1.77	0.52	1.99	0.43	2.50
29	0.80	1.25	0.73	1.39	0.69	1.48	0.58	1.76	0.53	1.97	0.43	2.47
30	0.80	1.25	0.73	1.38	0.69	1.47	0.59	1.74	0.53	1.95	0.44	2.44
35	0.81	1.24	0.74	1.36	0.70	1.44	0.60	1.69	0.54	1.88	0.45	2.31
40	0.82	1.23	0.75	1.34	0.71	1.42	0.61	1.65	0.55	1.83	0.46	2.22
45	0.82	1.22	0.75	1.33	0.71	1.40	0.62	1.62	0.56	1.78	0.47	2.15
50	0.83	1.21	0.76	1.32	0.72	1.39	0.63	1.60	0.57	1.75	0.48	2.10

(*continued*)

Table A.5—continued
($\nu_1 = 55$)

ν_2	50%		67%		75%		90%		95%		99%	
3	0.58	1.93	0.47	2.67	0.41	3.33	0.30	6.44	0.25	10.38	0.18	30.49
4	0.62	1.74	0.51	2.29	0.46	2.74	0.34	4.66	0.29	6.80	0.21	15.67
5	0.65	1.64	0.55	2.07	0.49	2.42	0.38	3.81	0.32	5.24	0.24	10.50
6	0.67	1.56	0.57	1.93	0.52	2.22	0.40	3.32	0.35	4.39	0.26	8.00
7	0.69	1.51	0.59	1.84	0.54	2.08	0.42	3.00	0.37	3.85	0.28	6.57
8	0.70	1.47	0.61	1.76	0.56	1.98	0.44	2.77	0.38	3.48	0.30	5.66
9	0.71	1.44	0.62	1.71	0.57	1.90	0.46	2.60	0.40	3.21	0.31	5.02
10	0.72	1.42	0.63	1.66	0.59	1.84	0.47	2.47	0.41	3.01	0.32	4.56
11	0.73	1.39	0.65	1.62	0.60	1.79	0.49	2.36	0.43	2.85	0.33	4.21
12	0.74	1.38	0.65	1.59	0.61	1.75	0.50	2.28	0.44	2.72	0.35	3.93
13	0.75	1.36	0.66	1.57	0.62	1.71	0.51	2.21	0.45	2.61	0.36	3.71
14	0.75	1.35	0.67	1.54	0.62	1.68	0.52	2.14	0.46	2.52	0.36	3.52
15	0.76	1.34	0.68	1.52	0.63	1.66	0.52	2.09	0.47	2.44	0.37	3.37
16	0.77	1.32	0.68	1.50	0.64	1.63	0.53	2.05	0.47	2.38	0.38	3.24
17	0.77	1.32	0.69	1.49	0.64	1.61	0.54	2.01	0.48	2.32	0.39	3.13
18	0.77	1.31	0.69	1.47	0.65	1.59	0.54	1.97	0.49	2.27	0.39	3.03
19	0.78	1.30	0.70	1.46	0.65	1.58	0.55	1.94	0.49	2.22	0.40	2.94
20	0.78	1.29	0.70	1.45	0.66	1.56	0.56	1.91	0.50	2.18	0.40	2.87
21	0.78	1.29	0.71	1.44	0.66	1.55	0.56	1.89	0.50	2.15	0.41	2.80
22	0.79	1.28	0.71	1.43	0.67	1.53	0.56	1.86	0.51	2.11	0.41	2.74
23	0.79	1.28	0.72	1.42	0.67	1.52	0.57	1.84	0.51	2.08	0.42	2.68
24	0.79	1.27	0.72	1.41	0.68	1.51	0.57	1.82	0.52	2.06	0.42	2.63
25	0.80	1.27	0.72	1.41	0.68	1.50	0.58	1.80	0.52	2.03	0.43	2.59
26	0.80	1.26	0.72	1.40	0.68	1.49	0.58	1.79	0.53	2.01	0.43	2.55
27	0.80	1.26	0.73	1.39	0.69	1.48	0.58	1.77	0.53	1.99	0.43	2.51
28	0.80	1.25	0.73	1.39	0.69	1.48	0.59	1.76	0.53	1.97	0.44	2.47
29	0.80	1.25	0.73	1.38	0.69	1.47	0.59	1.74	0.54	1.95	0.44	2.44
30	0.81	1.25	0.73	1.37	0.69	1.46	0.59	1.73	0.54	1.93	0.44	2.41
35	0.81	1.23	0.74	1.35	0.70	1.43	0.61	1.68	0.55	1.86	0.46	2.28
40	0.82	1.22	0.75	1.33	0.71	1.41	0.62	1.64	0.56	1.81	0.47	2.19
45	0.83	1.21	0.76	1.32	0.72	1.39	0.63	1.61	0.57	1.77	0.48	2.12
50	0.83	1.21	0.76	1.31	0.73	1.38	0.63	1.58	0.58	1.73	0.49	2.07
55	0.83	1.20	0.77	1.30	0.73	1.37	0.64	1.56	0.59	1.71	0.49	2.02

(*continued*)

Table A.5—continued
($\nu_1 = 60$)

ν_2	50%		67%		75%		90%		95%		99%	
3	0.58	1.93	0.47	2.67	0.41	3.33	0.30	6.43	0.25	10.36	0.18	30.43
4	0.62	1.74	0.51	2.29	0.46	2.73	0.34	4.65	0.29	6.78	0.21	15.63
5	0.65	1.63	0.55	2.07	0.49	2.42	0.38	3.80	0.32	5.22	0.24	10.45
6	0.67	1.56	0.57	1.93	0.52	2.22	0.40	3.31	0.35	4.37	0.26	7.97
7	0.69	1.51	0.59	1.83	0.54	2.08	0.43	2.99	0.37	3.83	0.28	6.54
8	0.70	1.47	0.61	1.76	0.56	1.98	0.45	2.76	0.39	3.47	0.30	5.63
9	0.72	1.44	0.62	1.70	0.57	1.90	0.46	2.59	0.40	3.20	0.31	5.00
10	0.73	1.41	0.64	1.66	0.59	1.84	0.48	2.46	0.42	3.00	0.33	4.53
11	0.74	1.39	0.65	1.62	0.60	1.79	0.49	2.35	0.43	2.84	0.34	4.18
12	0.74	1.37	0.66	1.59	0.61	1.74	0.50	2.27	0.44	2.71	0.35	3.91
13	0.75	1.36	0.67	1.56	0.62	1.71	0.51	2.20	0.45	2.60	0.36	3.68
14	0.76	1.34	0.67	1.54	0.63	1.68	0.52	2.13	0.46	2.51	0.37	3.50
15	0.76	1.33	0.68	1.52	0.63	1.65	0.53	2.08	0.47	2.43	0.38	3.35
16	0.77	1.32	0.69	1.50	0.64	1.63	0.53	2.04	0.48	2.36	0.38	3.22
17	0.77	1.31	0.69	1.48	0.65	1.60	0.54	2.00	0.48	2.31	0.39	3.10
18	0.78	1.30	0.70	1.47	0.65	1.59	0.55	1.96	0.49	2.26	0.40	3.01
19	0.78	1.30	0.70	1.46	0.66	1.57	0.55	1.93	0.50	2.21	0.40	2.92
20	0.78	1.29	0.71	1.44	0.66	1.55	0.56	1.90	0.50	2.17	0.41	2.84
21	0.79	1.28	0.71	1.43	0.67	1.54	0.56	1.87	0.51	2.13	0.42	2.78
22	0.79	1.28	0.71	1.42	0.67	1.53	0.57	1.85	0.51	2.10	0.42	2.71
23	0.79	1.27	0.72	1.42	0.68	1.52	0.57	1.83	0.52	2.07	0.43	2.66
24	0.80	1.27	0.72	1.41	0.68	1.50	0.58	1.81	0.52	2.04	0.43	2.61
25	0.80	1.26	0.73	1.40	0.68	1.49	0.58	1.79	0.53	2.02	0.43	2.56
26	0.80	1.26	0.73	1.39	0.69	1.49	0.59	1.78	0.53	1.99	0.44	2.52
27	0.80	1.25	0.73	1.39	0.69	1.48	0.59	1.76	0.53	1.97	0.44	2.48
28	0.80	1.25	0.73	1.38	0.69	1.47	0.59	1.75	0.54	1.95	0.45	2.45
29	0.81	1.25	0.74	1.37	0.70	1.46	0.60	1.73	0.54	1.93	0.45	2.42
30	0.81	1.24	0.74	1.37	0.70	1.45	0.60	1.72	0.54	1.92	0.45	2.39
35	0.82	1.23	0.75	1.35	0.71	1.43	0.61	1.67	0.56	1.85	0.47	2.26
40	0.82	1.22	0.76	1.33	0.72	1.40	0.62	1.63	0.57	1.79	0.48	2.17
45	0.83	1.21	0.76	1.31	0.73	1.38	0.63	1.60	0.58	1.75	0.49	2.10
50	0.83	1.20	0.77	1.30	0.73	1.37	0.64	1.57	0.59	1.72	0.50	2.05
55	0.84	1.20	0.77	1.29	0.74	1.36	0.65	1.55	0.59	1.69	0.50	2.00
60	0.84	1.19	0.78	1.29	0.74	1.35	0.65	1.53	0.60	1.67	0.51	1.96

Further Reading

(a) *Hierarchical methods*

It has been suggested that while most of the justifications that have been advanced for subjective theories of probability assume that such probabilities are clear-cut, there is always some vagueness in your judgements and that one way of coping with this is to say that your beliefs are expressed by a distribution over a family of prior densities. Such an approach has connections with empirical Bayes theory. Some useful references on this area are Good (1965, 1983), Berger (1985, Section 4.6) and, for the empirical Bayes aspect, Morris (1983). A hierarchical approach has also been applied with considerable success to the analysis of the general linear model; see Lindley and Smith (1971).

(b) *Robustness*

Although the importance of robustness was mentioned at the end of Section 2.3 on "Several normal observations with a normal prior", not much attention has been devoted to this topic in the rest of the book. Some useful references are Berger (1985, Section 4.7), Box and Tiao (1973, Section 3.2 and *passim.*), Hartigan (1983, Chapter 12) and Kadane (1984).

(c) *Nonparametric statistics*

Throughout this book it is assumed that the data we are analysing come from some parametric family, so that the density $p(x|\theta)$ of any observation x depends on one or more parameters θ (for example, x is normal of mean θ and known variance). In classical statistics, much attention has been devoted to developing methods which do not make any such assumption, so that we can, for example, say something about the median of a set of observations without assuming that they come from a normal distribution. Some attempts have been made to develop a Bayesian form of nonparametric theory, though this is not easy as it involves setting up a prior distribution over a very large class of densities for the observations. Useful references are Florens *et al.* (1983) and Dalal (1980).

(d) Multivariate estimation

In order to provide a reasonably simple introduction to Bayesian statistics, avoiding matrix theory as far as possible, the coverage of this book has been restricted largely to cases where only one measurement is taken at a time. Useful references for multivariate Bayesian statistics are Box and Tiao (1973, Chapter 8), Zellner (1971, Chapter 8) and Press (1972, 1982).

(e) Sequential methods

Some idea on how to apply Bayesian methods in cases where observations are collected sequentially through time can be obtained from Berger (1985, Chapter 7).

(f) General reading

Apart from Jeffreys (1939, 1948, 1961), Berger (1985) and Box and Tiao (1973), which have been frequently referred to, some useful references are Lindley (1971), DeGroot (1970) and Raiffa and Schlaifer (1961). Anyone interested in Bayesian statistics will gain a great deal by reading de Finetti (1972, 1974–75) and Savage (1954, 1972, 1981). A useful collection of essays on the foundations of Bayesian statistics is Kyburg and Smokler (1964, 1980). A valuable treatment of probability from a personalistic standpoint is provided by O'Hagan (1988). A comparison of Bayesian and other approaches to statistical inference is provided by Barnett (1973, 1982).

References

Abramowitz, M. and Stegun, M. A., *Handbook of Mathematical Functions*. Washington, DC: National Bureau of Standards (1964); New York: Dover (1965).

Altham, P. M. E., "Exact Bayesian analysis of a 2×2 contingency table and Fisher's 'exact' significance test". *J. Roy. Statist. Soc.*, **B, 31** (1969) 261–9.

Arbuthnot, J., "An argument for Divine Providence taken from the constant Regularity of the Births of Both Sexes". *Phil. Trans. Roy. Soc. London*, **23** (1710) 186–90 [reprinted in Kendall and Plackett (1977)].

Armitage, P. and Berry, G., *Statistical Methods in Medical Research*, 2nd edn. Oxford: Blackwells (1987) [1st edn, by Armitage alone (1971)].

Arnold, B. C., *Pareto Distributions*, Fairland, MD: International Co-operative Publishing House (1983).

Aykaç A. and Brumat, C., eds, *New Methods in the Applications of Bayesian Methods*. Amsterdam: North-Holland (1977).

Baird, R. D., *Experimentation: An Introduction to Measurement Theory and Experiment Design*. Englewood Cliffs, NJ: Prentice-Hall (1962).

Barnard, G. A., "Thomas Bayes's essay towards solving a problem in the doctrine of chances". *Biometrika*, **45** (1958) 293–315 [reprinted in Pearson and Kendall (1970)].

Barnett, V., "Evaluation of the maximum-likelihood estimator where the likelihood equation has multiple roots". *Biometrika* **53** (1966) 151–65.

Barnett, V., *Comparative Statistical Inference*. New York: Wiley (1973, 1982).

Bartlett, M. S., "The information available in small samples". *Proc. Cambridge Philos. Soc.*, **32** (1936) 560–6.

Bartlett, M. S., "A comment on D. V. Lindley's statistical paradox". *Biometrika*, **44** (1957) 533–4.

Batschelet, E., *Circular Statistics in Biology*. New York: Academic Press (1981).

Bayes, T. R., "An essay towards solving a problem in the doctrine of chances". *Phil. Trans. Roy. Soc. London*, **53** (1763) 370–418 [reprinted as part of Barnard (1958) and Pearson and Kendall (1970)].

Bayes, T. R., "A demonstration of the second rule in the essay towards the solution of a problem in the doctrine of chances". *Phil. Trans. Roy. Soc. London*, **54** (1764) 296–325.

Behrens, W. A., "Ein Betrag zur Fehlenberechnung bei wenigen Beobachtungen". *Landwirtschaftliche Jahrbücher*, **68** (1929) 807–37.

Benford, F., "The law of anomalous numbers". *Proc. Amer. Philos. Soc.*, **78** (1938) 551–72.

Berger, J. O., *Statistical Decision Theory and Bayesian Analysis*, 2nd edn. Berlin: Springer-Verlag (1985) [1st edn published as *Statistical Decision Theory: Foundations, Concepts and Methods*. Berlin: Springer-Verlag (1980)].

Berger, J. O. and Delampady, M., "Testing precise hypotheses" (with discussion). *Statistical Science*, **2** (1987) 317–52.

Berger, J. O. and Wolpert, R. L., *The Likelihood Principle*. Hayward, CA: Institute of Mathematical Statistics (1984).

Bernardo, J. M., DeGroot, M. H., Lindley, D. V. and Smith, A. F. M., eds, *Bayesian Statistics*, Valencia: Valencia University Press (1980).

Bernardo, J. M., DeGroot, M. H., Lindley, D. V. and Smith, A. F. M., eds, *Bayesian Statistics 2*, Amsterdam: North-Holland and Valencia: Valencia University Press (1985).

Birnbaum, A., "On the foundations of statistical inference" (with discussion). *J. Amer. Statist. Assoc.*, **57** (1962) 269–306.

Box, G. E. P. and Cox, D. R., "An analysis of transformations" (with discussion). *J. Roy. Statist. Soc.*, **B**, **26** (1964) 211–52.

Box, G. E. P. and Tiao, G. C., *Bayesian Inference in Statistical Analysis*. Reading, MA: Addison-Wesley (1973).

British Association for the Advancement of Science, *Mathematical Tables, Vol. VI: Bessel Functions Part I, Functions of Order Zero and Unity*. Cambridge: University Press (1937).

Calvin, T. W., *The ASQC Basic References in Quality Control: Statistical Techniques, Volume 7: How and When to Perform Bayesian Acceptance Sampling*, Milwaukee, WI: American Society for Quality Control (1984).

Cochran, W. G. and Cox, G. M., *Experimental Designs*. New York: Wiley (1950) [2nd edn (1957)].

Cornish, E. A., "The multivariate t-distribution associated with a set of normal sample deviates". *Austral. J. Phys.*, **7** (1954) 531–42.

Cornish, E. A., "The sampling distribution of statistics derived from the multivariate t-distribution". *Austral. J. Phys.*, **8** (1955) 193–9.

Cornish, E. A., *Published Papers of E. A. Cornish*. Adelaide: E. A. Cornish Memorial Appeal (1974).

Dalal, S. R., "Non-parametric Bayes decision theory" (with discussion). In Bernardo *et al.* (1980).

David, F. N., *Tables of the Correlation Coefficient*. Cambridge: University Press, for Biometrika (1954).

DeGroot, M. H., *Optimal Statistical Decisions*. New York: McGraw-Hill (1970).

DeGroot, M. H., Fienberg, S. E. and Kadane, J. B., eds, *Statistics and the Law*. New York: Wiley (1986).

Di Raimondo, F., "*In Vitro* and *in vivo* antagonism between vitamins and antibiotics". *Int. Rev. Vitamin Res.*, **23** (1951) 1–12.

Diaconis, P. and Ylvisaker, D., "Quantifying prior opinion". In Bernardo *et al.* (1985).

Diaconis, P. and Ylvisaker, D., "Conjugate priors for exponential families". *Ann. Statist.*, **7** (1979) 269–81.

Dunnett, C. W. and Sobel, M., "A bivariate generalization of Student's t distribution with tables for special cases". *Biometrika*, **41** (1954) 153–76.

Edwards, A. W. F., "The measure of association in a 2×2 table". *J. Roy. Statist. Soc.*, **A**, **126** (1963) 109–13.

Edwards, A. W. F., *Likelihood*. Cambridge: University Press (1972).
Edwards, J., *A Treatise on the Integral Calculus* (2 vols). London: Macmillan (1921) [reprinted New York: Chelsea (1955)].
Edwards, W., Lindman, H. and Savage, L. J., "Bayesian statistical inference for statistical research". *Psychological Review*, **70** (1963) 193–242 [reprinted in Luce *et al.* (1965), Kadane (1984) and Savage (1981)].
Eisenhart, C., Hastay, M. W. and Wallis, W. A., eds, *(Selected) Techniques of Statistical Analysis* by the Statistical Research Group of Columbia University. New York: McGraw-Hill (1947).
Feller, W., *A Course in Probability Theory and its Applications* (2 vols). New York: Wiley, Vol. **1**, 1950, 1957, 1968; Vol. **2**, 1966, 1971.
Ferguson, T. S., *Mathematical Statistics: A Decision Theoretic Approach*. New York: Academic Press (1967).
de Finetti, B., *Probability, Induction and Statistics: The Art of Guessing*. New York: Wiley (1972).
de Finetti, B., *Theory of Probability: A Critical Introductory Treatment* (2 vols). New York: Wiley (1974–75).
Fisher, R. A., "Frequency distribution of the values of the correlation coefficient in samples from an indefinitely large population". *Biometrika*, **10** (1915) 507–21.
Fisher, R. A., "On the 'probable error' of a coefficient of correlation deduced from a small sample". *Metron*, **1** (1921) 3–32.
Fisher, R. A., "On the mathematical foundations of theoretical statistics". *Phil. Trans. Roy. Soc. London*, **A**, **222** (1922) 309–68.
Fisher, R. A., "On a distribution yielding the error function of several well-known statistics". *Proc. Internat. Congress of Math.*, Toronto: University of Toronto Press, Vol. 2, pp. 805–13 (1924).
Fisher, R. A., "Theory of statistical information". *Proc. Cambridge Philos. Soc.*, **22** (1925a) 700–25.
Fisher, R. A., *Statistical Methods for Research Workers*. Edinburgh: Oliver & Boyd (1925b) (many subsequent editions).
Fisher, R. A., "The fiducial argument in statistical inference". *Ann. Eugenics,* **6** (1935) 391–8.
Fisher, R. A., "On a point raised by M. S. Bartlett in fiducial probability". *Ann. Eugenics*, **7** (1937) 370–5.
Fisher, R. A., "The comparison of samples with possibly unequal variances". *Ann Eugenics*, **9** (1939) 174–80.
Fisher, R. A., "The analysis of variance with various binomial transformations". *Biometics*, **10** (1954) 130–9.
Fisher, R. A., *Statistical Methods and Scientific Inference*. Edinburgh: Oliver & Boyd (1956) [2nd edn (1959)].
Fisher, R. A., *Collected Papers of R. A. Fisher* (5 vols), ed. J. H. Bennett, Adelaide: University of Adelaide Press (1971–74).
Florens, J. P., Mouchart, M., Raoult, J. P., Simar, L. and Smith, A. F. M., *Specifying Statistical Models from Parametric to Non-parametric Using Bayesian or Non-Bayesian Approaches* (Lecture Notes in Statistics No. 16). Berlin: Springer-Verlag (1983).
Good, I. J., *Probability and the Weighing of Evidence*. London: Griffin (1950).
Good, I. J., *The Estimation of Probabilities: An Essay on Modern Bayesian Methods*. Cambridge, MA: M.I.T. Press (1965).
Good, I. J., *Good Thinking: The Foundations of Probability and its Applications*. Minneapolis, MN: University of Minnesota Press (1983).

Haldane, J. B. S., "A note on inverse probability". *Proc. Cambridge Philos. Soc.*, **28** (1931) 55–61.

Harter, H. L., "The method of least squares and some alternatives". *Internat. Statist. Review*, **42** (1974) 147–74, 235–64; **43** (1975) 1–44, 125–90, 269–78; **44** (1976) 113–59.

Hartigan, J. A. *Bayes Theory*. Berlin: Springer-Verlag (1983).

Hill, B., "On statistical paradoxes and non-conglomerability". In Bernardo *et al.* (1980).

Holland, G. D., "The Reverend Thomas Bayes, F.R.S. (1702–1761)". *J. Roy. Statist. Soc.*, **A**, **125** (1962) 451–61.

Huzurbazar, V. S., *Sufficient Statistics*. New York: Marcel Dekker (1976).

Isaacs, G. L., Christ, D. E., Novick, M. R. and Jackson, P. H., *Tables for Bayesian Statisticians*. Ames, IO: Iowa University Press (1974).

Jackson, P. H., "Formulae for generating highest density credibility regions". *ACT Technical Bulletin*, **20**. Iowa City, IO: American College Testing Program (1974).

Jeffreys, H. S., *Theory of Probability*. Oxford: University Press (1939) [2nd edn (1948), 3rd edn (1961)].

Johnson, N. L. and Kotz, S., *Distribution in Statistics* (4 vols). New York: Houghton-Mifflin/Wiley (1969–72).

Kadane, J. B, ed., *Robustness of Bayesian Analyses*. Amsterdam: North-Holland (1984).

Kale, B. K., "On the solution of the likelihood equation by iteration processess". *Biometrica*, **48** (1961) 452–6.

Kelley, T. L., *Interpretation of Educational Measurements*. Yonkers-on-Hudson, NY: World Book Co. (1927).

Kendall, M. G. and Plackett, R. L., *Studies in the History of Probability and Statistics*, Vol. II. London: Griffin (1977).

Kendall, M. G., Stuart, A. and Ord, J. K., *The Advanced Theory of Statistics*, Vol. I, 5th edn. London: Griffin (1987).

Kennett, P. and Ross, C. A., *Geochronology*. London: Longmans (1983).

Knuth, D. E., *The Art of Computer Programming, Vol. 2: Seminumerical Algorithms*. Reading, MA: Addison-Wesley (1969).

Kyburg, H. E. and Smokler, H. E., eds, *Studies in Subjective Probability*. New York: Wiley (1964) [2nd edn (much altered) Melbourne, FL: Krieger (1980)].

Laplace, P. S., "Mémoire sur la probabilité des causes par les évenements". *Mém. de math. et phys. presenté à l'Acad. roy. des sci.*, **6** (1774) 621–86; reprinted in his *Œuvres complètes*, **8**, 27–65. An English translation is to be found in Stigler (1986).

Lehmann, E. L., *Testing Statistical Hypotheses*. New York: Wiley (1959) [2nd edn (1986)].

Lieberman, G. J. and Owen, D. B., *Tables of the Hypergeometric Probability Distribution*. Stanford, CA: Stanford University Press (1961).

Lindley, D. V., *Introduction to Probability and Statistics from a Bayesian Viewpoint* (2 vols—*Part I: Probability* and *Part II: Inference*). Cambridge University Press (1965).

Lindley, D. V., "A statistical paradox". *Biometrika*, **44** (1957) 187–92.

Lindley, D. V., *Bayesian Statistics: A Review*. Philadelphia, PA: S.I.A.M.—Society for Industrial and Applied Mathematics (1971).

Lindley, D. V., "A problem in forensic science". *Biometrika*, **64** (1977) 207–13.

Lindley, D. V. and Smith, A. F. M., "Bayes estimates for the linear model" (with discussion). *J. Roy. Statist. Soc.*, **A**, **34** (1971) 1–41.

Luce, R. D., Bush, R. B. and Galanter, E., *Readings in Mathematical Psychology*, Vol. 2. New York: Wiley (1965).

Mardia, K. V., *Statistics of Directional Data*. New York: Academic Press (1972).

Maritz, J. S., *Empirical Bayes Methods*. London: Methuen (1970).

Meyer, D. L. and Collier, R. O., eds, *Bayesian Statistics*. Itasca, IL: F. E. Peacock (1970).

Morris, C., "Parametric empirical Bayes confidence sets: theory and applications". *J. Amer. Statist. Soc.*, **78** (1983) 47–65.

Nagel, E., *Principles of the Theory of Probability*. Chicago: University of Chicago Press (1939) [reprinted in Neurath *et al.* (1955)].

Neave, H. R., *Statistics Tables for Mathematicians, Engineers, Economists and the Behavioural and Management Sciences*. London: George Allen & Unwin (1978).

Neurath, O., Carnap, R. and Morris, C., eds, *Foundations of the Unity of Science*, Vol. I. Chicago: Chicago University Press (1955).

Newcomb, S., "Note on the frequency of use of the different digits in natural numbers". *Amer. J. Math.*, **4** (1881) 39–40 [reprinted in Stigler (1980)].

Novick, M. R. and Jackson, P. H., *Statistical Methods for Educational and Psychological Research*. New York: McGraw-Hill (1974).

O'Hagan, A., *Probability: Methods and Measurements*. London: Chapman & Hall (1988).

Patil, V. H., "Approximations to the Behrens–Fisher distribution". *Biometrika*, **52** (1965) 267–71.

Pearson, E. S. and Hartley, H. O., *Biometrika Tables for Statisticians* (2 vols). Cambridge: University Press, for Biometrika (Vol. I—1954, 1958, 1966; Vol. II—1972).

Pearson, E. S. and Kendall, M. G., *Studies in the History of Probability and Statistics*. London: Griffin (1970).

Pearson, K., *Tables of the Incomplete Gamma Function*. Cambridge: University Press (1922, 1924).

Pearson, K., *Tables of the Incomplete Beta Function*. Cambridge: University Press (1934, 1968).

Peirce, C. S., "The probability of induction". *Popular Science Monthly*, **12** (1878) 705–18; reprinted as Arts. 2.669–2.693 of Peirce (1931–58).

Peirce, C. S., *Collected Papers*. Cambridge, MA: Harvard University Press (1931–58).

Press, S. J., *Applied Multivariate Analysis: Using Bayesian and Frequentist Measures of Inference*, 2nd edn. Melbourne FL: Krieger (1982) [1st edn published as *Applied Multivariate Analysis*. New York: Holt, Rinehart, Winston (1972)].

Raiffa, H. and Schlaifer, R., *Applied Statistical Decision Theory*. Cambridge, MA: Harvard University Press (1961).

Raimi, R. A., "The first digit problem". *Amer. Math. Monthly*, **83** (1976) 521–38.

Rényi, A., *Foundations of Probability*. San Francisco, CA: Holden-Day (1970).

Roberts, H. V., "Informative stopping rules and inference about population size". *J. Amer. Statist. Soc.*, **62** (1967) 763–75.

Robinson, G. K., "Properties of Student's t and of the Behrens–Fisher solution to the two means problem". *Ann. Statist.*, **4** (1976) 963–71; **10** (1982) 321.

Rothschild, V. and Logothetis, N., *Probability Distributions*. New York: Wiley (1986).

Savage, L. J., *The Foundations of Statistics*. New York: Wiley (1954); New York: Dover (1972).

Savage, L. J., *The Writings of Leonard Jimmie Savage: A Memorial Selection*. Washington, DC: American Statistical Association/Institute of Mathematical Statistics (1981).

Savage, L. J., *et al.*, *The Foundations of Statistics: A Discussion*. London: Methuen (1962).

Scheffé, H., *The Analysis of Variance*. New York: Wiley (1959).

Schlaifer, R., *Introduction to Statistics for Business Decisions*. New York: McGraw-Hill (1961).

Schmitt, S. A., *Measuring Uncertainty: An Elementary Introduction to Bayesian Statistics*. Reading, MA: Addison-Wesley (1969).

Shafer, G., "Lindley's paradox". *J. Amer. Statist. Assoc.*, **77** (1982) 325–51.

Silcock, A., *Verse and Worse*. London: Faber & Faber (1952).

Sprent, P., *Models in Regression and Related Topics*. London: Methuen (1969).

Stigler, S., *American Contributions to Mathematical Statistics in the Nineteenth Century* (2 vols). New York: Arno Press (1980).

Stigler, S., *The History of Statistics: The Measurement of Uncertainty before 1900*. Cambridge, MA: Harvard University Press (1986).

Stigler, S., "Laplace's 1774 memoir on inverse probability". *Statistical Science*, **1** (1986) 359–78.

"Student" (W. S. Gosset), "The probable error of a correlation coefficient". *Biometrika*, **6** (1908) 1–25.

"Student" (W. S. Gosset), *Student's Collected Papers*. Cambridge University Press, for Biometrika (1942).

Todhunter, I., *A History of the Mathematical History of Probability from the Time of Pascal to that of Laplace*. London: Macmillan (1865) [reprinted New York: Chelsea (1949)].

Turner, P. S., "The distribution of l.s.d. and its implications for computer design", *Math. Gazette*, **71** (1987) 26–31.

von Bortkiewicz, L., *Das Gesetz der Kleinen Zahlenen*. Leipzig: Teubner (1898).

von Mises, R., "Über die 'Ganz-zahligkeit' der Atomgewicht und verwandte Fragen". *Physikal. Z.*, **19** (1918) 490–500.

von Mises, R., "On the correct use of Bayes' formula". *Ann. Math. Statist.*, **13** (1942) 156–65.

Von Neumann, J. and Morgenstern, O., *Theory of Games and Economic Behaviour*. Princeton, NJ: Princeton University Press (1944, 1947, 1953).

Watkins, P., *Story of the W and the Z*. Cambridge: University Press (1986).

Weisbers, H. L., "Bayesian comparison of two ordered multinomial populations". *Biometrics*, **23** (1972) 859–67.

Whitaker's Almanack. London: Whitaker (annual).

Whittaker, E. T. and Robinson, G., *The Calculus of Observations*. Edinburgh: Blackie (1924) [2nd edn (1926); 3rd edn (1940)].

Williams, J. D., *The Compleat Strategyst*. New York: McGraw-Hill (1954).

Wilson, E. B., *An Introduction to Scientific Research*. New York: McGraw-Hill (1952).

Young, A. S., "A Bayesian approach to prediction using polynomials", *Biometrika*, **64** (1977), 309–317.

Zellner, A., *An Introduction to Bayesian Inference in Econometrics*. New York: Wiley (1971).

Zellner, A., *Basic Issues in Econometrics*. Chicago: University of Chicago Press (1974).

Zellner, A., "Maximal data information prior distributions". In Aykaç and Brumat (1977).

Index

Absolute error loss 221
Action 218
Addition law, generalized 9
Alternative hypothesis 123
Analysis of variance 188, 190, 196
Ancillary statistic 58, 176
Antilogarithms 122
Arc-sine distribution 88, 93, 217

Bayes, Rev. Thomas *viii*, 22, 47, 86
  estimator 219
  factor 125, 142, 149
  postulate 47, 86
  risk 239
  rule 219, 231
  theorem 8, 9, 10, 20
  theorem, sequential use of 35
Bayesian confidence interval 51
Behrens' distribution 154, 157, 166, 255
Behrens-Fisher problem *ix*, 153, 156, 166
Benford's law 109
Bernoulli trials 14, 21, 99, 211, 216, 230
Beta distribution 22, 80, 120, 162, 231, 241, 252
Beta-binomial distribution 86, 227
Beta-Pascal distribution 86, 215
Bilateral bivariate Pareto
  distribution 101, 246
Binomial distribution 14, 21, 25, 32, 64, 67, 80, 93, 120, 145, 161, 227, 231, 241
Bivariate normal distribution 27, 32, 169, 175
Bivariate probability density 17

Cardioid distribution 248
Cauchy distribution 62, 118, 138, 145, 252
Central limit theorem 17, 37, 62
Change of variable rule 15, 47
Characteristic function 99
Chest measurements 42
Chi-squared distribution 31, 53, 164, 234, 235, 257
Circular normal distribution 110, 247
Classical statistics *viii*, 51, 69, 87, 90, 123, 129, 173, 210, 212, 216, 224
Components of variance model 19
Composite hypothesis 126, 225
Conditional
  density 17
  distribution function 18, 19
  expectation 28
  variance 28
Conditionality principle 205, 206
Confidence interval, Bayesian 51
Conjugate priors 63, 67, 73, 78, 80, 95, 100, 101
  mixtures of 65
Continuous random variables 14, 18, 21
Contrast 192
Correlation coefficient 26, 168, 174, 201, 203, 253
  sample 169, 174, 192, 201, 202
Covariance 26
Cramèr-Rao bound 90
Credible interval 51
Cumulative distribution function 13

Data-translated likelihood 48, 103, 217
Decision rule (function) 219, 224
Density 13, 233
  bivariate 17
  conditional 18, 19
  improper 45
  joint 17, 19
  marginal 18, 69, 71
  proper 44

Discrete random variable 11, 14, 17, 21
Discrete uniform distribution 105, 245
Discrimination 11
Distribution function 13, 14, 233
  conditional 18, 19
  joint 17, 19
Dominant likelihood 44
Doog, K. Caj 142
Drunken soldier 229

ESP 131
Elementary event 2
Empirical Bayes method 226
Event 2
  elementary 2
  independent 7
Evidence 206
  weight of 127
Exact significance level 129
Expectation 22, 23
  conditional 28, 32
Experiment 205
Exponential distribution 78, 236
Exponential family 66, 68, 78, 101, 111
Extensive form 220
Extra-sensory perception 131

F distribution 81, 159, 162, 187, 196, 198, 250, 263
Factorization theorem 58, 66
First digit 106
Fish 214
Fisher, Sir Ronald Aylmer 158
Fisher-Behrens problem 153, 156, 165, 166
Fisher's information 90

Game 218
Gamma distribution 78, 235
General linear model 197
Generalized addition law 9
Generalized multiplication law 9
Genetics 30, 87, 143
Geology 38, 50
Glass 144
Good, I. J. 142

HDR 50
Haar priors 109
Haldane's prior 88
Hay 158
Hierarchical methods 282
Highest density region 50
Homoscedastic 177
Horse-kicks 121
Hyperbolic tangent 169, 172, 174, 254
Hypergeometric distribution 214, 244
Hypothesis 123
  composite 126, 225
  simple 125, 225
  testing 123, 224
  testing, one-sided 128

Improper prior 44, 214, 220
Independent events 7
Independent random variables 22, 176
Indicator function 99
Information 90
Informative stopping rule 214
Intercept 177
Invariant prior 109
Inverse chi distribution 237
Inverse chi-squared distribution 53, 178, 236, 259
Inverse root-sine transformation 27, 89, 163, 166
Iterated logarithm 211

Jacobian 94, 166, 174
Jeffreys, Sir Harold *viii*, 87, 105, 124
  prior 91, 96
  rule 88, 90, 92, 96, 105, 121, 174, 203, 205, 217, 231
Joint distribution function 17, 19
Joint density 17, 19

King's Arms 60

Laplace, Pierre-Simon, Marquis de 86, 120
Layout, one way 185
Layout, two way 193
Least squares 178
Likelihood 34
  data-translated 48, 103, 217
  dominant 44
  nearly constant 132
  principle *ix*, 205, 212, 216
  ratio 126, 225
  standardized 35, 42
Limericks 85
Lindley, D. V. *viii*, 145, 182
Lindley's method 130, 189
Lindley's paradox 137, 144, 145
Line, regression 177
Line of best fit 179
Linear model 177, 197, 200
Log chi-squared distribution 56, 238, 261
Log-likelihood 34, 96, 115, 201

Log-odds 81, 88, 162, 166
Loss 218, 224
absolute error 222
quadratic 220
weighted quadratic 221
zero-one 222, 231

MAD 222
MVUE 87
Marginal density 18, 69, 71
Maximum 31, 100, 102
Maximum likelihood 87, 115, 215
Mean 22, 233
absolute deviation 222
square error 220
Median 29, 222, 233
Method of scoring 116
Mice 30, 163
Minimum sufficient statistic 61
Minimum 31, 102
Minimum variance unbiased estimator 87
Misprints 97
Mixed experiment 206
Mixed random variables 17
Mixtures of conjugate priors 65
Modal interval 223
Mode 30, 223, 233
Model
components of variance 19
general linear 197
linear 177, 197, 200
one way 185
random-effects 191
two way 193
Multiple regression 181, 198
Multiplication law, generalized 9
Multivariate estimation 283
Multivariate t distribution 187

Negative binomial distribution 31, 98, 121, 215, 217, 230, 243
Negative exponential distribution 78, 236
Newton-Raphson method 116
Neyman-Pearson theory 123, 225
Neyman's factorization theorem 58, 66
Nitrogen 160
Nonparametric methods *ix*, 282
Normal distribution 16, 25, 31, 36, 234, 235 and *passim*.
bivariate 31, 53, 164, 234, 235, 257
wrapped 111
Normal form 219
Normal variance 52, 55, 60, 62, 64, 67, 73, 77, 92, 116
Normal/chi-squared distribution 72, 73, 78, 240
Nuisance parameter 69, 186, 214
Null hypothesis 123, 205, 224
Null hypothesis, point (or sharp) 131, 134

Odds 7, 81, 125
One-sided test 128
One way model 185
Optional (optimal) stopping 212
Order statistics 61

P-value 128
Paired comparisons 147
Parallel axes theorem 24
Parameter space 218
Pareto distribution 99, 101, 121, 246
Patil's approximation 154, 250
Point estimator 220, 223, 232
Point null hypothesis *ix*, 131, 134
Poisson distribution 32, 60, 62, 64, 67, 95, 117, 121, 164, 167, 220, 230, 242, 243, 244, 342
Polynomial regression 181
Pólya distribution 86
Posterior 9, 34
odds 125
Precision 24, 38
Predictive distribution 36, 40, 43, 86, 98, 180, 226
Prior 9, 34
conjugate 63, 67, 73, 78, 80, 95, 100, 101
diffuse 131
Haar 109
Haldane's 88
improper 44, 214, 220
invariant 109
Jeffrey's 91, 96
odds 125
reference 46, 49, 54, 69, 86, 103, 111, 173, 203
uniform 47, 111, 173, 217, 230, 231
vague 131
Probability 2
density 13, 17, 19, 233
element 15
posterior 9, 34
prior 9, 34
unconditional 6
Proper prior 44

Quadratic loss 230
weighted 221
Quartile 252

Rainfall 179, 182
Random digits 103
Random variables 11, 14, 17, 18, 20, 21, 233
  continuous 14, 18, 21
  discrete 11, 17, 21
  independent 22, 176
  mixed 17
  uncorrelated 26
Rats 56, 71, 130
Rectangular distribution 99
Recursive construction 183
Reference prior 46, 49, 54, 69, 86, 96, 103, 111, 173, 203
Regression 175, 182, 191, 197, 203
  multiple 181, 198
  polynomial 181
Regression line 177
  recursive construction 183
Rejection region 123, 225
Roberts, H. V. 214
Robustness 43, 55, 282

Sample correlation coefficient 169, 174, 192, 201, 202
Sampling theory statistics 51, 69, 87, 90, 123, 129, 173, 210, 212, 216, 224
Savage, L. J. 209, 210, 213
Scab disease 189
Schlaiffer, R. 128
Scoring, method of 116
Semi-interquartile range 252
Sequential analysis 283
Sequential use of Bayes' theorem 35
Sharp null hypotheses *ix*, 131, 134
Significance level 129
Simply hypothesis 125, 225
Simpson's rule 113
Size 124, 225
Skewness 31, 248
Slope 177
Soldier, drunken 229
Standard deviation 24
Standardized likelihood 35, 42
Statistic 57
  minimum sufficient 61
  order 61
  sufficient 48, 57, 78, 134
Statistical decision problem (game) 219
Stopping rule
  informative 214
  principle 210
Stopping time 210, 230
"Student" (W. S. Gosset) 148
Student's t distribution 70, 140, 146, 178, 180, 193, 239, 252, 253
Sufficiency principle 59, 205, 207, 228
Sufficient statistic *x*, 48, 57, 78, 134, 176
Support 99

t distribution 70, 140, 146, 178, 180, 193, 239, 252, 253
  multivariate 187
tanh 169, 172, 174, 254
Taylor's theorem 27, 115
Traffic accidents 112
Tramcars 105
Trinomial distribution 78
Twins 4, 10, 31
Two by two table 161
Two-sample problem 147
Two way layout 193
Type I and Type II errors 123, 124

Unbiased 87, 212, 213
Unconditional probability 6
Uncorrelated random variables 26
Uniform distribution 21, 31, 99, 103, 121, 138, 145, 245
  discrete 105, 245
Uniform prior 47, 111, 173, 217, 230, 231

Variable
  dependent 175
  explanatory 175
  independent 22, 175
  random 11, 14, 17, 18, 20, 21, 233
Variance 24, 26, 27, 28
  analysis of 188, 190, 196
  conditional 28
  normal 52, 55, 60, 62, 64, 67, 73, 77, 92, 116
Variance/covariance matrix 174
Variance-stabilizing transformation 28
Variance ratio 159
von Mises' distribution 110, 247

W and Z particles 226
Water 226
Weak conditionality principle 207
Weak sufficiency principle 207
Weight of evidence 127
Wheat 75
Window 144
Wrapped normal distribution 111

z distribution 81, 162, 251, 263
Zero-one loss 223, 231